Springer Series in Computational Physics

Editors

H. Cabannes   M. Holt   H. B. Keller
J. Killeen   S. A. Orszag

Springer Series in Computational Physics

Editors: H. Cabannes, M. Holt, H. B. Keller, J. Killeen, S. A. Orszag

Numerical Methods in Fluid Dynamics
M. Holt
1977. viii, 253 pages. 107 illustrations. 2 tables.

A Computational Method in Plasma Physics
P. Garabedian, F. Bauer, and O. Betancourt
1978. vi, 144 pages. 22 figures.

Unsteady Viscous Flows
D. Telionis
1981. 406 pages. 127 figures.

Finite-Difference Techniques for Vectorized Fluid Dynamics Calculations
D. Book
1981. 240 pages. 60 figures.

Implementation of Finite Element Methods for Navier-Stokes Equations
F. Thomasset
1981. 176 pages. 86 figures.

Optimal Shape Design for Elliptic Systems
O. Pironneau
1983. xiii, 192 pages. 57 figures.

Computational Methods in Bifurcation Theory and Dissipative Structures
M. Kubicek, M. Marek
1983. xi, 243 pages. 91 figures.

Numerical Methods for Nonlinear Variational Problems
R. Glowinski
1984. xvii, approx. 462 pages. 80 figures.

Computational Galerkin Methods
C. Fletcher
1984. xvi, approx. 320 pages. 107 figures.

C. A. J. Fletcher

Computational Galerkin Methods

With 107 Figures

Springer-Verlag
New York Berlin Heidelberg Tokyo

C.A.J. Fletcher
Department of Mechanical Engineering
University of Sydney
New South Wales 2006
Australia

Editors

H. Cabannes
Mécanique Théorique
Université Pierre et Marie Curie
F-75005 Paris
France

H. B. Keller
California Institute of Technology
Pasadena, CA 91125
USA

S. A. Orszag
Massachusetts Institute of Technology
Cambridge, MA 02139
USA

M. Holt
College of Engineering
Department of Mechanical Engineering
University of California
Berkeley, CA 94720
USA

J. Killeen
Lawrence Livermore Laboratory
Livermore, CA 94551
USA

Library of Congress Cataloging in Publication Data
Fletcher, C. A. J.
    Computational Galerkin methods.
    (Springer series in computational physics)
    Bibliography: p.
    Includes index.
    1. Galerkin methods. 2. Differential equations—Numerical solutions.
3. Differential equations, Partial—Numerical solutions. I. Title. II. Series.
QA372.F635   1984   519.4   83-17086

Typeset by Asco Trade Typesetting Ltd, Chai Wan, Hong Kong.
Printed and bound by R. R. Donnelley & Sons, Harrisonburg, Virginia.
Printed in the United States of America.

9 8 7 6 5 4 3 2 1

ISBN 0-387-12633-3 Springer-Verlag New York Heidelberg Berlin Tokyo
ISBN 3-540-12633-3 Springer-Verlag Berlin Heidelberg New York Tokyo

Preface

In the wake of the computer revolution, a large number of apparently unconnected computational techniques have emerged. Also, particular methods have assumed prominent positions in certain areas of application. Finite element methods, for example, are used almost exclusively for solving structural problems; spectral methods are becoming the preferred approach to global atmospheric modelling and weather prediction; and the use of finite difference methods is nearly universal in predicting the flow around aircraft wings and fuselages.

These apparently unrelated techniques are firmly entrenched in computer codes used every day by practicing scientists and engineers. Many of these scientists and engineers have been drawn into the computational area without the benefit of formal computational training. Often the formal computational training we do provide reinforces the arbitrary divisions between the various computational methods available.

One of the purposes of this monograph is to show that many computational techniques are, indeed, closely related. The Galerkin formulation, which is being used in many subject areas, provides the connection. Within the Galerkin frame-work we can generate finite element, finite difference, and spectral methods.

Not only does a study of Galerkin methods draw attention to the connections, it also makes it easier to compare the various methods and to identify the features of the various methods that are well suited to particular fields of study. For example, the use of low-order polynomial interpolation with isoparametric mapping makes the finite element method a powerful technique for solving problems with complex boundary shapes. When modelling meteorological flows, the Galerkin spectral formulation has the important property of automatically conserving many physically conserved parameters like kinetic energy, enstrophy, etc.

The connections and comparisons of the different Galerkin methods are emphasized by using the same model problems to illustrate the different methods. The model problems are presented in sufficient depth as to encourage the reader to program these problems on a computer and to check and extend the results for himself (or herself).

To provide guidance in this area, computer programs for the solution of the "propagating shock" problem are described and listed in the Appendices. The programs are written in FORTRAN IV. Using these programs solutions to the propagating shock problem can be obtained by the traditional Galerkin (section 1.2.5), spectral (section 5.2.2), finite element (section 3.2.5), and finite difference methods.

The emphasis on solving problems to the stage of generating numerical results is deliberate. First, this encourages the viewpoint that computation is the total process of problem formulation, mathematical analysis, algorithm construction, and the execution of a computer program to obtain the results. Second, it has been found, in presenting this material in the classroom, that upper-level undergraduates, graduates, and practicing engineers and scientists can more quickly understand a new technique and its limitations through the solution of well-chosen model problems carried to completion.

Roughly the same approach has been taken to introduce each computational Galerkin method. A brief statement of the technique is made or the key features are discussed; then a series of worked examples are presented to illustrate the properties and subtleties of the method. Next the mathematical aspects, particularly *a priori* error estimates, are examined. Extensions of the method are described and, finally, specific applications are provided.

Examples have been chosen to indicate the breadth of application and to encourage wider dissemination. Thus no attempt has been made to treat the major areas of application comprehensively. However, sufficient references (over 250) have been provided to permit the reader to get within reach of the computational techniques currently being developed.

Although this monograph has been written with engineers and scientists in mind, it is anticipated that the applied numerical perspective and inclusion of specific applications will be of interest to applied mathematics students, by providing a balance to more analytic numerical courses. This should assist them to more easily adjust to the research and development working environment.

It is assumed that the reader is familiar with the basic computational techniques as provided by Dahlquist, Bjorck and Anderson in *Numerical Methods* or by Forsythe, Malcolm, and Moler in *Computer Methods for Mathematical Computation*. The reader should also have some awareness of the application of finite difference techniques to the solution of partial differential equations, as provided by Carnaghan, Luther, and Wilkes in *Applied Numerical Analysis*, for example.

In preparing this monograph I have been helped by many people in many different ways. In particular I would like to thank Dr. K. Srinivas for carefully reading the text and making many helpful suggestions. I am grateful to June and Dennis Jeffery for producing illustrations of a consistently high standard. I acknowledge with gratitude the efforts of Tessie Santos and Meriel Knight in typing the manuscript and revisions. They also deserve special thanks for meeting the most unreasonable of deadlines with cheerful forebearance.

I am grateful to Professors M. Holt and R. Temam who provided the ideal environment for writing a monograph during my 1981 sabbatical at the University of California, Berkeley and the Universite de Paris-Sud, Orsay. I am also grateful to the staff of Springer-Verlag, both for including this monograph in the Springer Series in Computational Physics, and for its efficient production.

Finally I am particularly appreciative of my family, Mary, Paul, and Samantha, who both helped in the preparation and forewent their share of my attention while this monograph was being written.

Contents

Chapter 1

Traditional Galerkin Methods

1.1. Introduction

Galerkin methods have been used to solve problems in structural mechanics, dynamics, fluid flow, hydrodynamic stability, magnetohydrodynamics, heat and mass transfer, acoustics, microwave theory, neutron transport, etc. Problems governed by ordinary differential equations, partial differential equations, and integral equations have been investigated via a Galerkin formulation. Steady, unsteady, and eigenvalue problems have proved to be equally amenable to a Galerkin treatment. Essentially, *any problem for which governing equations can be written down is a candidate for a Galerkin method.*

The origin of the method is generally associated with a paper published by Galerkin (1915) on the elastic equilibrium of rods and thin plates. Galerkin, a Russian engineer and applied mechanician, was born in 1871, graduated from the St. Petersburg (Leningrad) Technological Institute in 1899, and gained early engineering experience in the Kharkov locomotive works. It is known (Anon., 1941) that Galerkin began his very fruitful research career while in prison during 1906 and 1907 for his anti-Tsarist views. Galerkin's first academic appointment was at the St. Petersburg Polytechnical Institute in 1909. Galerkin became Head of the Department of Applied Mechanics at the same establishment in 1920 and held various academic positions, mainly in Leningrad, throughout the rest of his distinguished career.

Galerkin methods were well known in the Russian literature from the work of Galerkin and his colleagues. The method first attracted attention in the Western literature as a result of the work of Duncan (1937, 1938) on the dynamics of aeronautical structures. Subsequently Bickley (1941) used a Galerkin method to solve an unsteady heat-conduction problem through consideration of an equivalent electrical circuit. Bickley obtained solutions with a Galerkin method and compared them with solutions obtained with the collocation method and method of least squares. Use of Galerkin methods increased rapidly during the 1950s.

An essential requirement of the method from the beginning was that it should provide solutions of significant accuracy with minimal manual effort. Clearly, at the time of the conception of the method and for perhaps the next 50 years, the calculations were all done by hand or on primitive calculating machines.

The ready availability of computers during the 1960s produced a change in emphasis in the use of Galerkin methods. Since large amounts of computation were now almost inconsequential, the method was required to produce solutions of greater accuracy than before with minimal computer execution time rather than minimal manual effort.

The link between the Galerkin method and the Fourier representation led to the development of *Galerkin spectral methods*, with special treatment of nonlinear terms, for the direct simulation of turbulence (Orszag, 1977) and for global weather prediction (Bourke et al., 1977). These are problems characterized by complex physics and simple boundaries.

It was well known (Kantorovich and Krylov, 1958) that a close connection exists between the Galerkin method and the variational methods such as the Rayleigh–Ritz method. This connection assumed considerable importance during the late 1960s in connection with the *finite-element method.*

Initially, the finite-element method was interpreted as an ad hoc engineering procedure for constructing matrix solutions to stress and displacement calculations in structural analysis. It soon became apparent that this procedure could be given a variational interpretation by considering the potential energy of the system. The link with the Galerkin method permitted finite-element techniques to be extended into areas where no variational principle was obvious; for example, many aspects of fluid mechanics and heat transfer (Oden, 1972). Thus at the present time the Galerkin finite-element formulation is by far the most popular finite-element method (based upon papers published).

The structure of this monograph reflects the chronological development of the various Galerkin methods. First, the traditional Galerkin method is illustrated by some simple examples. The interpretation of the Galerkin method as a method of weighted residuals is examined, and a brief detour is made to look at the other common methods of weighted residuals. This permits a tentative comparison to be made of the various methods of weighted residuals. There are close links between the traditional Galerkin method and other classical approximate methods, particularly the variational methods. Some of these connections are highlighted. This leads into a consideration of convergence and error estimates for the traditional method. That method is still being applied to contemporary problems; some of these modern applications are reviewed to illustrate the breadth of the method.

Attempts to apply the traditional Galerkin method to problems with many unknowns (e.g., the typical computer situation) exposes some inherent limitations in the method. These are discussed along with improvements that lead to the Galerkin spectral and the Galerkin finite-element methods.

Galerkin finite-element methods are introduced with some typical illustrative examples. The Galerkin concept is sufficiently flexible that it is straightforward to introduce additional techniques to improve the efficiency of the complete algorithm. Some of the modified Galerkin finite-element formulations are described. It is possible to interpret many of the well-known finite-difference algorithms as finite-element methods; this aspect is explored. In parallel with the essentially engineering development of the Galerkin finite-element method, considerable theoretical analysis of Galerkin schemes has been undertaken in relation to establishing convergence and estimating errors. Some of this work is described. A representative collection of modern Galerkin finite-element applications are included.

Galerkin spectral methods are described with emphasis on choosing the form of the approximating functions and the treatment of nonlinear terms. Methods other than traditional Galerkin methods can be upgraded to the status of spectral methods. This is illustrated in relation to the orthonormal method of integral relations. Applications of the spectral method, particularly in the area of turbulence and global atmospheric modeling, are described.

By applying the various Galerkin techniques to model problems, such as Burgers' equations, it is possible to make some comparative statements about the various Galerkin methods. These comparisons are also made in the context of more sophisticated problems such as ocean modeling.

A major difficulty in computing viscous fluid flows is how to handle the convection–dissipation interaction correctly. Consideration of this problem in a Galerkin framework is facilitated if the Galerkin method is generalized. The use of generalized Galerkin methods (or Petrov–Galerkin methods) to avoid stability problems in convection-dominated applications is emphasized.

The key features of the Galerkin method can be stated quite concisely. It will be assumed that a two-dimensional problem is governed by a linear differential equation

$$L(u) = 0 \qquad (1.1.1)$$

in a domain $D(x, y)$, with boundary conditions

$$S(u) = 0 \qquad (1.1.2)$$

on ∂D, the boundary of D. The Galerkin method assumes that u can be accurately represented by an *approximate* solution

$$u_a = u_0(x, y) + \sum_{j=1}^{N} a_j \phi_j(x, y), \qquad (1.1.3)$$

where the ϕ_j's are known *analytic functions*, u_0 is introduced to satisfy the boundary conditions, and the a_j's are coefficients to be determined. Substitution of eq. (1.1.3) into eq. (1.1.1) produces a nonzero *residual*, R, given by

$$R(a_0, a_1 \cdots a_N, x, y) = L(u_a) = L(u_0) + \sum_{j=1}^{N} a_j L(\phi_j). \qquad (1.1.4)$$

It is convenient to define an inner product (f, g) in the following manner:

$$(f, g) = \iint_D fg \, dx \, dy. \qquad (1.1.5)$$

In the Galerkin method the unknown coefficients, a_j in eq. (1.1.3) are obtained by solving the following system of equations:

$$(R, \phi_k) = 0, \qquad k = 1, \ldots, N. \tag{1.1.6}$$

Here R is just the equation residual, and the ϕ_k's are the *same* analytic functions that appear in eq. (1.1.3). Since this example is based on a linear differential equation, eq. (1.1.6) can be written directly as a matrix equation for the coefficients a_j as

$$\sum_{j=1}^{N} a_j(L(\phi_j), \phi_k) = -(L(u_0), \phi_k). \tag{1.1.7}$$

Substitution of the a_j's resulting from the solution of eq. (1.1.7) into eq. (1.1.3) gives the required approximate solution u_a.

Some of the ramifications of the Galerkin method will become apparent after considering the simple examples of section 1.2. The main features of the method are described in section 1.3.5.

1.2. Simple Examples

Here we examine the mechanics of applying the Galerkin method to some representative problems of escalating complexity. In the first example, which is due to Duncan (1937), the Galerkin method is used to reduce an ordinary differential equation to a system of algebraic equations. In the second example an elliptic partial differential equation is reduce to a system of algebraic equations. The third and fourth examples illustrate the use of the Galerkin method to reduce parabolic partial differential equations to systems of ordinary differential equations in time. The third example considers a linear partial differential equation; the fourth example considers a nonlinear partial differential equation, Burgers' equation. A computer program, BURG1, to solve Burgers' equation by the traditional Galerkin method is provided in appendix 1.

1.2.1. An ordinary differential equation

Consider the ordinary differential equation

$$\frac{dy}{dx} - y = 0 \tag{1.2.1}$$

with boundary condition $y = 1$ at $x = 0$. An approximate solution is sought in the domain $0 \leq x \leq 1$. The exact solution is $y = e^x$. An approximate (often called *trial*) solution is introduced by

$$y_a = 1 + \sum_{j=1}^{N} a_j x^j, \qquad (1.2.2)$$

where the leading term is included to satisfy the boundary condition. The trial functions x^j then satisfy homogeneous boundary conditions. Deliberately *structuring the trial solution to satisfy the boundary conditions* is a common practice in applying the traditional Galerkin method. Usually this technique produces a more accurate solution for a given number of unknowns, N. Eq. (1.2.2) can also be written in the form

$$y_a = \sum_{j=0}^{N} a_j x^j, \qquad (1.2.3)$$

where a_0 is chosen to satisfy the boundary condition, that is, $a_0 = 1$. Substitution of eq. (1.2.2) into eq. (1.2.1) produces an equation residual

$$R = -1 + \sum_{j=1}^{N} a_j (jx^{j-1} - x^j). \qquad (1.2.4)$$

Evaluating the inner product defined by eq. (1.1.5),

$$(R, x^{k-1}) = 0, \qquad (1.2.5)$$

for $k = 1, \ldots, N$ produces a system of equations that can be written

$$\mathbf{MA} = \mathbf{D}, \qquad (1.2.6)$$

where $\mathbf{A}$ is the vector of the unknown coefficients a_j. An element of $\mathbf{D}$ is given by

$$d_k = (1, x^{k-1}) = \frac{1}{k}.$$

An element of $\mathbf{M}$ is given by

$$m_{kj} = (jx^{j-1} - x^j, x^{k-1}) = \frac{j}{j+k-1} - \frac{1}{j+k}.$$

For $N = 3$ the solution of Eq. (1.2.6) is

$$A^t = \{1.0141, 0.4225, 0.2817\}, \qquad (1.2.7)$$

and substitution into the trial solution, eq. (1.2.2), gives

$$y_a = 1 + 1.0141x + 0.4225x^2 + 0.2817x^3.$$

An indication of the rapid improvement in accuracy with increasing N is given in Table 1.1. This is a very important feature of the traditional Galerkin method; namely that *high accuracy can be achieved with a relatively modest algebraic effort*. This can be quantified by evaluating the discrete L_2 error between the approximate and exact solutions, $\| y - y_a \|_{2,d}$.

We will make frequent use of the following error definitions. The *error in the L_2 norm* (or the L_2 error) is defined by

Table 1.1. Traditional Galerkin Solutions of $dy/dx - y = 0$

	Approximate solution, y_a			Exact solution
x	Linear ($N = 1$)	Quadratic ($N = 2$)	Cubic ($N = 3$)	$y = e^x$
0	1	1	1	1
0.2	1.4	1.2057	1.2220	1.2214
0.4	1.8	1.4800	1.4913	1.4918
0.6	2.2	1.8229	1.8214	1.8221
0.8	2.6	2.2349	2.2259	2.2251
1.0	3.0	2.7143	2.7183	2.7183
$\|y - y_a\|_{2,d}$	0.6997	0.0217	0.0013	
$\|R\|_{2,d}$	1.2911	0.1429	0.0119	

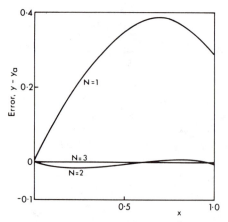

Figure 1.1. Error distribution of N parameter solution of $dy/dx - y = 0$

$$\|y - y_a\|_2 = \left[\int_0^1 (y - y_a)^2 \, dx \right]^{1/2}. \tag{1.2.8}$$

The *discrete L_2 error* is defined by

$$\|y - y_a\|_{2,d} = \left[\sum_{l=1}^{L} (y - y_a)_l^2 \right]^{1/2}. \tag{1.2.9}$$

The discrete L_2 error is related to the *rms error* σ by

$$\|y - y_a\|_{2,d} = \sigma L^{1/2}.$$

With reference to Table 1.1, the discrete L_2 error has been computed from the exact and solution values at the discrete x values 0, 0.2, 0.4, etc. In Fig. 1.1 the solution error for various values of the order of representation, N, is shown. It can be seen how rapidly the error reduces as N is increased. It may

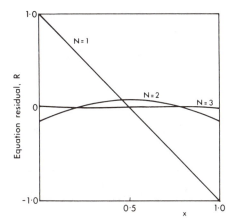

Figure 1.2. Equation residual distribution for N parameter solution of $dy/dx - y = 0$

also be noted from Table 1.1 and Fig. 1.2 that the discrete L_2 norm of the residual, $\|R\|_{2,d}$, also diminishes rapidly with increasing N. Since the boundary condition is satisfied exactly, one would expect that $\|R\|_2 \to 0$ as $\|y - y_a\|_2 \to 0$. For the general problem an exact solution is not available and $\|y - y_a\|_2$ cannot be evaluated. However, $\|R\|_2$ can be evaluated without difficulty. Finlayson (1972) has explored the use of $\|R\|_2$ to estimate the error $\|y - y_a\|_2$.

It was pointed out in section 1.1 that in evaluating the inner product in eq. (1.1.6), the *weight* (often called *test*) functions ϕ_k are chosen from the same functions as the trial functions ϕ_j. It is important that the ϕ's be chosen as the lowest-order members of a complete set of functions. This is the reason for setting $\phi_k = x^{k-1}$ in eq. (1.2.5), and this is consistent with the trial solution written in the form of eq. (1.2.3).

Less accurate solutions, for the same number of unknowns, are obtained if the weight functions are chosen as higher-order members of ϕ_j. This is illustrated by the results shown in Table 1.2. Each solution is a cubic, but in columns 3 and 4 larger values of k have been used in the weight function x^{k-1}. An examination of $\|y - y_a\|_{2,d}$ indicates a dramatic improvement in accuracy when using test and trial functions as the lowest-order members of a complete set. It is interesting to note that when Duncan (1937) considered this example, he presented results with weight functions x^{k-1}, $k = 2, 3, 4$, and in so doing missed the significantly more accurate solution corresponding to $k = 1, 2, 3$ in Table 1.2.

Another possibility exists for improving the accuracy of the solution. If the boundary condition at $x = 0$ is combined with the governing equation, (1.2.1), another constraint or "boundary condition" can be obtained:

$$\frac{dy}{dx} = 1 \quad \text{at } x = 0. \tag{1.2.10}$$

Table 1.2. Different weight functions for Galerkin solution of $dy/dx - y = 0$

| | Cubic solution with weight function x^{k-1} | | | Exact solution |
x	$k = 1,2,3$	$k = 2,3,4$	$k = 3,4,5$	$y = e(x)$
0	1	1	1	1
0.2	1.2220	1.2248	1.2282	1.2214
0.4	1.4913	1.4952	1.5004	1.4918
0.6	1.8214	1.8255	1.8317	1.8221
0.8	2.2259	2.2303	2.2375	2.2251
1.0	2.7183	2.7241	2.7330	2.7183
$\|y - y_a\|_{2,d}$	0.0013	0.0087	0.0215	
$\|R\|_{2,d}$	0.0119	0.0192	0.0335	

Table 1.3. Traditional Galerkin solution of $dy/dx - y = 0$ with extra "boundary condition" (1.2.10)

| | Approximate solution | | | |
x	$N = 2$	$N = 3$	$N = 4$	$y = e^x$
0	1.0000	1.0000	1.0000	1.0000
0.2	1.2300	1.2205	1.2219	1.2214
0.4	1.5200	1.4903	1.4926	1.4918
0.6	1.8700	1.8215	1.8223	1.8221
0.8	2.2800	2.2267	2.2250	2.2251
1.0	2.7500	2.7179	2.7184	2.7183
$\|y - y_a\|_{2,d}$	0.0774	0.0022	0.0009	

To satisfy this condition we must put $a_1 = 1$ in eq. (1.2.2). It can be seen from eq: (1.2.7) that previously this "boundary condition" was not satisfied exactly. Results of applying the Galerkin method with this additional condition are indicated in Table 1.3. N refers to the largest power of x in eq. (1.2.2). If the discrete L_2 error is compared for the same N between Tables 1.3 and 1.1 it can be seen that imposing the extra condition has reduced the accuracy of the solution. This is not completely unexpected, since the Galerkin criterion (1.1.6) is reducing the *global* error, whereas eq. (1.2.10) uses up one of the disposable parameters to improve the solution *locally*. In obtaining the solution most of the effort is required to solve the system of equations (1.2.6). Thus a comparison based on equal effort should compare a particular value of N in Table 1.3 with the corresponding value of $N - 1$ in Table 1.1. In this comparison it is clear that imposition of the extra "boundary condition" leads to a more accurate solution for almost the same effort.

1.2.2. An eigenvalue problem

Here we consider the model eigenvalue problem governed by

$$\frac{d^2p}{dx^2} + \lambda P = 0 \tag{1.2.11}$$

with boundary conditions

$$P(0) = P(1) = 0. \tag{1.2.12}$$

This problem has an exact solution. The exact eigenvalues are given by

$$\lambda_j = (j\pi)^2, \qquad j = 1, 2, 3, \ldots, \tag{1.2.13}$$

and the corresponding eigenfunctions are

$$P_j(x) = \sqrt{2}\sin(j\pi x). \tag{1.2.14}$$

An approximate solution is introduced as

$$P_a(x) = \sum_{j=1}^{N} a_j(x - x^{j+1}), \tag{1.2.15}$$

where the form of the trial functions, $x - x^{j+1}$, has been chosen to satisfy the boundary conditions exactly., Substitution of eq. (1.2.15) into the governing eq. (1.2.11) creates a residual

$$R = \sum_{j=1}^{N} a_j(-(j + 1)jx^{j-1} + \lambda(x - x^{j+1})). \tag{1.2.16}$$

Evaluation of the inner product, that is (using eq. (1.1.5)),

$$(R, x - x^{k+1}) = 0, \tag{1.2.17}$$

produces the matrix eigenvalue equation

$$\mathbf{LA} = \lambda\mathbf{MA}, \tag{1.2.18}$$

where the elements of $\mathbf{L}$ and $\mathbf{M}$ are given by

$$l_{kj} = \frac{jk}{k + j + 1} \tag{1.2.19}$$

and

$$m_{kj} = \frac{jk(j + k + 6)}{3(j + 3)(k + 3)(j + k + 3)}. \tag{1.2.20}$$

The solution for the eigenvalues are obtained by solving

$$\det|\mathbf{L} - \lambda\mathbf{M}| = 0. \tag{1.2.21}$$

For $N = 1$ in eq. (1.2.15), eq. (1.2.18) becomes

$$\tfrac{1}{3}a_1 = \lambda\tfrac{1}{30}a_1, \tag{1.2.22}$$

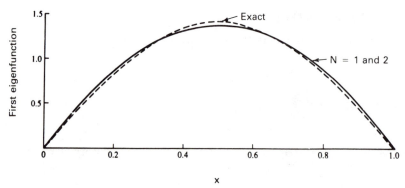

Figure 1.3. First eigenfunction comparison, Galerkin and exact (after Harrington, 1968; reprinted with permission of Krieger Publishing Co.)

or $\lambda = 10$. The corresponding eigenfunction, $a_1(x - x^2)$, can be compared with the exact value by noting that the exact solution has been *orthonormalized*, that is, using eq. (1.1.5),

$$(P_j, P_k) = \begin{cases} 1 & \text{if } j = k, \\ 0 & \text{if } j \neq k. \end{cases} \tag{1.2.23}$$

Requiring the same of the approximate eigenfunction gives

$$1 = a_1^2 \tfrac{1}{30},$$

or

$$a_1(x - x^2) = \sqrt{30}(x - x^2). \tag{1.2.24}$$

Eq. (1.2.24) is compared with $P_1 = \sqrt{2} \sin \pi x$ in Fig. 1.3.

For $N = 2$ in eq. (1.2.15), eq. (1.2.18) gives

$$\begin{aligned} \tfrac{1}{3}a_1 + \tfrac{1}{2}a_2 &= \lambda(\tfrac{1}{30}a_1 + \tfrac{1}{20}a_2), \\ \tfrac{1}{2}a_1 + \tfrac{4}{5}a_2 &= \lambda(\tfrac{1}{20}a_1 + \tfrac{8}{105}a_2). \end{aligned} \tag{1.2.25}$$

The corresponding eigenvalues are $\lambda_1 = 10$, $\lambda_2 = 42$. Substituting the values of λ into eq. (1.2.25) permits a_2 to be eliminated in terms of a_1. Imposition of the orthonormal condition allows a_1 (and hence a_2) to be obtained. The first two eigenfunctions are then

$$\sqrt{30}(x - x^2) \quad \text{and} \quad 3\sqrt{210}(x - x^2) - 2\sqrt{210}(x - x^3).$$

The first eigenfunction is the same as for the $N = 1$ approximation; this is plotted in Fig. 1.3. The second eigenfunction is plotted in Fig. 1.4. Comparing the results plotted in Figs. 1.3 and 1.4 indicates that the agreement for the first eigenfunction is better than for the second eigenfunction. This result is consistent with the better agreement for the lower-order eigenvalues shown in Table 1.4.

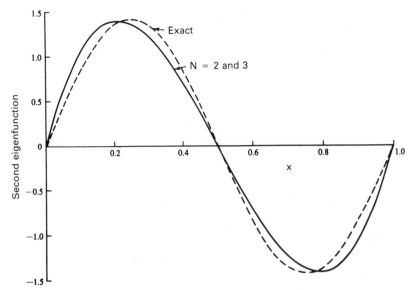

Figure 1.4. Second eigenfunction comparison, Galerkin and exact (after Harrington, 1968; reprinted with permission of Krieger Publishing Co.)

Table 1.4. Comparison of eigenvalues of $d^2p/dx^2 + \lambda p = 0$ for N-term Galerkin method, $x - x^{j+1}$ as trial functions

N	1	2	3	4	Exact
λ_1	10.0000	10.0000	9.8697	9.8697	9.8696
λ_2		42.0000	42.000	39.497	39.478
λ_3			102.133	102.133	88.826
λ_4				200.583	157.914

As N, the number of terms in the approximate solution, increases, the accuracy of the lower-order eigenvalues rapidly increases whereas the accuracy of the higher-order eigenvalues rapidly decreases. Since the governing equation is self-adjoint and positive definite, the Galerkin method will always produce eigenvalues that are greater than the exact eigenvalues.

An indication of the effect of using different trial functions in eq. (1.2.15) can be obtained by considering the piecewise linear function shown in Fig. (1.5).

Replacement of eq. (1.2.15) with

$$P_a(x) = \sum_{j=1}^{N} a_j \phi_j(x)$$

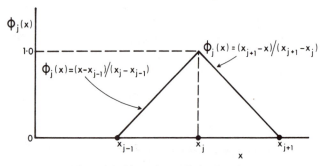

Figure 1.5. Linear interpolating function

indicates that the a_j's become the *nodal* values of P_a, since $\phi_j(x)$ is an *interpolating* function. Such functions find widespread use in the finite-element method (see section 2.2 and chapter 3).

Following the same steps as for the previous case, eqs. (1.2.15) to (1.2.20) lead to

$$(R, \phi_k(x)) = 0 \qquad (1.2.26)$$

instead of eq. (1.2.17) and

$$l_{kk} = 2(k + 1),$$
$$l_{kk-1} = l_{kk+1} = -(k + 1), \qquad (1.2.27)$$
$$l_{kj} = 0 \quad \text{for } j < k - 1 \text{ or } j > k + 1$$

and

$$m_{kk} = \frac{2}{3(k + 1)},$$
$$m_{kk-1} = m_{kk+1} = \frac{1}{6(k + 1)}, \qquad (1.2.28)$$
$$m_{kj} = 0 \quad \text{for } j < k - 1 \text{ and } j > k + 1$$

instead of eqs. (1.2.19) and (1.2.20).

The form of eqs. (1.2.27) and (1.2.28) indicates that the combination of piecewise polynomials of small support and the Galerkin method produce matrices, like **L** and **M** in eq. (1.2.18), that have nonzero entries adjacent to the main diagonal only. When N is large this leads to considerable economy in operating on the matrices.

However there is a price to pay for this economy; and this can be seen by considering the eigenvalues obtained using the piecewise linear interpolating functions $\phi_j(x)$ (Table 1.5). Although the accuracy of the evaluation of the eigenvalues improves with N, the accuracy, for a particular value of N, is less

Table 1.5. Comparison of eigenvalues of $d^2P/dx^2 + \lambda P = 0$
for N-term Galerkin method, $\phi_j(x)$ as trial function

N	1	2	3	4	Exact
λ_1	12.000	10.800	10.386	10.198	9.870
λ_2		54.000	48.000	44.903	39.478
λ_3			128.868	116.118	88.826
λ_4				227.838	157.914

than the accuracy indicated in Table 1.4. The results shown in Table 1.4 were
obtained with trial functions that spanned the domain $0 \le x \le 1$. Thus the
trial functions are global in character. By comparison the trial functions $\phi_j(x)$
are local in character. The reduced accuracy of the Galerkin method when
using local rather than global trial functions, *for the same N*, is typical.

Harrington (1968), also uses the above example to illustrate the effect of
not choosing the approximate solution to satisfy the boundary conditions
exactly. For this problem a spurious negative eigenvalue is generated and the
accuracy of the remaining eigenvalues, for the same N, is reduced.

1.2.3. Viscous flow in a channel

Consider a channel of square cross section shown in Fig. 1.6. The steady flow
of a viscous fluid is governed by the z momentum equation

$$u\frac{\partial w}{\partial x} + v\frac{\partial w}{\partial y} + w\frac{\partial w}{\partial z} + \frac{1}{\rho}\frac{\partial p}{\partial z} = v\left\{\frac{\partial^2 w}{\partial x^2} + \frac{\partial^2 w}{\partial y^2} + \frac{\partial^2 w}{\partial z^2}\right\}. \qquad (1.2.29)$$

Far from the exit or entrance the flow does not vary in the z direction and
u and v are zero. Then eq. (1.2.29) becomes

$$\frac{1}{\mu}\frac{\partial p}{\partial z} = \frac{\partial^2 w}{\partial x^2} + \frac{\partial^2 w}{\partial y^2}. \qquad (1.2.30)$$

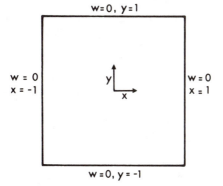

Figure 1.6. Flow in a square channel

For such a flow $\partial p/\partial z$ is a constant. With an appropriate nondimensionaliza-tion eq. (1.2.30) becomes

$$\frac{\partial^2 w}{\partial x^2} + \frac{\partial^2 w}{\partial y^2} + 1 = 0, \tag{1.2.31}$$

that is, a Poisson equation for w. The boundary conditions are

$$w = 0 \quad \text{on } x = \pm 1 \text{ and } y = \pm 1. \tag{1.2.32}$$

If a trial solution is based on trigonometric functions, it is possible for each trial function to satisfy the boundary conditions. Thus let

$$w_a = \sum_{i=1,3,5,\dots}^{N} \sum_{j=1,3,5,\dots}^{N} a_{ij} \cos i\frac{\pi}{2}x \cos j\frac{\pi}{2}y. \tag{1.2.33}$$

Substituting eq. (1.2.33) into eq. (1.2.31) produces the residual

$$R = -\left[\sum_{i=1,3,5,\dots}^{N} \sum_{j=1,3,5,\dots}^{N} a_{ij} \cos i\frac{\pi}{2}x \cos j\frac{\pi}{2}y \left\{ \left(i\frac{\pi}{2}\right)^2 + \left(j\frac{\pi}{2}\right)^2 \right\} - 1 \right]. \tag{1.2.34}$$

The Galerkin method obtains algebraic equations for the a_{ij}'s from evaluating

$$\left(R, \cos i\frac{\pi}{2}x \cos j\frac{\pi}{2}y \right) = 0, \quad i = 1, 3, 5, \dots, \quad j = 1, 3, 5, \dots \tag{1.2.35}$$

For the example considered in section 1.2.1, evaluation of equations like (1.2.35) produced a system of simultaneous equations that could be solved for the unknown coefficients. In this example the particular choice of the trial functions permits a_{ij} to be evaluated directly. Namely,

$$a_{ij} = +\left(\frac{8}{\pi^2}\right)^2 \frac{(-1)^{(i+j)/2-1}}{ij(i^2+j^2)}. \tag{1.2.36}$$

The avoidance of solving a system of simultaneous equations is brought about by the choice of trial (and test) functions that are *orthogonal* over the domain considered. Such a choice has important advantages; these are discussed in chapter 2. Applying the Galerkin method with *trigonometric* trial (and test) functions produces the same equations and solutions as applying the Fourier method with a finite number of terms in the expansion (Kantorovich and Krylov, 1958, p. 70).

Substitution of eq. (1.2.36) into eq. (1.2.33) gives the required solution:

$$w_a = \left(\frac{8}{\pi^2}\right)^2 \sum_{i=1,3,5,\dots}^{N} \sum_{j=1,3,5,\dots}^{N} \frac{(-1)^{(i+j)/2-1}}{ij(i^2+j^2)} \cos i\frac{\pi}{2}x \cos j\frac{\pi}{2}y. \tag{1.2.37}$$

Exact solution for the centerline velocity w_{CL} and the nondimensional flow rate $\dot{q}$ have been given by Dryden et al. (1956). The flow rate, $\dot{q}$, is related to w_a by

$$\dot{q} = \int_{-1}^{1} \int_{-1}^{1} w_a(x,y)\, dx\, dy. \tag{1.2.38}$$

Table 1.6. Solutions to viscous flow in a channel using the traditional Galerkin method

Trial solution	N	Approximate solution						Exact solution
		1	2	3	4	5	6	
Eq. (1.2.33)	w_{CL}	0.3285	0.2888	0.2968	0.2938	0.2952	0.2944	0.2947
	$\dot{q}$	0.5326	0.5570	0.5606	0.5615	0.5619	0.5621	0.5623
Eq. (1.2.40)	w_{CL}	0.3125	0.2927					
	$\dot{q}$	0.5556	0.5607					

Evaluating eq. (1.2.38) gives

$$\dot{q} = 2\left(\frac{8}{\pi^2}\right)^3 \sum_{i=1,3,5,\dots}^{N} \sum_{j=1,3,5,\dots}^{N} \frac{1}{i^2 j^2 (i^2 + j^2)}. \qquad (1.2.39)$$

The solutions for w_{CL} and $\dot{q}$ with increasing number of terms, N, converge rapidly. Results with increasing N are presented in Table 1.6. Solutions could also have been obtained with other appropriate trial solutions. Thus Finlayson (1969) used

$$w_a = \sum_{j=1}^{N} a_j (1 - x^2)^j (1 - y^2)^j. \qquad (1.2.40)$$

Results with $N = 1$ and 2 using eq. (1.2.40) are shown in Table 1.6. It can be seen that the solution accuracy is greater for the same N than using the trial solution (1.2.33). However, use of such a trial solution would require the solution of a system of equations like (1.2.6) for the unknown coefficients a_j. Increasing N requires a complete new set of unknown coefficients a_j. In contrast, use of orthogonal functions permits the calculation of the coefficients a_{ij} *independently of the preceding coefficients.* Consequently the magnitude of the higher-order coefficients in the series gives an indication of the error in the solution.

1.2.4. Unsteady heat conduction

This example is included to illustrate the use of the Galerkin procedure to reduce a partial differential equation to a system of ordinary differential equations. This procedure is particularly useful because algorithms for the efficient integration of ordinary differential equations have been developed further (Gear, 1981) than algorithms for the direct solution of partial differential equations.

A typical unsteady heat-condition problem is illustrated in Fig. 1.7. The ends of the bar are maintained at temperatures T_1 and T_2, respectively. At $t = 0$ the temperature of the bar is

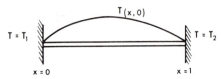

Figure 1.7. Unsteady heat conduction

$$T(x,0) = T_1 + (\sin \pi x + x)(T_2 - T_1). \qquad (1.2.41)$$

The subsequent temperature distribution in the bar is governed by the parabolic partial differential equation

$$\frac{\partial T}{\partial t} - \alpha \frac{\partial^2 T}{\partial x^2} = 0. \qquad (1.2.42)$$

It is convenient to define a nondimensional temperature

$$\theta = \frac{T - T_1}{T_2 - T_1}. \qquad (1.2.43)$$

If α is absorbed into a nondimensional time, the problem becomes one of solving the equation

$$\frac{\partial \theta}{\partial t} - \frac{\partial^2 \theta}{\partial x^2} = 0 \qquad (1.2.44)$$

subject to the initial condition

$$\theta(x,0) = \sin \pi x + x \qquad (1.2.45)$$

and the boundary conditions

$$\theta(0,t) = 0 \quad \text{and} \quad \theta(1,t) = 1. \qquad (1.2.46)$$

An approximate solution of the following form is introduced:

$$\theta_a(x,t) = \theta_0(x) + \sum_{j=1}^{N} a_j(t)\phi_j(x), \qquad (1.2.47)$$

where

$$\theta_0(x) = \sin \pi x + x$$

and

$$\phi_j(x) = x^j - x^{j+1}.$$

It may be noted that θ_0 satisfies both the initial and boundary conditions and that the ϕ_j's satisfy homogeneous boundary conditions.

Substitution of eq. (1.2.47) into the governing equation (1.2.44) produces a residual

$$R = -\frac{d^2\theta_0}{dx^2} + \sum_{j=1}^{N} \left(\frac{da_j}{dt}\phi_j - \frac{a_j d^2\phi_j}{dx^2} \right). \qquad (1.2.48)$$

Repeated evaluation of the inner product

$$(R, \phi_k) = 0$$

produces a system of ordinary differential equations that can be written

$$\mathbf{M\dot{A}} + \mathbf{BA} + \mathbf{C} = 0, \qquad (1.2.49)$$

where an element of $\mathbf{\dot{A}}$ is da_j/dt and an element of $\mathbf{M}$ is given by

$$m_{kj} = (\phi_j, \phi_k). \qquad (1.2.50)$$

The elements of $\mathbf{B}$ and $\mathbf{C}$ are respectively

$$b_{kj} = -\left(\frac{d^2\phi_j}{dx^2}, \phi_k\right) \qquad (1.2.51)$$

and

$$c_k = -\left(\frac{d^2\theta_0}{dx^2}, \phi_k\right). \qquad (1.2.52)$$

For a small number of unknown coefficients a_j, it is convenient to manipulate eq. (1.2.49) into an explicit form

$$\mathbf{\dot{A}} = \mathbf{SA} + \mathbf{T} \qquad (1.2.53)$$

where

$$\mathbf{S} = -\mathbf{M}^{-1}\mathbf{B} \quad \text{and} \quad \mathbf{T} = -\mathbf{M}^{-1}\mathbf{C}. \qquad (1.2.54)$$

Since the system of ordinary differential equations (1.2.53) is linear, it would be possible to obtain an analytic solution. However, in order to illustrate a typical procedure for more complex problems, a numerical integration will be carried out. In general, starting values $\mathbf{A}_0$ are required to integrate eq. (1.2.53). In this instance $\theta_0(x)$ satisfies the initial conditions exactly, so that $\mathbf{A}_0 = 0$.

For N up to 5 in eq. (1.2.47), eq. (1.2.53) has been integrated with a first-order (*Euler*) scheme

$$a_j^{n+1} = a_j^n + \frac{da_j}{dt}\bigg|_j^n \Delta t, \qquad (1.2.55)$$

where n indicates the time level. At each time level, substitution into eq. (1.2.47) gives the approximate solution, $\theta_a(x, t)$. Some results obtained with $\Delta t = 0.001$ are shown in Table 1.7. The exact solution of equation (1.2.44) to (1.2.46) is

$$\theta = \sin \pi x \exp(-\pi^2 t) + x. \qquad (1.2.56)$$

Thus the initial conditions may be interpreted as a perturbation, $\sin \pi x$, on a linear variation x. With increasing t the perturbation dies away, leaving the linear temperature distribution as the steady-state solution.

Examination of the results in Table 1.7 indicates that the results for $N = 3$ and 4 are the same. This is due to the particular trial functions used in eq. (1.2.47). It can be seen (Table 1.7) that the solution for $N = 3, 4$ agree more closely with the exact solution than the solutions for $N = 5$. However, this

Table 1.7. Numerical solutions of eq. (1.2.44)

	Approximate solution $\theta_a(0.5, t)$				Exact solution	Approx. solution
t	$N = 2$	$N = 3$	$N = 4$	$N = 5$	θ	θ_b
0	1.50000	1.50000	1.50000	1.50000	1.50000	1.50000
0.02	1.32611	1.32020	1.32020	1.32007	1.32087	1.32006
0.04	1.18389	1.17278	1.17278	1.17251	1.17383	1.17251
0.06	1.06757	1.05188	1.05188	1.05150	1.05312	1.05150
0.08	0.97242	0.95274	0.95274	0.95227	0.95404	0.95226
0.10	0.89461	0.87144	0.87144	0.87089	0.87271	0.87089
0.12	0.83096	0.80477	0.80477	0.80416	0.80594	0.80415
0.14	0.77890	0.75009	0.75009	0.74943	0.75114	0.74942
0.16	0.73632	0.70526	0.70526	0.70455	0.70615	0.70454
0.18	0.70150	0.66849	0.66849	0.66775	0.66923	0.66774
0.20	0.67301	0.63833	0.63833	0.63756	0.63891	0.63756

superior agreement is deceptive. The approximate results shown in Table 1.7 contain errors from two sources: the error in the trial solution (1.2.47) and the error in the numerical integration of eq. (1.2.53).

We can separate the two sources of errors in the following manner. Suppose a solution of eq. (1.2.44) to (1.2.46) were sought via a separation-of-variables technique. Thus let

$$\theta = \chi(t) \sin \pi x + x. \qquad (1.2.57)$$

Then $\chi(t)$ satisfies the equation

$$\frac{d\chi}{dt} + \pi^2 \chi = 0, \qquad (1.2.58)$$

which has the solution $\chi = \exp(-\pi^2 t)$. However, if eq. (1.2.58) were integrated using the numerical scheme (1.2.55), the solution χ_a would be different from χ. If χ_a is substituted into eq. (1.2.57), the result, θ_b, only differs from the exact solution due to the errors associated with the numerical integration.

Thus the errors due to the Galerkin formulation can be isolated by comparing the approximate solution θ_a of Table 1.7 with the approximate solution θ_b, which is also shown in Table 1.7. It is apparent that the accuracy of the Galerkin formulation with $N = 5$ is substantial.

1.2.5. Burgers' equation

Burgers' equation, in one dimension, is

$$\frac{du}{dt} + u\frac{du}{dx} - \frac{1}{Re}\frac{d^2u}{dx^2} = 0. \qquad (1.2.59)$$

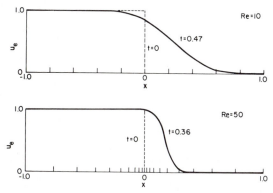

Figure 1.8. Exact solution of Burgers' equation

The equation is a very good model for the Navier–Stokes equations, since it represents, in the simplest manner possible, the balance between the nonlinear convective process ($u\,du/dx$) and the dissipative process ($(1/\text{Re})\,d^2u/dx^2$). Although Burgers' equation is nonlinear, it does possess an *exact solution* for many combinations of initial and boundary conditions. For this reason Burgers' equation has often been used to assess the accuracy of computational algorithms. A broad discussion of Burgers' equation may be found in Fletcher (1982). Here we apply the traditional Galerkin method to Burgers' equation to see how a nonlinear term is dealt with.

A particularly concise form of the exact solution is obtained if a propagating shock wave is considered. Typical situations are shown in Fig. 1.8 for different values of the Reynolds number Re. The shock wave is initially discontinuous and located at $x = 0$. Subsequently the shock propagates to the right, but its initially discontinuous profile is smoothed by the dissipative viscous process ($(1/\text{Re})\,d^2u/dx^2$).

Solutions to eq. (1.2.59) will be sought in the region $-1 \le x \le 1$ for $t \ge 0$. Initial and boundary conditions are taken to be

$$u_0(x) = u(x,0) = \begin{cases} 1 & \text{if } -1 \le x \le 0, \\ 0 & \text{if } 0 < x \le 1, \end{cases} \tag{1.2.60}$$

$$u(-1,t) = 1, \qquad u(1,t) = 0.$$

The exact solution of eqs. (1.2.59) and (1.2.60) is

$$u_e = \frac{\displaystyle\int_{-\infty}^{\infty} \frac{x-\xi}{t} \exp\{-0.5\,\text{Re}\,F\}\,d\xi}{\displaystyle\int_{-\infty}^{\infty} \exp\{-0.5\,\text{Re}\,F\}\,d\xi} \tag{1.2.61}$$

where

$$F(\xi; x, t) = \int_0^\xi u_0(\xi')\, d\xi' + \frac{0.5(x - \xi)^2}{t}.$$

This equation is plotted in Fig. 1.8. In order to compare the approximate solutions of eq. (1.2.59) and (1.2.60) with the exact solution, it is necessary to limit the time and the smallest value of Re so that the boundary conditions (1.2.60) still correspond to the exact solution.

It would be possible to utilize a polynomial trial solution as in the examples of sections 1.2.1 and 1.2.4. However, here we introduce Chebyshev polynomials as the trial functions.

It is well known that Chebyshev polynomials are more robust for curve fitting than conventional polynomials. Interpolating polynomials of high order are susceptible to gross errors near the edges of the domain. If the function $f(x) = 1/(1 + 25x^2)$ is fitted over the interval $(-1, 1)$ it is found (Forsythe et al., 1977) that a high-order interpolating polynomial will deviate markedly from the underlying function between the interpolating points adjacent to the boundaries $x = \pm 1$.

Chebyshev polynomials are orthogonal over the interval $(-1, 1)$ with respect to the weight $(1 - x^2)^{-1/2}$. Thus more emphasis is placed on the points *close to the boundaries* of the interval. The three lowest-order Chebyshev polynomials are

$$T_0(x) = 1, \qquad T_1(x) = x, \qquad T_2(x) = 2x^2 - 1, \qquad (1.2.62)$$

and a recurrence relationship (Dahlquist et al., 1974) is available for the higher-order polynomials:

$$T_{n+1}(x) = 2xT_n(x) - T_{n-1}(x). \qquad (1.2.63)$$

The robustness of the Chebyshev polynomials is equally relevant to the Galerkin method, since there is a strong connection between curve fitting and solving differential equations via methods like the Galerkin method.

In the present application it will be seen that the initial values of the unknown coefficients, a_j, in eq. (1.2.64) are chosen by *curve-fitting* the initial data (eq. (1.2.71)). Subsequently the a_j's change according to eq. (1.2.67) to keep the approximate solution a reasonable fit to the exact solution.

An approximate solution of the following form is assumed for the present problem:

$$u_a(x, t) = \sum_{j=0}^N a_j(t)\, T_j(x). \qquad (1.2.64)$$

Substituting eq. (1.2.64) into eq. (1.2.59) produces a residual

$$R = \sum_j \dot{a}_j T_j + \sum_j a_j \sum_i a_i T_j \frac{dT_i}{dx} - \frac{1}{\text{Re}} \sum_j a_j \frac{d^2 T_j}{dx^2}. \qquad (1.2.65)$$

A system of ordinary differential equations for the unknown coefficients a_j is obtained from

$$(R, T_k) = 0, \qquad k = 0, \ldots, N - 2. \qquad (1.2.66)$$

Two coefficients in eq. (1.2.64) are obtained in terms of the other algebraic coefficients to satisfy the boundary conditions. The result of evaluating eq. (1.2.66) gives

$$\mathbf{M\dot{A}} + (\mathbf{B} + \mathbf{C})\mathbf{A} = 0, \qquad (1.2.67)$$

where an element of $\dot{\mathbf{A}}$ is $\dot{a}_j$, and elements of $\mathbf{M}$, $\mathbf{B}$, and $\mathbf{C}$ are given by

$$m_{kj} = (T_j, T_k), \qquad (1.2.68)$$

$$b_{kj} = \sum_i a_i \left(T_j \frac{dT_i}{dx}, T_k \right), \qquad (1.2.69)$$

$$c_{kj} = \frac{1}{\mathrm{Re}} \left(\frac{dT_j}{dx}, \frac{dT_k}{dx} \right). \qquad (1.2.70)$$

Table 1.8. Solution of Burgers' equation by traditional Galerkin method: $T = 0.92$, $\mathrm{Re} = 10$

	Approximate solution			Exact
x	$N = 7$	$N = 9$	$N = 11$	solution
−1.0	1.0000	1.0000	1.0000	1.0000
−0.9	0.9978	1.0054	1.0003	1.0000
−0.8	0.9863	1.0082	0.9987	0.9999
−0.7	0.9792	1.0029	0.9987	0.9998
−0.6	0.9818	0.9942	1.0002	0.9996
−0.5	0.9932	0.9881	1.0007	0.9990
−0.4	1.0088	0.9879	0.9985	0.9977
−0.3	1.0216	0.9924	0.9969	0.9948
−0.2	1.0241	0.9960	0.9869	0.9888
−0.1	1.0096	0.9903	0.9759	0.9768
0	0.9730	0.9661	0.9551	0.9539
0.1	0.9115	0.9159	0.9151	0.9131
0.2	0.8251	0.8359	0.8461	0.8451
0.3	0.7169	0.7272	0.7426	0.7428
0.4	0.5927	0.5962	0.6077	0.6074
0.5	0.4606	0.4543	0.4554	0.4545
0.6	0.3306	0.3162	0.3080	0.3092
0.7	0.2134	0.1969	0.1885	0.1926
0.8	0.1192	0.1084	0.1088	0.1110
0.9	0.0562	0.0550	0.0618	0.0602
1.0	0.0288	0.0288	0.0288	0.0309
$\|u_a - u_e\|_{\mathrm{rms}}$	0.0189	0.0083	0.0016	

The nonlinearity in the problem manifests itself in the dependence of b_{kj} on a summation of the unknown coefficients a_i. Green's theorem has been applied to eq. (1.2.66) to prevent second derivatives appearing in the expression for c_{kj}.

The initial values of a_j are obtained by applying the Galerkin method to the initial data

$$(u_a - u_0, T_k) = 0, \qquad k = 0, \ldots, N = 2. \tag{1.2.71}$$

This gives a system of algebraic equations

$$\mathbf{MA} = \mathbf{D}, \tag{1.2.72}$$

where an element of $\mathbf{D}$ is

$$d_k = (u_0, T_k) = \int_{-1}^{0} T_k \, dx. \tag{1.2.73}$$

The system of equations (1.2.67) has been integrated using a variable-step-size fourth-order Runge–Kutta scheme. The step size is kept sufficiently small that the errors associated with the numerical integration are negligible compared with the error in the approximation (1.2.64). After a fixed time, typically $t = 0.92$, the solution (1.2.64), is compared with the exact solution (1.2.61). Results for Re $= 10$ and various orders of representation are shown in Table 1.8. It can be seen that at the time of comparison the value of $u_e(1, t) \neq 0$. Therefore, to maintain the same interval $-1 \leq x \leq 1$, the boundary conditions at $x = \pm 1$, eq. (1.2.60), have been set equal to the exact solution after every five integration steps.

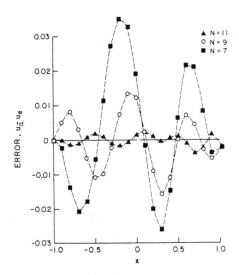

Figure 1.9. Error in the solution of Burgers' equation by the traditional Galerkin method for Re $= 10$

An examination of the rms error $\|u_a - u_e\|_{rms}$ indicates that the approximate solution converges rapidly to the exact solution as the order of the approximate representation is increased. The error $u_a - u_e$ is plotted in Fig. 1.9 for the same data presented in Table 1.8.

It can be seen from Fig. 1.9 that the error is distributed over the whole region and that the error diminishes rapidly with increasing order N. The main source of error for this problem is the difficulty in following the rapid change in the exact solution in the shock. However, the error manifests itself in the value of the a_j's, which, through eq. (1.2.64), influence the whole region.

For a value of $N = 9$ results have been obtained at various Reynolds numbers, Re = 1, 10, 100. These results are presented in Table 1.9 and Fig. 1.10. The results presented in Table 1.9 indicate that the error is large at higher Reynolds number Re. This follows directly from the steeper shock profile when Re is large (Fig. 1.8) and the difficulty that a truncated approximate solution has in resolving the profile.

This can be seen in Fig. 1.10. At Re = 1 and 10 a nine-term approximate solution is able to represent the exact solution reasonably accurately. How-

Table 1.9. Solution of Burgers' equation by traditional Galerkin method with $N = 9$

x	Re = 1.0, $t = 0.23$		Re = 10, $t = 0.92$		Re = 100, $t = 0.92$	
	Approx.	Exact	Approx.	Exact	Approx.	Exact
−1.0	0.9609	0.9599	1.0000	1.0000	1.0000	1.0000
−0.9	0.9443	0.9447	1.0054	1.0000	0.9956	1.0000
−0.8	0.9263	0.9253	1.0082	0.9999	1.0456	1.0000
−0.7	0.9029	0.9008	1.0029	0.9998	1.0672	1.0000
−0.6	0.8731	0.8709	0.9942	0.9996	1.0402	1.0000
−0.5	0.8369	0.8350	0.9881	0.9990	0.9831	1.0000
−0.4	0.7947	0.7931	0.9879	0.9977	0.9303	1.0000
−0.3	0.7469	0.7452	0.9924	0.9948	0.9128	1.0000
−0.2	0.6938	0.6919	0.9960	0.9889	0.9444	1.0000
−0.1	0.6359	0.6340	0.9903	0.9768	1.0159	1.0000
0	0.5742	0.5728	0.9661	0.9539	1.0963	1.0000
0.1	0.5099	0.5097	0.9159	0.9131	1.1411	1.0000
0.2	0.4449	0.4464	0.8359	0.8451	1.1057	1.0000
0.3	0.3814	0.3844	0.7272	0.7428	0.9613	0.9998
0.4	0.3213	0.3254	0.5962	0.6074	0.7099	0.9714
0.5	0.2665	0.2706	0.4543	0.4545	0.3933	0.1861
0.6	0.2179	0.2210	0.3162	0.3092	0.0905	0.0015
0.7	0.1756	0.1772	0.1969	0.1925	−0.1017	0.0000
0.8	0.1388	0.1395	0.1084	0.1110	−0.1154	0.0000
0.9	0.1066	0.1077	0.0550	0.0601	−0.0091	0.0000
1.0	0.0793	0.0817	0.0288	0.0309	0.0000	0.0000
$\|u_a - u_e\|_{rms}$	0.0022		0.0083		0.1049	

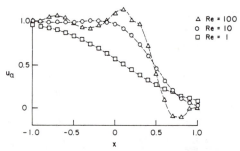

Figure 1.10. Solution of Burgers' equation by the traditional Galerkin method for various Reynolds numbers

ever, at Re = 100 the inability of the approximate solution to represent the sharp profile of the shock is apparent in the unphysical wiggles contained in the approximate solution. An obvious remedy is to increase the number of terms in the approximate solution (1.2.64). But this increases the computation time.

A computer program, BURG1, that applies the traditional Galerkin method to Burgers' equation is provided in appendix 1.

1.3. Method of Weighted Residuals

In section 1.1 the key ideas of the Galerkin method were given. In section 1.2 some simple example were considered to emphasize the practical aspects of implementing the Galerkin method and to draw attention to the influence of various features of the method. Here we intend to examine where the Galerkin method fits in the broad spectrum of computational methods. This will lead naturally to the connections between the Galerkin method and a number of other methods.

The Galerkin method is a member of the larger class of methods known as the *methods of weighted residuals* (MWR). The name appears to have been introduced by Crandall (1956). However a similar idea was considered by Collatz (1960) under the name "error distribution principles". More recently MWRs as a class have been considered by Finlayson and Scriven (1966), Vichnevetsky (1969), and Finlayson (1972). Harrington (1968) has looked at a similar concept under the title of "method of moments", but restricted himself to linear field problems arising in electromagnetic theory.

The method of weighted residuals can be described in the following way. It will be assumed that a differential equation

$$L(u) = 0 \qquad\qquad (1.3.1)$$

is to be solved subject to initial conditions $I(u) = 0$ and boundary conditions $S(u) = 0$. An approximate solution u_a is introduced so that

$$L(u_a) = R,$$
$$I(u_a) = R_I, \qquad (1.3.2)$$
$$S(u_a) = R_b.$$

The approximate solution u_a can be structured (Collatz, 1960) so that:

(i) A differential equation is satisfied exactly, that is, $R = 0$. This is called a *boundary method*.

(ii) The boundary conditions are satisfied exactly, that is, $R_b = 0$. This is called an *interior method*.

(iii) Neither differential equation nor boundary conditions is satisfied exactly. This is called a *mixed method*.

Here MWRs will be described in relation to interior methods, but the formulation extends directly to boundary and mixed methods. Probably the best example of a boundary method is the *method of singularities* (Hess and Smith, 1964), or *panel method* (Rubbert and Saaris, 1972). *Boundary-element methods* have become an important subclass of finite-element methods (Brebbia, 1978). The mixed method has been proposed by Shuleshko (1959).

For an interior method the approximate solution u_a might be written in the form

$$u_a(\mathbf{x}, t) = u_0(\mathbf{x}, t) + \sum_{j=1}^{N} a_j(t)\phi_j(\mathbf{x}), \qquad (1.3.3)$$

where the ϕ_j's are known analytic functions. These are often called trial functions, and eq. (1.3.3) the trial solution. The coefficients a_j are to be determined. In eq. (1.3.3) $u_0(\mathbf{x}, t)$ is chosen to satisfy the initial and boundary conditions, *exactly if possible*.

As written, eq. (1.3.3) implies that eq. (1.3.1) is to be reduced to an ordinary differential equation in t. If $\phi_j = \phi_j(t)$ and $a_j = a_j(\mathbf{x})$, a partial differential equation in $\mathbf{x}$ is obtained. This approach has been used by Wadia and Payne (1979), but it is unusual. If $\phi_j = \phi_j(\mathbf{x}, t)$ or the problem is steady, then the a_j's are constants and eq. (1.3.1) is reduced to a system of algebraic equations.

To obtain equations for the a_j's the inner product of the weighted residual is set equal to zero:

$$(R, w_k(\mathbf{x})) = 0, \qquad k = 1, \ldots, N, \qquad (1.3.4)$$

and it is from this relationship that the method takes its name. w_k is referred to as the weight function or test function.

Since independent relationships are needed to solve for the unknown coefficients a_j, it is clear that the $w_k(\mathbf{x})$, where they have an obvious analytic form, must be independent functions. If the w_k are members of a complete set of functions, then as $N \to \infty$, eq. (1.3.4) indicates that the equation residual R must be orthogonal to every member of a complete set of functions. However, this implies that R converges to zero *in the mean* (in the limit $N \to \infty$). If R converges to zero in the mean and eq. (1.3.3) satisfies the boundary conditions exactly, then we would expect the approximate solution u_a to converge to the exact solution of eq. (1.3.1) in the mean, that is,

$$\lim_{N \to \infty} \|u_a - u_e\|_2 = 0.$$

This may be contrasted with *uniform convergence*, which we define by

$$\lim_{N \to \infty} \|u_a - u_e\|_\infty = 0,$$

where

$$\|u_a - u_e\|_\infty \equiv \max |u_a - u_e|.$$

The form of eq. (1.3.4) is analogous to the *weak form* of eq. (1.3.1), namely

$$(L(u), w) = 0, \tag{1.3.5}$$

where w is a general test function.

The inner product used in eq. (1.3.4) was defined by eq. (1.1.5) to be continuous over the domain of interest. However, the inner product could equally well have been defined in a discrete sense, that is,

$$(f, g) = \sum_{i=1}^{N} f_i g_i.$$

This would lead to the *discrete method of weighted residuals*. This has been considered by Neuman (1974) as an appropriate technique for handling discrete boundary-value problems. In practice the evaluation of eq. (1.3.4) by numerical quadrature is strictly a discrete method of weighted residuals.

The importance of the general nature of eq. (1.3.4) is that many apparently unconnected methods can be made to conform to it. The different methods correspond to different choices of the weight function. The more common of these are described in the following subsections.

1.3.1. Subdomain method

The domain is split up into n subdomains D_j, which may overlap. Then

$$w_k = \begin{cases} 1 & \text{in } D_k, \\ 0 & \text{outside } D_k. \end{cases} \tag{1.3.6}$$

This formulation has the attractive feature that it reverses the normal way that partial differential equations are derived from conservation principles. In fact, the method is the same as the *finite-volume* method, which is a popular method in fluid mechanics and heat transfer. For example, in considering the conservation of mass associated with some finite volume of compressible fluid, the net flux of mass through the surface is equal to the time rate of decrease of mass in the volume. In a finite-volume formulation this statement would be written

$$\int_s \rho u_n \, ds = -\frac{\partial}{\partial t} \iint_v \rho \, dv. \tag{1.3.7}$$

But application of Green's theorem gives

$$\iint_v \left\{ \frac{\partial \rho}{\partial t} + \operatorname{div}(\rho \mathbf{u}) \right\} 1 \, dv = 0,$$

which is, of course, just a subdomain method applied over a finite-volume v.

The subdomain method originated with the work of Biezeno and Koch (1923). In a finite-difference context, finite-volume methods have been presented recently by Jameson and Caughey (1977), Rizzi and Inouye (1973), and Demirdzic et al. (1981). However, in the finite-difference context the equivalent of the trial solution (1.3.3) is implicit.

Examples of the subdomain method that fit into the definition (1.3.6) can be found in the methods of Pallone (1961) and Bartlett and Kendall (1968) to solve laminar boundary layers, and Murphy (1977) to solve the incompressible Navier–Stokes equations. The *method of integral relations* (MIR) has been used extensively in connection with the blunt-body problem in supersonic flow. As used for this problem, MIR is a subdomain method. MIR for the supersonic blunt-body problem is described by Belotserkovskii and Chushkin (1965). MIR in a more general context is described by Holt (1977).

1.3.2. Collocation method

If the weight function is given by

$$w_k(\mathbf{x}) = \delta(\mathbf{x} - \mathbf{x}_k), \tag{1.3.8}$$

where δ is the Dirac delta function, the evaluation of eq. (1.3.4) reduces to setting $R(\mathbf{x}_k) = 0$. This property is shared by most finite-difference methods. The first example (and the name) appears to have been provided by Frazer et al. (1937). Bickley (1941) applied the collocation method (along with the least-squares and Galerkin methods) to eq. (1.2.44), but expressed as an equivalent electric-circuit problem.

Jain (1962) introduced an *extremal-point collocation* by sampling the residual at the zeros of Chebyshev polynomials and requiring that

$$R(x_i^k) - (-1)^i R(x_0^k) = 0.$$

This is essentially exploiting the Chebyshev property of minimizing the maximum error (Hamming, 1973, p. 477). Jain combined the procedure with Newton's method to obtain solutions for viscous flow near a stagnation point and viscous flow near a rotating disc.

Villadsen and Stewart (1967) introduced a method called *orthogonal colloca-tion* by setting the equation residual to zero at the zeros of Jacobi polynomials and using Jacobi polynomials as trial functions. An additional attractive feature of the method is that solutions are sought in terms of the dependent variables u at the collocation points. Many examples of orthogonal colloca-tion, drawn from the area of chemical engineering, are described in Finlayson (1972).

Schetz (1963) has applied a low-order collocation method to a number of boundary-layer flows. The assumed form of the trial solution was chosen to conform to the exact solution of a related problem. In this way results of acceptable accuracy were achieved with just one unknown coefficient a_1.

More recently Panton and Sallee (1975) have applied a collocation method to problems of unsteady heat conduction (like the example in section 1.2.4) and to boundary-layer flows. They used *B-splines* as trial functions and compared results with a finite-difference method and with a Galerkin method also based on *B*-splines. Results were more accurate than those achieved from the finite-difference method, but also required more computer time. A comparison of the collocation and Galerkin methods applied to the problem is given in section 1.3.7.

Viviand and Ghazzi (1974) apply a collocation method to the problem of computing the inviscid flow about an inclined three-dimensional wing.

1.3.3. Least-squares method

In this method the weight function can be expressed as

$$w_k = \frac{\partial R}{\partial a_k},$$
(1.3.9)

where the a_k's are the unknown coefficients in eq. (1.3.3). Thus eq. (1.3.9) is equivalent to choosing the a_k's so that

$$(R, R) \text{ is a minimum.}$$

The method has a natural appeal for steady problems, where it might be expected that a minimization of the square of the equation residual would imply a small value of $\|u_a - u_e\|_2$. For transient problems, such as the general situation expressed by eq. (1.3.1) to (1.3.3), the implementation of the least-squares method is not strictly valid. To make the method correct it is necessary to include trial functions for the time behavior as well as the spatial behavior (Citron, 1960). Thus the a_j's in eq. (1.3.3) are then constants and the inner product (1.3.4) includes an integration *over the time domain* of the solution.

Bickley (1941) used this approach in solving eq. (1.2.44). However, Bickley found that by taking the integration over an infinite time domain, insufficient weight was ascribed to early time and the early solution was poor. In principle a weighted least-squares method could be considered with the additional weight chosen to favor the early-time behavior. The precise criterion for doing this is not obvious.

The least-squares method is by far the oldest of the methods of weighted residuals. Crandall (1956) states that the method was discovered by Gauss in 1975. The least-squares method has been applied to the example in section 1.2.1 by Frazer et al. (1937), and to the one in section 1.2.4 with different boundary conditions by Bickley (1941). Narasimha and Deshpande (1969) used a least-squares formulation to examine shock structure via the Boltzmann equation. More recently the least-squares formulation has been used with the finite-element method. We discuss the Galerkin finite-element method in chapter 3. In particular, Lynn (1974) applied a least-squares finite-element formulation to boundary-layer flow. Fletcher (1979) and Chattot et al. (1981) have obtained solutions to problems of compressible, inviscid flow using a least-squares finite-element formulation. Milthorpe and Steven (1978) have considered viscous, incompressible flow.

1.3.4. Method of moments

In this case the weight function becomes

$$w_k(x) = x^k, \qquad k = 0, \ldots, N. \tag{1.3.10}$$

The first application of the method of moments appears to be due to Yamada (1947), who applied it to a problem of nonlinear diffusion. The governing equation is the nonlinear counterpart of the equation considered in section 1.2.4. Yamada's formulation is described in Ames (1965). Yamada (1948) applied the same general method to the solution of the laminar boundary-layer equations. The method has proved to be very effective when used to represent the velocity variation across the boundary layer analytically. Subsequently Truckenbrodt (1952) developed a moment-of-momentum method that was applied to both laminar and turbulent boundary-layer flow. The application of the method of integral relations to boundary-layer flow by Dorodnitsyn (1960) in (x, u) space is closely related to the method of moments. In this method the weight function is given by

$$w_k(u) = (1 - u)^k, \qquad k = 1, \ldots, N. \tag{1.3.11}$$

It is interesting that a development of the Dorodnitsyn scheme, called *orthonormal method of integral relations* by Fletcher and Holt (1975), can be interpreted as a *spectral* scheme (Fletcher, 1978). This interpretation is set out in section 5.5. Many of the subsequent applications and developments of the Dorodnitsyn method of integral relations are described in Holt (1977).

1.3.5. Galerkin method

In this case the weight function is chosen from the same family of functions as the trial functions, that is,

$$w_k(\mathbf{x}) = \phi_k(\mathbf{x}), \qquad k = 1, \ldots, N. \tag{1.3.12}$$

The historical development of this method has already been described.

It may be noted that Mikhlin (1964) also associates the name of Bubnov with the Galerkin method. However, as noted by Mikhlin, Bubnov put forward the formulation in connection with a variational formulation for solving eigenvalue problems. In addition, Bubnov required that the trial and test functions be orthogonal. Since this is not a requirement of the Galerkin method and since the Galerkin method does not require a corresponding variational formulation, we prefer to consider the Galerkin method to be a separate method.

The test and trial functions, $\phi_j(\bar{x})$, should be chosen from the first N functions of a complete set of functions (Kantorovich and Krylov, 1958, p. 262). This is a necessary condition for convergence to the exact solution as $N \to \infty$.

Many applications of the Galerkin method are given in Kantorovich and Krylov (1958), Collatz (1960), Finlayson and Scriven (1966), Finlayson (1972), and Ames (1972). Some more recent applications are given in section 1.6.

It is useful here to reiterate the conditions required in applying the traditional Galerkin method:

(i) The test functions w_k are chosen from the *same family* as the trial functions, ϕ_j.

(ii) The trial and test functions must be *linearly independent*.

(iii) The trial and test functions should be the *first N members of a complete set of functions*.

(iv) The trial functions should *satisfy the boundary conditions* (and initial conditions if applicable) *exactly*.

Condition (i) defines the Galerkin method, and condition (ii) is necessary to ensure that independent equations are available to obtain the unknown coefficients a_j. Conditions (iii) and (iv) relate to the efficiency of the method. Violation of these conditions reduces the efficiency.

As the traditional Galerkin method was originally devised and used, it was essentially a hand method or at best a desk-calculator method. In this framework efficiency would be measured in terms of solution accuracy per unit of manual effort. This clearly emphasizes using as few unknown coefficients a_j as possible while still achieving acceptable accuracy. An indication of the loss of accuracy due to not satisfying condition (iii) is given in the results shown in Table 1.2 in relation to the problem of section 1.2.1. Some indication of the effect of not satisfying the boundary conditions exactly can be obtained from the imposition of extra boundary conditions (see Table 1.3).

Assuming that conditions (i) to (iv) are met, one may ask what else can be done to improve efficiency. First we might expect *the choice of trial functions* to influence the accuracy. In section 1.2.2 it was seen that the use of purely local, linear polynomials, as trial and test functions, produced less accurate eigenvalues than using a global, higher-order polynomial. For the viscous-flow problem considered in section 1.2.3 it was seen that using suitable polynomials as the trial functions led to a more accurate solution than a Fourier series for the same N.

However the use of trial functions orthogonal over the domain greatly reduced the effort of obtaining a solution at larger N. For this example the reduction in effort more than compensated for the reduction in accuracy. The use of orthogonal test and trial functions is an important feature of the spectral method (chapter 5). In general, one would expect that choosing the trial and test functions based on knowledge of the form of the exact solution of a closely related problem would enhance the efficiency of the method.

The problem in section 1.2.4 demonstrated that there is no point reducing the error in the Galerkin solution beyond the level at which other errors in the overall algorithm dominate the solution accuracy. From a consideration of the problem in section 1.2.5 it is clear that the inclusion of nonlinear terms puts an ever greater premium on obtaining accurate solutions with N as small as possible. Specific techniques to reduce the computational effort of manipulating nonlinear terms are considered in section 5.3.2.

A final question on efficiency would be whether some other method might be preferred to the traditional Galerkin method. This question will be considered in section 1.3.7.

For the modern Galerkin methods—that is the spectral and the Galerkin finite-element method—efficiency can be defined in terms of solution accuracy per unit of computer execution time. The required solution accuracies are perhaps around 0.1 to 1% for the modern Galerkin methods and around 1 to 10% for the traditional Galerkin method. Typically in modern applications the number of unknowns would be anything up to 10,000 to 20,000. At this level particular choices for the trial functions are less important, at least for the Galerkin finite-element method. For the spectral method, the choice of the trial functions affects the convergence rate significantly (section 5.1).

1.3.6. Generalized Galerkin method

In this category the weight function in eq. (1.3.4) is represented by

$$w_k(\mathbf{x}) = P_k(\mathbf{x}), \tag{1.3.13}$$

where $P_k(\mathbf{x})$ is an analytic function, similar to the test function ϕ_k used by the Galerkin method, but with additional terms or factors to impose some additional requirement on the solution. The motivation for this category is that in applying a conventional Galerkin finite-element method with linear elements to convection-domained flows, the resulting algebraic equations have undesirable stability properties. These aspects are described further in chapter 7. Sometimes the generalized Galerkin method is referred to as the Petrov–Galerkin method (e.g. Anderssen and Mitchell, 1979).

The term "generalized Galerkin method" has also been used in a sense equivalent to the method of weighted residuals as a whole by Murphy (1973). What is more, Harrington (1968) uses the term "method of moments" in a sense equivalent to the method of weighted residuals as a whole.

1.3.7. Comparison of weighted-residual methods

Here we consider some simple examples to which different methods of weighted residuals have been applied to see if particular methods are consistently more accurate or converge faster or are easier to apply.

We begin by considering the problem in section 1.2.1 and apply to it the various methods of weighted residuals considered above. Because of the choice of the trial functions in this example, the method of moments is identical with the Galerkin method.

Application of the least-squares method (1.3.4), (1.3.9), with a three-term trial solution like eq. (1.2.2), to the equation

$$\frac{dy}{dx} - y = 0 \qquad (1.3.14)$$

and boundary condition $y = 1$ at $x = 0$, produces the following system of equations:

$$\begin{bmatrix} \frac{1}{3} & \frac{1}{4} & \frac{1}{5} \\ \frac{1}{4} & \frac{8}{15} & \frac{2}{3} \\ \frac{1}{5} & \frac{2}{3} & \frac{33}{35} \end{bmatrix} \begin{bmatrix} a_1 \\ a_2 \\ a_3 \end{bmatrix} = \begin{bmatrix} \frac{1}{2} \\ \frac{2}{3} \\ \frac{3}{4} \end{bmatrix}. \qquad (1.3.15)$$

Application of the Galerkin method, which was used in section 1.2.1, gives the following system of equations:

$$\begin{bmatrix} \frac{1}{2} & \frac{2}{3} & \frac{3}{4} \\ \frac{1}{6} & \frac{5}{12} & \frac{11}{20} \\ \frac{1}{12} & \frac{3}{10} & \frac{13}{30} \end{bmatrix} \begin{bmatrix} a_1 \\ a_2 \\ a_3 \end{bmatrix} = \begin{bmatrix} 1 \\ \frac{1}{2} \\ \frac{1}{3} \end{bmatrix}. \qquad (1.3.16)$$

A subdomain method with intervals 0 to $\frac{1}{3}$, $\frac{1}{3}$ to $\frac{2}{3}$, and $\frac{2}{3}$ to 1 produces the following system of equations:

$$\begin{bmatrix} \frac{5}{18} & \frac{8}{81} & \frac{11}{324} \\ \frac{3}{18} & \frac{20}{81} & \frac{69}{324} \\ \frac{1}{18} & \frac{26}{81} & \frac{163}{324} \end{bmatrix} \begin{bmatrix} a_1 \\ a_2 \\ a_3 \end{bmatrix} = \begin{bmatrix} \frac{1}{3} \\ \frac{1}{3} \\ \frac{1}{3} \end{bmatrix}. \qquad (1.3.17)$$

A collocation method with the residual evaluated at the points 0, 0.5, and 1.0 gives the following system of equations:

$$\begin{bmatrix} 1 & 0 & 0 \\ 0.5 & 0.75 & 0.625 \\ 0 & 1 & 2 \end{bmatrix} \begin{bmatrix} a_1 \\ a_2 \\ a_3 \end{bmatrix} = \begin{bmatrix} 1 \\ 1 \\ 1 \end{bmatrix}. \qquad (1.3.18)$$

The values of a_1, a_2, a_3 for the various methods are shown in Table 1.10. The coefficients there for the optimal $L_{2,d}$ case are obtained by choosing a_1, a_2, a_3 so that the error in the discrete L_2 norm between the approximate solution and the exact solution, $y = e^x$, is a minimum. The corresponding solutions over the interval $0 \leq x \leq 1$ are shown in Table 1.11.

The optimal $L_{2,d}$ solution is the best possible solution (in the discrete L_2 norm) given an approximate solution (1.2.2) with only three disposable constants. Thus it is more appropriate to compare the various methods of weighted residuals with this solution rather than the exact solution. It is clear, from the errors $\| y_a - y \|_{2,d}$, that the least-squares, Galerkin, and sub-domain methods are all close to optimal for this example. The collocation method is producing solutions of lower accuracy. However, the accuracy of the collocation method depends on the sample points. Choosing Gauss points $x = 0.1127, 0.5$, and 0.8873 for the sample points produces an equation

Table 1.10. Comparison of coefficients for approximate solution of $dy/dx - y = 0$

Scheme	a_1	a_2	a_3
Least squares	1.0131	0.4255	0.2797
Galerkin	1.0141	0.4225	0.2817
Subdomain	1.0156	0.4219	0.2813
Collocation	1.0000	0.4286	0.2857
Taylor series	1.0000	0.5000	0.1667
Optimal $L_{2,d}$	1.0138	0.4264	0.2781

Table 1.11. Comparison of approximate solutions of $dy/dx - y = 0$

x	Least squares	Galerkin	Sub-domain	Collo-cation	Taylor series	Optimal $L_{2,d}$	Exact
0	1.0000	1.0000	1.0000	1.0000	1.0000	1.0000	1.0000
0.2	1.2219	1.2220	1.2223	1.2194	1.2213	1.2220	1.2214
0.4	1.4912	1.4913	1.4917	1.4869	1.4907	1.4915	1.4918
0.6	1.8214	1.8214	1.8220	1.8160	1.8160	1.8219	1.8221
0.8	2.2260	2.2259	2.2265	2.2206	2.2053	2.2263	2.2255
1.0	2.7183	2.7183	2.7187	2.7143	2.6667	2.7183	2.7183
$\| y_a - y \|_{2,d}$	0.00105	0.00103	0.00127	0.0094	0.0512	0.00101	

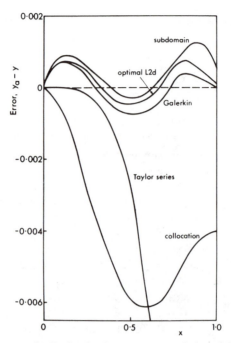

Figure 1.11. Error distribution for three-parameter solution of $dy/dx - y = 0$

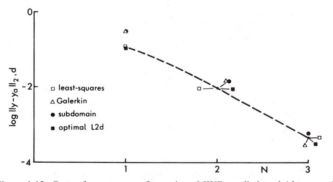

Figure 1.12. Rate of convergence for various MWR applied to $dy/dx - y = 0$

and solution identical with the Galerkin method. The Taylor-series solution, which only seeks to match the exact solution at $x = 0$, is included for comparison.

It can be confirmed from Fig. 1.11 that the Galerkin solution is close to optimal in magnitude and distribution. In contrast, the errors of the Taylor series and the collocation solutions are always negative.

An indication of the convergence with N of the least-squares, Galerkin, and subdomain methods is given in Fig. 1.12. For $N = 1$ and 2 the Galerkin and subdomain methods coincide. For $N = 1$ the Galerkin and subdomain methods are suboptimal. Above $N = 1$ all methods are close to optimal.

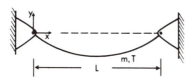

Figure 1.13. Vibrating-string problem

In evaluating the coefficients using eq. (1.3.4), the integrands for the least-squares method are of higher order than those for the Galerkin method. The integrands for the Galerkin method are of higher order than those for the subdomain method. Thus there is a small advantage in *computational effort* to the subdomain method.

The problem of a vibrating string (Fig. 1.13) has been considered by Frazer et al. (1937). A stretched string of length L has a mass per unit length of m and a tension T. The lateral displacement of the string is governed by the wave equation

$$\frac{\partial^2 y}{\partial x^2} = \frac{m}{T}\frac{\partial^2 y}{\partial t^2}.\tag{1.3.19}$$

If $a^2 = m/T$ and it is assumed that the wave motion can be represented by

$$y = P\sin\frac{wt}{a}\tag{1.3.20}$$

with $P = 0$ at $\xi = 0$ and 1, then eq. (1.3.19) can be replaced with

$$\frac{d^2 P}{d\xi^2} + \lambda P = 0,\tag{1.3.21}$$

where $\xi = x/L$ and $\lambda = w^2$. Eq. (1.3.21) is the same as eq. (1.2.11). Solutions obtained via the collocation, Galerkin, and least-squares methods are presented in Table 1.12. These solutions have assumed a trial solution of the form

$$P_a(\xi) = \xi(\xi - 1)\sum_{j=1}^{N} a_j\xi^{j-1}.\tag{1.3.22}$$

The collocation results have been obtained by setting the residual equal to zero at equal intervals in ξ. It can be seen that the least-squares formulation generates a complex eigenvalue. This arises because in the least-squares

Table 1.12. MWR comparison for eigenvalues λ of eq. (1.3.21)

Method	Collocation			Least-squares		Galerkin		Exact
N	3	4	5	2	3	2	3	
λ_1	9.6	9.82	9.872	$10 \pm 4.47i$	$9.87 \pm 0.23i$	10	9.870	9.8696
λ_2		36	38.4	$42 \pm 27.5i$	$42 \pm 27.5i$	42	42	39.478
λ_3			75.3		$102 \pm 8.3i$		102	88.826

formulation λ^2 appears. However, the real part of the least-squares solution coincides with the Galerkin solution for this problem. For self-adjoint eigenvalue problems like eq. (1.3.21), only the Galerkin method *guarantees* real eigenvalues. Because of its link with the variational methods (see section 1.4), the Galerkin method also provides an upper bound on the eigenvalue. The least-squares method is generally not suitable for eigenvalue problems, since the effort required to obtain the solution is substantially greater than for the other methods of weighted residuals (Crandall, 1956). It can be seen from Table 1.12 that the collocation method is rather less accurate (for the same N) than the Galerkin method.

Finlayson (1969) has compared the Galerkin, collocation, and subdomain methods for the eigenvalue problem

$$\frac{d^2y}{dx^2} + \lambda(1 - x^2)y = 0$$

with boundary conditions $y(0) = dy/dx(1) = 0$. Using a trial solution $y_a = a_1 \sin(\pi x/2)$, Finlayson found that the Galerkin method gave the most accurate and the collocation method the least accurate solution for the first eigenvalue.

Bickley (1941) solved the unsteady heat-conduction equation (1.2.42) by considering the equivalent transient electric-circuit problem. After appropriate nondimensionalization the governing equation becomes

$$\frac{\partial \theta}{\partial \tau} = \frac{\partial^2 \theta}{\partial \xi^2} \tag{1.3.23}$$

subject to the initial and boundary conditions,

$$\theta(\xi, 0) = 0, \qquad 0 \leq \xi \leq 1,$$

$$\theta(0, \tau) = 1, \qquad \tau > 0, \tag{1.3.24}$$

$$\frac{\partial \theta}{\partial \xi}(1, \tau) = 0, \qquad \tau > 0.$$

Solutions were sought using the collocation, least-squares, and Galerkin methods. An approximate solution of the following form was introduced:

$$\theta = a_0 + a_1\xi + a_2\xi^2 + \cdots + a_n\xi^n. \tag{1.3.25}$$

After applying the boundary conditions (to determine a_0 and a_1), an approximate solution was obtained in the form

$$\theta = 1 - a_2(2\xi - \xi^2) - a_3(3\xi - \xi^3) - \cdots - a_n(n\xi - \xi^n). \tag{1.3.26}$$

To impose the least-squares criterion the following integral was minimized:

$$\int_0^\infty \int_0^1 R \, d\xi \, d\tau,$$

Table 1.13. Error in solution for θ at $\xi = 1$ for $\partial\theta/\partial\tau = \partial^2\theta/\partial\xi^2$

τ	1-parameter Galerkin	1-parameter least squares	2-parameter Galerkin	2-parameter collocation	3-parameter collocation
0	0	0	0.0385	0	0
0.2	0.1535	0.1940	0.0089	0.0061	0.0041
0.4	0.0916	0.1402	0.0018	0.0025	0.0063
0.6	0.0528	0.0964	−0.0013	0.0009	0.0047
0.8	0.0303	0.0651	−0.0025	0.0002	0.0034
1.0	0.0173	0.0433	−0.0021	−0.0001	0.0023
1.5	0.0042	0.0151	−0.0012	−0.0002	0.0009
2.0	0.0009	0.0049	−0.0006	−0.0002	0.0003
2.5	0.0002	0.0015	−0.0002	0.0002	0.0001
3.0	0	0.0005	0	0	0.0001

where R is the equation residual. The error in the solution for a one-parameter least-squares method is shown in Table 1.13. The approximate solution was compared with the exact solution for increasing time at one location, $\xi = 1$. It is clear that the least-squares solution is poor at small time and progressively improves. Bickley attributes this to the large weight associated with large time inherent in the integration over time having an upper limit of infinity. Bickley concludes that the least-squares formulation is not well suited to evolutionary problems.

Solutions for a two-parameter collocation formulation are shown in Table 1.13. For the collocation points Bickley takes the points $\xi = 0.5$ and 1.0. The result for θ at $\xi = 1$ is more accurate than for a two-parameter Galerkin formulation. However, the strength of the method of weighted residuals is its ability to control the error *globally*. Since one of the collocation points was imposed at $\xi = 1$, this is probably contributing to the higher accuracy at this point. The three-parameter collocation solution used $\xi = 0$, 0.5, and 1.0 as collocation points, and it can be seen that this has produced an inferior solution. The main conclusion from Bickley's data is that the least-squares criterion is unsuitable for evolutionary problems and that the collocation method can be effective if solutions at one point in the domain only are required.

Thorsen and Landis (1965) have considered unsteady heat-conduction problems (governed by eq. (1.2.42)) and sought solutions using a Galerkin (unnamed) and a subdomain method. The results indicated that the Galerkin method was slightly more accurate.

Panton and Sallee (1975) have considered unsteady heat conduction (eq. (1.3.23)) with initial and boundary conditions

$$\theta(\xi, 0) = 0, \qquad 0 \le \xi \le 1,$$

$$\frac{d\theta}{d\xi}(0, \tau) = -1, \qquad \tau > 0,$$

$$\frac{d\theta}{d\xi}(1, \tau) = 0, \qquad \tau > 0. \tag{1.3.27}$$

Table 1.14. Comparison of Galerkin, collocation, and finite-difference methods for unsteady heat conduction

	Galerkin		Collocation		Finite Difference
	Piecewise cubic	B-spline	Piecewise cubic	B-spline	
Number of unknowns	14	7	14	7	11
Average error (%):					
at $\tau = 0.5$	0.008	0.065	2.20	1.14	12.2
at $\tau = 1.0$	0.00006	0.030	1.03	0.53	11.4
Comp. time (sec)					
per Δt	0.69	0.14	0.79	0.20	0.03

They used a piecewise cubic and B-splines (Ahlberg et al., 1967) as trial functions and compared solutions from the collocation and Galerkin method with solutions from a finite-difference formulation. The results are summarized in Table 1.14.

It can be seen that the Galerkin method is more accurate than collocation for the same number of unknowns, which is in agreement with the earlier comparisons. Also it appears that the Galerkin method is more economical than the collocation method, which is a rather surprising result. Both weighted residual methods are considerably more accurate than the finite-difference method, but less economical.

The Galerkin and subdomain methods have been compared by Fuller et al. (1970) for problems in space-dependent nuclear-reactor dynamics, and by Pomraning (1966) for a diffusion problem and an eigenvalue problem. The Galerkin method gave superior results to the subdomain method, particularly when eigenvalue problems were considered.

This comparative section is concluded with the observation that *the Galerkin method produces results of consistently high accuracy and has a breadth of application as wide as any method of weighted residuals*. The least-squares formulation is of comparable accuracy to the Galerkin method when applied to equilibrium (elliptic) problems. However, it is more complex to apply and is unsuited to evolutionary and eigenvalue problems. The subdomain method is almost as accurate as the Galerkin method, is typically simpler to apply, and has an important physical interpretation via the conservation principles. The collocation method is generally less accurate than the Galerkin method, but simpler to set up. However the orthogonal collocation method of Villadsen and Stewart (1967) can give accuracies as high as the Galerkin method in

Table 1.15. Subjective comparison of different methods of weighted residuals

MWR	Galerkin	Least-squares	Subdomain	Collocation
Accuracy	Very high	Very high	High	Moderate
Ease of formulation	Moderate	Poor	Good	Very good
Additional remarks	Equivalent to Ritz method where applicable	Not suited to eigenvalue or evolutionary problems	Equivalent to finite-volume method; suited to conservation formulation	Orthogonal collocation gives high accuracy

particular situations. This perhaps subjective comparison is summarized in Table 1.15.

In choosing to use a method of weighted residuals in circumstances where only a few unknowns will be permitted (to minimize the manual effort), two aspects are important. First, what is the best choice of MWR for the particular problem? The comparisons above and Table 1.14 should be of assistance.

Secondly, the accuracy is often dependent on the particular choice for the trial functions. Following Crandall (1956) and Finlayson and Scriven (1966), the following advice can be given. The trial functions should satisfy the boundary and initial conditions exactly and be the lowest-order members of a complete set of functions. Any natural symmetry in the problem should be exploited. The exact solutions of closely related problems may suggest an appropriate form for the trial solution.

Orthonormal functions are recommended particularly for the Galerkin formulation. This will be discussed further in chapter 2. Polynomials have been used far more than any other class of trial functions. On the assumption that, even for small N, the problem will probably end up on a computer, Chebyshev polynomials are recommended if there is no good reason for choosing something else.

Clearly, choosing the best trial functions is something of an art. Fortunately the ready availability of cheap computing has made this particular art less important than previously, as long as N is not too large.

1.4. Connection with Other Methods

In obtaining solutions to the ordinary differential equation in section 1.2.1, it was found that for $N = 1, 2$ the Galerkin and the subdomain method produced identical equations and hence solutions. For the eigenvalue problem considered in section 1.2.2 the real part of the eigenvalue obtained from the least-squares solution coincided with the Galerkin solution for the eigenvalue. In this section we wish to emphasize situations in which the Galerkin method is related to methods outside the class of methods of weighted residuals.

First, for problems that can be solved by the separation-of-variables method, a connection is possible. This arises if the trial functions used in the Galerkin formulation are *eigenfunctions* of the problem—for example, as obtained from a separation-of-variables solution. Since such a solution is obtained with N unknowns in practice, application of an N-term Galerkin method will produce a solution that coincides with the separation-of-variables solution. For viscous flow in a channel (section 1.2.3) the same solution would have been obtained if the problem had been solved by the separation-of-variables method (Kantorovich and Krylov, 1958).

The orthogonal collocation method of Villadsen and Stewart (1967) was mentioned in section 1.3.2. The trial functions are chosen to satisfy the orthogonality relationship

$$\int_a^b w(x)P_i(x)P_j(x)\,dx = \begin{cases} 1 & \text{if } i = j, \\ 0 & \text{if } i \neq j, \end{cases} \tag{1.4.1}$$

and the zeros of $P_N(x)$ are used as the collocation points. If a Galerkin method, with test and trial functions $w(x)P_j(x)$, were used to solve a linear differential equation $L(u) = 0$, the evaluation of the inner product (1.3.4) would give

$$\int w(x)P_k(x)L(u)\,dx = 0. \tag{1.4.2}$$

A quadrature scheme could be devised in which the zeros of $P_N(x)$ were the sample points and which evaluated eq. (1.4.2) exactly. Thus eq. (1.4.2) is replaced by

$$\sum_i H_i w(x_i)P_k(x_i)L(u(x_i)) = 0, \tag{1.4.3}$$

where H_i are the weights in the quadrature scheme. It may be noted that the solution of $L(u) = 0$ by the orthogonal collocation scheme is just

$$L(u(x_i)) = 0. \tag{1.4.4}$$

Clearly the solution obtained using the orthogonal collocation method coincides with the solution using the Galerkin method.

The most important connection is between the Galerkin method and the Rayleigh-Ritz method. As stated by Finlayson (1972), "There is always a Galerkin method that corresponds to the variational method . . . . The variational method is the Rayleigh–Ritz method." The converse is not true, of course. There are many problems for which a corresponding variational principle cannot be obtained. However, the Galerkin method is still applicable. An important consequence of the equivalence between the Galerkin method and the variational method arises from the fact that approximate solution to the variational problem using the Rayleigh–Ritz method is either an upper or a lower bound on the exact solution. Clearly the corresponding Galerkin solution is also an upper or a lower bound. This is exploited in obtaining error estimates. Variational methods are discussed in Finlayson (1972), Mikhlin (1964, 1971), and Kantorovich and Krylov (1958).

The solution of the differential equation

$$A(u) = f, \tag{1.4.5}$$

where A is a positive definite operator, is equivalent (Mikhlin, 1964) to finding the minimum of

$$F(u) = (A(u), u) - 2(u, f). \tag{1.4.6}$$

In the Rayleigh–Ritz method an approximate solution of eq. (1.4.6) is sought as

$$u_a(\mathbf{x}) = \sum_{j=1}^{N} a_j \phi_j(\mathbf{x}). \tag{1.4.7}$$

The trial functions $\phi_j(\mathbf{x})$ must be the lowest-order members of a complete set and linearly independent. Substitution into eq. (1.4.6) causes the function F to depend on the N unknown coefficients $a_1, a_2, \ldots, a_N$:

$$F(u_a) = \left\{ \sum_{j=1}^{N} a_j A(\phi_j), \sum_{k=1}^{N} a_k \phi_k \right\} - 2 \left\{ \sum_{j=1}^{N} a_j \phi_j, f \right\}$$
$$= \sum_{j=1}^{N} \sum_{k=1}^{N} (A(\phi_j), \phi_k) a_j a_k - 2 \sum_{j=1}^{N} (\phi_j, f) a_j. \tag{1.4.8}$$

The condition that $F(u_a)$ is a minimum is given by

$$\frac{\partial F(u_a)}{\partial a_k} = 0, \qquad k = 1, 2, \ldots, N. \tag{1.4.9}$$

From eq. (1.4.8), which can be written

$$F(u_a) = (A(\phi_k), \phi_k) a_k^2 + 2 \sum_{j \neq i} (A(\phi_j), \phi_k) a_j a_k - 2(\phi_k, f) a_k,$$

one obtains

$$\frac{\partial F(u_a)}{\partial a_k} = 2 \sum_{j=1}^{N} (A(\phi_j), \phi_k) a_j - 2(\phi_k, f) = 0,$$

or

$$\sum_{j=1}^{N} (A(\phi_j), \phi_k) a_j = (\phi_k, f), \qquad k = 1, 2, \ldots, N, \tag{1.4.10}$$

which is a linear system of equation for the coefficients a_j. However, application of the Galerkin method to eq. (1.4.5), using eq. (1.4.7) as the trial solution, would give eq. (1.4.10) directly.

The importance of the equivalence of the Galerkin and Rayleigh–Ritz methods is that the convergence of the Rayleigh–Ritz solution to the exact solution as N in eq. (1.4.7) approaches infinity is well established (Kantorovich and Krylov, 1958) as long as the ϕ_j's in eq. (1.4.7) are members of a complete set of functions.

A specific example of the above equivalence is provided by considering the Poisson equation,

$$\frac{\partial^2 u}{\partial x^2} + \frac{\partial^2 u}{\partial y^2} - f = 0. \tag{1.4.11}$$

Since eq. (1.4.11) is linear in u, boundary conditions of the form $u = \phi(s)$ on the boundary ∂D can always be reduced to homogeneous boundary conditions by changing the dependent variable u. Therefore we consider the homogeneous boundary-condition case here.

Obtaining a solution to eq. (1.4.11) is equivalent to finding the minimum of

$$I = \iint_D \left\{ \left(\frac{\partial u}{\partial x} \right)^2 + \left(\frac{\partial u}{\partial y} \right)^2 + 2fu \right\} dx\,dy. \tag{1.4.12}$$

To obtain a solution via the Rayleigh–Ritz method the following trial solution is introduced:

$$u_a = \sum_{j=1}^{N} a_j \phi_j(x, y), \tag{1.4.13}$$

where the trial functions ϕ_j are chosen so that they satisfy the boundary condition on ∂D and in addition, for $\varepsilon > 0$, satisfy the following:

$$\left| u - \sum_{j=1}^{N} a_j \phi_j \right| < \varepsilon,$$

$$\iint_D \left(\frac{\partial u}{\partial x} - \sum_{j=1}^{N} a_j \frac{\partial \phi_j}{\partial x} \right)^2 dx\,dy < \varepsilon, \tag{1.4.14}$$

$$\iint_D \left(\frac{\partial u}{\partial y} - \sum_{j=1}^{N} a_j \frac{\partial \phi_j}{\partial y} \right)^2 dx\,dy < \varepsilon.$$

If eq. (1.4.13) is substituted into eq. (1.4.12) and $\partial I(u_a)/\partial a_k = 0$, the result is

$$2 \iint_D \left(\frac{\partial u_a}{\partial x} \frac{\partial \phi_k}{\partial x} + \frac{\partial u_a}{\partial y} \frac{\partial \phi_k}{\partial y} + f\phi_k \right) dx\,dy = 0. \tag{1.4.15}$$

Application of Green's theorem and noting the homogeneous boundary condition gives

$$\iint_D \left(\frac{\partial^2 u_a}{\partial x^2} + \frac{\partial^2 u_a}{\partial y^2} - f \right) \phi_k \, dx\,dy = 0, \tag{1.4.16}$$

which is just the Galerkin method applied to eq. (1.4.11).

An interesting connection with a Fourier representation can be obtained as follows. For the equation

$$A(u) = f, \tag{1.4.17}$$

where A is a symmetric, positive definite operator, a Galerkin solution can be sought with a trial solution

$$u_a = \sum_{j=1}^{N} \alpha_j \phi_j(x) \tag{1.4.18}$$

based on trial functions that are *energy orthonormal* (Mikhlin, 1964, p. 67), that is,

$$(A(\phi_j), \phi_k) = \begin{cases} 0 & \text{if } j \neq k, \\ 1 & \text{if } j = k. \end{cases} \tag{1.4.19}$$

Then application of the Galerkin method to eq. (1.4.17) gives

$$\sum_{j=1}^{N} (A(\phi_j), \phi_k)\alpha_j = (f, \phi_k),$$

or, using eq. (1.4.19),

$$\alpha_k = (f, \phi_k),$$

so that the Galerkin solution (1.4.18) becomes

$$u_a = \sum_{j=1}^{N} (f, \phi_j)\phi_j(x). \tag{1.4.20}$$

However if an energy-orthonormal set of functions ϕ_k is complete in energy, then the Fourier series of any function u (i.e. the solution of eq. (1.4.17)) is

$$u_f = \sum_{k=1}^{\infty} (A(u), \phi_k)\phi_k(x) \tag{1.4.21}$$

and converges in energy to u. Here *convergence in energy* means

$$(A(u - u_f), (u - u_f)) \xrightarrow[N \to \infty]{} \varepsilon,$$

where ε is an arbitrarily small positive constant. But from eq. (1.4.19) it is clear that the Galerkin solution (1.4.20) is just the Fourier solution (1.4.21) truncated to finite N. Clearly this equivalence to a truncated Fourier series implies convergence in energy for the Galerkin solution as well. However although the approach of expanding in terms of energy-orthonormal functions is conceptually useful, the labor of creating the energy-orthonormal functions prevents this from being a practical scheme.

1.5. Theoretical Properties

For problems that can be cast in the equivalent variational form, the convergence properties associated with the Rayleigh–Ritz method carry over to the Galerkin method. Suppose the following operator equation is to be solved:

$$A(u) = f, \tag{1.5.1}$$

where the operator A is a symmetric positive definite operator. It can be shown (Mikhlin, 1964, p. 74) that eq. (1.5.1) has a unique solution. Further, the problem of solving eq. (1.5.1) can be replaced by the problem of finding the function u which minimizes the functional

$$F(u) = (A(u), u) - 2(u, f). \tag{1.5.2}$$

In a manner analogous to the inner product, an *energy product* associated with the operator A can be defined as

$$[u, v] \equiv (A(u), v). \tag{1.5.3}$$

Then if u_e is the exact solution of eq. (1.5.1), eq. (1.5.2) can be written

$$F(u) = [u, u] - 2[u_e, u] \tag{1.5.4}$$

or

$$F(u) = [u - u_e, u - u_e] - [u_e, u_e] \tag{1.5.5}$$

or

$$F(u) = \|u - u_e\|_A^2 - \|u_e\|_A^2, \tag{1.5.6}$$

where $\| \ \|_A$ is the *energy norm* and is defined by

$$\|u\|_A = (A(u), u)^{1/2}. \tag{1.5.7}$$

It is clear that when $u = u_e$, $F(u)$ has a minimum value, and that the minimum value is proportional to the energy.

The energy norm $\|u\|_A$ is finite if A is positive-bounded-below and f has a finite norm (Mikhlin, 1964, p. 121). A is positive-bounded-below if

$$(Au, u) \geq \gamma^2 \|u\|_2^2, \tag{1.5.8}$$

where γ is a positive constant. For this case a sequence u_k, $k = 1, 2, \ldots,$ of functions is a minimizing sequence if

$$\lim_{k \to \infty} F(u_k) = d, \tag{1.5.9}$$

where d is $\inf F(u)$, the greatest lower bound of $F(u)$. Any sequence u_k which satisfies eq. (1.5.9) converges in energy to the solution of eq. (1.5.1). Convergence in energy implies that $u_k(x)$ converges to u_e if

$$\|u_k - u_e\|_A \to \varepsilon \quad \text{as } k \to \infty, \tag{1.5.10}$$

where ε is an arbitrarily small positive constant.

It can be shown (Mikhlin, 1964, p. 88) that the Rayleigh–Ritz method provides a sequence of functions $u_k(x)$ that converge in energy to u_e, provided that u_e is a solution with finite energy. It would appear, in establishing convergence of the Rayleigh–Ritz method, that the trial functions ϕ_j in eq. (1.4.7) should satisfy two conditions:

(i) that the sequence $\phi_1, \phi_2, \ldots, \phi_j, \ldots, \phi_N(x)$ is complete in energy;
(ii) that the ϕ_j's are linearly independent.

Condition (i) is rather restrictive, but Mikhlin (1964, p. 93) indicates that it can be partially relaxed.

It was established in section 1.4 that the Rayleigh–Ritz solution minimizing eq. (1.5.2) coincides with the Galerkin solution of eq. (1.5.1). Thus for the class of problems represented by eq. (1.5.1), the convergence properties of the Rayleigh–Ritz solution also apply to the Galerkin solution.

For the Rayleigh–Ritz method (and hence the Galerkin method) Kantorovich and Krylov (1958) provide an estimate of the maximum error for an N-term trial solution of the self-adjoint ordinary differential equation

$$\frac{d}{dx}\left(p\frac{dy}{dx}\right) - qy = f \tag{1.5.11}$$

in the region $0 \le x \le 1$ with boundary conditions $y(0) = y(1) = 0$ and the subsidiary conditions

$$p(x) > 0, \qquad q(x) \ge 0.$$

The solution to this problem is equivalent (Kantorovich and Krylov, 1958, p. 243) to minimizing

$$I(y) = \int_0^1 \left[p\left(\frac{dy}{dx}\right)^2 + qy^2 + 2fy \right]. \tag{1.5.12}$$

Kantorovich and Krylov (1958, p. 262 and following) establish that a Rayleigh–Ritz solution (and hence a Galerkin solution) of eq. (1.5.11) will converge to the exact solution if the trial solution

$$y_a = \sum_{k=1}^{N} a_k\phi_k(x) \tag{1.5.13}$$

is made up of trial functions chosen from a complete set of functions. For this problem this requires that ϕ_k be chosen so that

$$\left| y(x) - \sum_{k=1}^{N} a_k\phi_k(x) \right|_{N\to\infty} < \varepsilon,$$

$$\left| \frac{dy}{dx}(x) - \sum_{k=1}^{N} a_k\frac{d\phi_k}{dx}(x) \right|_{N\to\infty} < \varepsilon, \tag{1.5.14}$$

where ε is an arbitrarily small positive constant.

Kantorovich and Krylov (1958, p. 263) consider the use of trial functions of the form,

$$\phi_k = \sin k\pi x \quad \text{and} \quad \phi_k = (1 - x)x^k, \qquad k = 1, 2, \ldots, N. \quad (1.5.15)$$

Using sine functions as trial functions, the following error bound is obtained:

$$\|y_a(x) - y(x)\|_\infty \le \frac{L}{N + 1},$$

where

$$L = \left\{ \frac{\|p\|_\infty + \|q\|_\infty}{(N+1)^2 \pi^2} \right\}^{1/2} \left[\left\| \frac{dp}{dx} \right\|_\infty + \frac{\|q\|_\infty}{\pi} + \pi p_{\min} \right] \left\{ \frac{\left(\int_0^1 f^2 \, dx \right)^{1/2}}{2\pi^4 p_{\min}} \right\} \quad (1.5.16)$$

and $\| \ \|_\infty \equiv \max | \ |$.

A more precise error bound is obtained for the particular case $p(x) = 1$. Then

$$\|y_a(x) - y(x)\|_\infty \le \left\{ 1 + 0.25\|q\|_\infty [6\pi(N+1)]^{-1/2} \left[1 + \frac{4\sqrt{2}}{N^{3/2}\pi^2} \right] \right.$$

$$\left. \times \left[1 - \frac{\|q\|_\infty}{\pi^2(N+1)^2} \right]^{-1} \right\} A^{1/2}$$

where

$$A = \frac{2\|q\|_\infty}{\pi^3(N+1)^3} \left[\int_0^1 \frac{f^2}{q} \, dx \right]. \qquad (1.5.17)$$

Comparable results can be obtained with the polynomial trial functions given in eq. (1.5.15).

For $x = 0.5$ and $q = f = 1$, Finlayson (1972, p. 363) has computed the error bound (1.5.17) and compared it with the actual error. The error bound is shown to be rather conservative. The ratio of the actual error to the error bound varies from 0.043 for $N = 1$ to 0.0073 for $N = 9$.

Mikhlin (1964, p. 455) considers essentially the same equation as (1.5.11) but weakens the requirement on the trial functions in eq. (1.5.13). Mikhlin requires that the ϕ_k's be complete in the sense that eq. (1.5.13) should be capable of representing any continuous function in the interval $0 \le x \le 1$ so that convergence in the mean of the derivative can be achieved, that is,

$$\lim_{N \to \infty} \int_0^1 \left| \frac{dy_a}{dx} - \frac{dy}{dx} \right|^2 dx < \varepsilon, \qquad (1.5.18)$$

where ε is an arbitrarily small positive constant. Applying the Galerkin method, Mikhlin shows that y_a converges uniformly to y.

Mikhlin (1964, p. 477) considers a more general ordinary differential equation

$$(-1)^m u^{(2m)} - \lambda K(u) = f(x),\qquad (1.5.19)$$

where $K(u)$ is a linear differential operator of order $2m - 1$ and $u^{(2m)}$ is a shorthand notation for d^{2m}/du^{2m}. A solution is sought in the region $a \le x \le b$ for boundary conditions

$$u(a) = u'(a) = \cdots = u^{(m-1)}(a) = 0,$$
$$u(b) = u'(b) = \cdots = u^{(m-1)}(b) = 0. \qquad (1.5.20)$$

An operator A_0 is introduced by

$$A_0(u) = (-1)^m u^{(2m)}. \qquad (1.5.21)$$

A Galerkin solution is sought with trial solutions that are complete in H_0, a Hilbert space with a finite energy norm. That is, we expect the Galerkin solution to converge in the energy of the operator A_0. Mikhlin establishes the convergence of the Galerkin solution *uniformly* for derivatives up to $m - 1$ and *in the mean* for the mth derivative as a consequence of demonstrating that the operator A_0 is positive-bounded-below.

For a second-order elliptic partial differential equation

$$-\sum_{i,k=1}^{N} \frac{\partial}{\partial x_i}\left(A_{ik}\frac{\partial u}{\partial x_k}\right) + \sum_{i=1}^{N} B_i\frac{\partial u}{\partial x_i} + Cu = f, \qquad (1.5.22)$$

it can be shown (Mikhlin, 1964, p. 480) that if a unique solution to eq. (1.5.22) exists, then a Galerkin solution will converge to it, under certain conditions.

First, it is assumed that A_{ik}, B_i, C, and f are functions of x, that A_{ik} and their first derivatives are continuous in the domain D, and that B_i and C are continuous in the domain D and on the boundary ∂D. Two new operators are introduced by

$$A_0(u) = -\sum_{i,k=1}^{N} \frac{\partial}{\partial x_i}\left(A_{ik}\frac{\partial u}{\partial x_k}\right),$$
$$K(u) = \sum_{i=1}^{N} B_i\frac{\partial u}{\partial x_i} + Cu. \qquad (1.5.23)$$

Mikhlin (1964, p. 165) establishes that A_0 is positive-bounded-below for functions u that vanish on the boundary ∂D. For trial functions that are complete in the energy of A_0 and satisfy $u = 0$ on ∂D, Mikhlin (1964, p. 483) demonstrates that the Galerkin solution converges in energy to the solution of eq. (1.5.22).

Temam (1973) considers the nonlinear elliptic problem

$$L(u) = -\nabla^2 u + \lambda u + u^3 - f = 0 \qquad (1.5.24)$$

in some domain D, with boundary condition $u = 0$ on ∂D. Temam (1973, p. 132) uses the Galerkin formulation to show that the weak form of eq. (1.5.24), namely eq. (1.3.5), has a unique solution and that this solution satisfies eq. (1.5.24). In addition Temam establishes that the Galerkin solution converges to the exact solution of eq. (1.5.24).

For eigenvalue problems, error estimates are more readily obtainable if the problem is amenable to the Rayleigh–Ritz method. Thus Kantorovich and Krylov (1958, p. 336) consider the problem of finding an error bound for the kth eigenvalue λ_k of

$$\frac{d^2 y}{dx^2} + \lambda q y = 0 \qquad (1.5.25)$$

in the domain $0 \le x \le 1$, if $q > 0$ and $y(0) = y(1) = 0$. For a trial functions of the form

$$\phi_l^{(x)} = x^l (1 - x), \qquad l = 1, 2, \ldots, N, \qquad (1.5.26)$$

the Rayleigh–Ritz (and hence the Galerkin) solution $\lambda_k^{(N)}$ obeys the following inequality:

$$\left| \frac{\lambda_k^{(N)} - \lambda_k}{\lambda_k} \right| < \frac{P \lambda_k^{(N)}}{N^2 (N+1)^2 (N+2)(N-1)}, \qquad (1.5.27)$$

where

$$P = \left[\left\| \frac{d^2 q / dx^2}{q^{1/2}} \right\|_\infty + 2 \left\| \frac{dq}{dx} \right\|_\infty \left\{ |\lambda_k^{(N)}| \frac{\|q^{1/2}\|_\infty}{q_{min}^{1/2}} \right\}^{1/2} + \|q^{3/2}\|_\infty |\lambda_k^{(N)}| \right]^2.$$

For the related problem (1.5.25) with $q = 1$ in the region $-1 \le x \le +1$ and $y(-1) = y(1) = 0$ and a trial solution,

$$y_a = (1 - x^2)(a_0 + a_1 x^2 + \cdots + a_{l-1} x^{2l-2}),$$

Kantorovich and Krylov obtain error bounds that are extremely conservative. The ratio of the actual error to the error bound was 0.02 for $l = 2$ and 0.00004 for $l = 3$. Mikhlin (1964, p. 483) establishes that the Galerkin solution to the eigenvalue problem corresponding to eq. (1.5.22) converges to the exact solution.

The stability of the flow between adjacent rotating cylindrical surfaces with a circumferential pressure gradient (Taylor–Dean problem) is governed by the following system of equations:

$$(D^2 - a^2)^2 u = f(x) v \qquad (1.5.28)$$

and

$$(D^2 - a^2) v = -a^2 T g(x) u \qquad (1.5.29)$$

with boundary conditions $u = v = Du = 0$ at $x = 0, 1$. In the above equations, D indicates differentiation with respect to x, a is a dimensionless disturbance wave number, T is related to the characteristic velocity of the system, and $f(x)$ and $g(x)$ are known continuous functions. u and v are the eigenfunctions of the problem.

Diprima and Sani (1965) demonstrate that eqs. (1.5.28) and (1.5.29) can be cast in the form

$$A(\mathbf{w}) - \lambda K(\mathbf{w}) = 0, \tag{1.5.30}$$

where A and K are linear operators with A a positive definite, self-adjoint operator. The eigenvalue λ is $aT^{1/2}$, and $\mathbf{w}$ is a vector with components u and v. Mikhlin (1964, p. 475) establishes that a Galerkin solution for the eigenvalues of eq. (1.5.30) converges to the exact eigenvalues. Consequently Galerkin solutions to eqs. (1.5.28) and (1.5.29), and many related equations governing hydrodynamic stability problems, are convergent.

Convergence and error estimates for the Galerkin method have been obtained for initial-value problems. Green (1953) considered the equation

$$L(\ ) = \frac{\partial u}{\partial t} - \frac{\partial^2 u}{\partial x^2} + g(x,t)u + f(x,t) = 0 \tag{1.5.31}$$

with boundary and initial conditions $u(0,t) = u(\pi,t) = u(x,0) = 0$. Green assumed that f, g, and their first derivations were continuous in the domain with $f(0,0) = f(\pi,0) = 0$. The following trial solution was assumed:

$$u_a(x,t) = \sum_{k=1}^{N} C_k(t) \sin kx. \tag{1.5.32}$$

Green established that the problem (1.5.31) possesses a unique solution and that the Galerkin solution u_a converges uniformly to the exact solution u. First derivatives with respect to x and t also converge uniformly, but $d^2(u_a)/dx^2$ converges in the mean to d^2u/dx^2. It was found that the L_2 error satisfies

$$\|u_a - u\|_2^2 \le t^2 F_a^2(t),$$

where

$$F_a \equiv \max_{\tau \le t} \left| \int_0^\pi [(L(u_a)]^2\, dx \right|^{1/2}. \tag{1.5.33}$$

Convergence for nonlinear initial-value problems is discussed by Finlayson (1972). In particular, the Navier–Stokes equations have received considerable attention (Ladyzhenskaya, 1963, 1975). Often the Galerkin method has been used to establish weak solutions of the Navier–Stokes equations, for example by Ladyzhenskaya (1963, p. 155) and Heywood (1970). As a consequence this implies weak convergence of the Galerkin method.

Clearly, the more complex the equation, the harder it is to obtain error bounds on the solution. However, the equation residual (after substitution of the trial solution) is readily available. Thus the possibility arises of relating an error estimate of the solution to an appropriate norm *on the residual*. Finlayson (1972) describes a number of applications where the L_2 norm of the residual is used to obtain a bound on the solution error. An example is provided by Sigilloto (1967) for the heat-conduction equation

$$\frac{\partial u}{\partial t} - \frac{\partial^2 u}{\partial x^2} = 0$$

with initial and boundary conditions

$$u(x, 0) = \cos x \quad \text{and} \quad u(0, t) = u(\pi, t) = e^{-t}.$$

The estimates are rather conservative. At $t = 4$ and $x = 0$ the ratio of the actual error to the error bound is 0.00025. However, Finlayson (1972) cites other examples for nonisothermal chemical reactions which lead to much more accurate error bounds.

Stephens (1976) considers the convergence of the residual $A(u_N) - f$ arising from solving the linear operator equation $A(u) = f$. Here u_N can be interpreted as an N-term Galerkin solution. Convergence is established if a bound can be placed on $\|A(u_N)\|$. This typically depends on the choice of functions. Stephens uses the convergence property of the residual to obtain an improved error estimate for a Galerkin solution, based on bicubic splines, of a Poisson equation with homogeneous Dirichlet (function-value) boundary conditions.

1.6. Applications

Most applications of the Galerkin method prior to 1972 were traditional Galerkin formulations. Most of these applications have been reviewed by Ames (1972) and Finlayson (1972). Here we describe some representative examples of the traditional Galerkin method that have appeared in the literature since 1972. It may be noted that in this same period the use of the traditional Galerkin method has been substantially superseded by the Galerkin finite-element method and to a lesser extent by the spectral method.

1.6.1. Natural convection in a rectangular slot

Steady two-dimensional natural convection in a rectangular slot, which is oriented arbitrarily with respect to the gravity vector, provides an interesting temperature–velocity interaction. The basic geometry is shown in Fig. 1.14. The walls at $x = \pm h$ are isothermal and maintained at nondimensional temperatures $T(\pm 1, Z) = \mp 1$. This temperature difference provides the driving mechanism for the problem. The two walls at $Z = \pm H$ are either isothermal or adiabatic. By varying θ the influence of slot inclination on the convection characteristics can be considered. The aspect ratio $A = H/h$ can also be varied and is found to have a significant effect on the temperature and velocity profiles within the slot.

The equations governing the temperature and velocity behavior can be

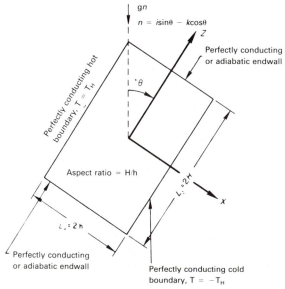

Figure 1.14. Geometry for natural-convection problem (after Catton et al., 1974; reprinted with permission of Pergamon Press)

written in nondimensional form as

$$\nabla \cdot \mathbf{v} = 0, \tag{1.6.1}$$

$$G^*(\mathbf{v} \cdot \nabla)\mathbf{v} = -\nabla p + \nabla^2 \mathbf{v} - T\mathbf{n}, \tag{1.6.2}$$

$$G^*(\mathbf{v} \cdot \nabla)T = \frac{1}{\text{Pr}}\nabla^2 T, \tag{1.6.3}$$

where G^* is the Grashof number and Pr is the Prandtl number. The two-dimensional velocity vector $\mathbf{v}$ is just

$$\mathbf{v} = \mathbf{i}u + \mathbf{k}w, \tag{1.6.4}$$

and $\mathbf{i}$ and $\mathbf{k}$ are unit vectors in the x and z directions. The boundary conditions for this problem are, for the velocity,

$$u(\pm 1, z) = w(\pm 1, z) = 0,$$
$$u(x, \pm A) = w(x, \pm A) = 0. \tag{1.6.5}$$

For the isothermal side walls (at $x = \pm 1$) the boundary conditions are

$$T(\pm 1, z) = \mp 1. \tag{1.6.6}$$

For the end walls the boundary conditions are

$$T(x, \pm A) = -x \tag{1.6.7}$$

or

$$\frac{\partial T}{\partial z}(x, \pm A) = 0. \tag{1.6.8}$$

Eq. (1.6.7) is an isothermal boundary condition that is compatible with eq. (1.6.6). Eq. (1.6.8) is an adiabatic boundary condition.

Trial solutions are introduced for the velocity and temperature distributions as follows:

$$u(x, z) = \sum_{j=1}^{N} B_j u_j(x, z), \tag{1.6.9}$$

$$w(x, z) = \sum_{j=1}^{N} B_j w_j(x, z), \tag{1.6.10}$$

$$T(x, z) = \left(\sum_{j=1}^{N} C_j f_j(x, z) \right) - x. \tag{1.6.11}$$

The trial functions $u_j(x, z)$ and $w_j(x, z)$ are chosen so that the continuity equation (1.6.1) is automatically satisfied. This is brought about by defining a stream function ψ_j such that

$$u_j = -\frac{\partial \psi_j}{\partial z} \quad \text{and} \quad w_j = \frac{\partial \psi_j}{\partial x} \tag{1.6.12}$$

and choosing a trial function for the stream function that will satisfy the boundary conditions (1.6.5). Because of the geometric symmetry it is convenient to define odd and even stream functions

$$\psi_j^{(e)} = C_{M_j}(x) C_{L_j}(z/A),$$
$$\psi_j^{(o)} = S_{M_j}(x) S_{L_j}(z/A), \tag{1.6.13}$$

where

$$C_{M_j}(x) = \frac{\cosh[\lambda_{M_j} x]}{\cosh \lambda_{M_j}} - \frac{\cos[\lambda_{M_j} x]}{\cos \lambda_{M_j}}$$

$$S_{M_j}(x) = \frac{\sinh[\mu_{M_j} x]}{\sinh \mu_{M_j}} - \frac{\sin[\mu_{M_j} x]}{\sin \mu_{M_j}}. \tag{1.6.14}$$

The parameters λ_m, μ_m must be chosen to satisfy the boundary conditions on u_j and v_j. The technique for doing this and the corresponding functions C_{M_j} and S_{M_j} are described in Chandrasekhar (1961, p. 634).

The trial functions f_j in eq. (1.6.11) are

$$f_j^{(e)}(x, j) = \sin[M_j \pi x] \cos[(2L_m - 1 - \alpha) 0.5\pi z/A]$$
$$f_j^{(o)}(x, j) = \cos[(2M_j - 1) 0.5\pi x] \sin[(2L_m - \alpha) 0.5\pi z/A] \tag{1.6.15}$$

where $\alpha = 0$ corresponds to the isothermal boundary condition (1.6.7) and $\alpha = 1$ corresponds to the adiabatic boundary condition (1.6.8).

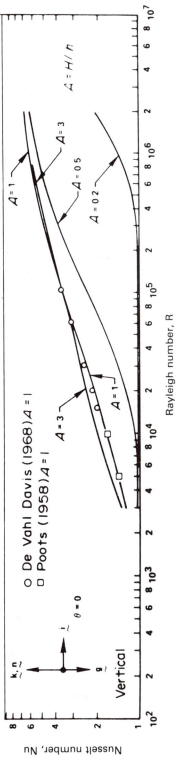

Figure 1.15. Heat transfer with isothermal end walls: zero tilt (after Catton et al., 1974; reprinted with permission of Pergamon Press)

Introduction of eqs. (1.6.9) to (1.6.11) into eqs. (1.6.2) and (1.6.3) produces equation residuals. Evaluating inner products of the residuals with the vector velocity trial function $\mathbf{v}_j = \mathbf{i}u_j + \mathbf{k}w_j$ and the temperature trial function f_j produces the following system of matrix equations for the unknown coefficients B_j and C_j in eqs. (1.6.9) to (1.6.11):

$$\mathbf{L}^{(1)}\mathbf{B} + \mathbf{L}^{(2)}\mathbf{C} + \frac{R^*}{\mathrm{Pr}}\mathbf{L}^{(6)}\mathbf{B} = \mathbf{L}^{(3)} \qquad (1.6.16)$$

and

$$R^*\mathbf{L}^{(4)}\mathbf{B} + \mathbf{L}^{(5)}\mathbf{C} - R^*\mathbf{L}^{(7)}\mathbf{C} = 0. \qquad (1.6.17)$$

Here the Rayleigh number $R^* = G^*\mathrm{Pr}$. The elements of the matrices $\mathbf{L}^{(1)}$ and so on are given by Catton et al. (1974). The matrices $\mathbf{L}^{(6)}$ and $\mathbf{L}^{(7)}$ depend on the unknown coefficients B_j, so that the equations are nonlinear.

However, in the limit of $\mathrm{Pr} \to \infty$, eq. (1.6.16) is linear and it is convenient to use this equation to eliminate $\mathbf{B}$ from eq. (1.6.17). The term dropped from eq. (1.6.16) implies that the convection-induced velocities will be small. The resulting equation for $\mathbf{C}$ is nonlinear and can be solved efficiently using the Newton–Raphson technique.

Catton et al. have obtained solutions for Rayleigh numbers up to 2×10^6,

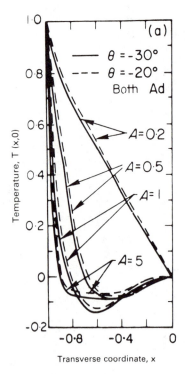

Figure 1.16. Temperature variation across slot at midheight, $Z = 0$, for $R^* = 300,000$. (after Catton et al., 1974; reprinted with permission of Pergamon Press)

Figure 1.17. Variation of longitudinal velocity across slot at midheight, $Z = 0$, for $R^* = 300,000$ (after Catton et al., 1974; reprinted with permission of Pergamon Press)

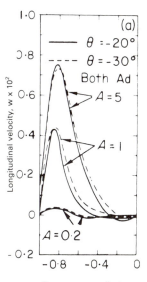

aspect ratios varying from 0.2 to 20, and tilt angles θ varying from -30 to $75°$. Galerkin solutions have been obtained with $N \leq 32$ in the trial solutions. For a vertical configuration, $\theta = 0°$, results for the Nusselt-number variation with Rayleigh number are shown in Fig. 1.15. Good agreement with the data of Poots (1958) and De Vahl Davis (1968) can be seen. The Nusselt number is a nondimensional measure of the heat transfer. Here it is defined by

$$\text{Nu} = -\frac{1}{2A} \int_{-A}^{+A} \frac{\partial T}{\partial x}\bigg|_{x=1} dz. \tag{1.6.18}$$

For adiabatic end walls the temperature and velocity components are shown in Figs. 1.16 and 1.17. These results correspond to $z = 0$ and a Rayleigh number of 3×10^5. It can be seen that the influence of aspect ratio is very pronounced. For small aspect ratio little motion in the z direction occurs, and the temperature profile is essentially linear in the x direction.

1.6.2. Hydrodynamic stability

For plane Poiseuille flow the critical Reynolds number R, above which the flow is no longer stable, is governed by the Orr–Sommerfeld equation

$$(D^2 - \alpha^2)\hat{v} = i\alpha R[(\bar{v} - \lambda)(D^2 - \alpha^2) - D^2\bar{v}]\hat{v} \tag{1.6.19}$$

with boundary conditions $\hat{v} = D\hat{v} = 0$ at $y = \pm 1$. Here D denotes differentiation, R is the Reynolds number, $\hat{v}$ is the amplitude of the disturbance velocity, $\bar{v}$ is the unperturbed velocity, α is a wave number, and λ is a complex velocity. If the imaginary part of λ is positive, the underlying motion is unstable.

Solving the Orr–Sommerfeld equation is equivalent to minimizing the local potential

$$\phi = \int_{-1}^{1} \mathscr{L}(\hat{v}, \hat{v}^{(0)})\, dy. \qquad (1.6.20)$$

During minimization, $\hat{v}^{(0)}$ is kept constant. After minimization $\hat{v}$ is set equal to $\hat{v}^{(0)}$. The function $\mathscr{L}$ has the following form (Platten et al., 1974):

$$\begin{aligned}
\mathscr{L}(\hat{v}, \hat{v}^{(0)}) =\ & -i\alpha R\lambda D\hat{v}^{(0)}D\hat{v} - i\alpha^3 R\lambda\hat{v}^{(0)}\hat{v} + \alpha^2 D\hat{v}^{(0)}D\hat{v} \\
& - 0.5(D^2\hat{v})^2 + \alpha^4\hat{v}^{(0)}\hat{v} + 0.5\alpha^2(D\hat{v})^2 + 2D^2\hat{v}^{(0)}D^2\hat{v} \\
& - i\alpha R\bar{v}\hat{v}^{(0)}D^2\hat{v} - 2i\alpha R(D\bar{v})\hat{v}^{(0)}D\hat{v} + i\alpha^3 R\bar{v}\hat{v}^{(0)}\hat{v}.
\end{aligned} \qquad (1.6.21)$$

Trial solutions for $\hat{v}$ and $\hat{v}^{(0)}$ are given by

$$\hat{v} = \sum_{j=1}^{N} a_j f_j,$$

$$\hat{v}^{(0)} = \sum_{j=1}^{N} a_j^{(0)} f_j. \qquad (1.6.22)$$

Platten et al. consider two different forms for f_j. First, they use

$$f_j(y) = (1 - y^2)^2 y^{2(j-1)}, \qquad (1.6.23)$$

which had been used by Lee and Reynolds (1964). However, the use of these f_j for the test functions in the Galerkin method leads to algebraic equations that are almost linearly dependent for large N (see section 2.1) and hence to ill-conditioned matrices. Alternatively Platten et al. use

$$f_j(y) = (1 - y^2)^2 T_{2j}(y), \qquad (1.6.24)$$

where T_{2j} is a Chebyshev polynomial. The factor $(1 - y^2)^2$ is retained to satisfy the boundary conditions. Yet another form for the trial function has been given by Orszag (1971) as

$$f_j(y) = T_{2j}(y) - j^2 T_2(y) + (j^2 - 1)T_0(y). \qquad (1.6.25)$$

Algebraic equations for the unknown coefficients a_j, $a_j^{(0)}$ in eq. (1.6.22) can be obtained by applying either the Rayleigh–Ritz method to eq. (1.6.20) or the Galerkin method to eq. (1.6.19). The resulting equations can be solved first for the eigenvalue λ, using

$$\det|\mathbf{A} - \lambda\mathbf{B}| = 0, \qquad (1.6.26)$$

where components of A and B are given by

$$A_{ij} = I_{ij}^{(2)} + 2\alpha^2 I_{ij}^{(1)} + \alpha^4 I_{ij}^{(0)} + i\alpha R[\alpha^2 J_{ij}^{(0)} + J_{ij}^{(1)} + J_{ij}^{(2)}],$$

$$B_{ij} = i\alpha R[I_{ij}^{(1)} + \alpha^2 I_{ij}^{(0)}],$$

$$(1.6.27)$$

in which

$$I_{ij}^{(0)} = \int_{-1}^{1} f_i f_j \, dy, \qquad J_{ij}^{(0)} = \int_{-1}^{1} \bar{v} f_i f_j \, dy,$$

$$I_{ij}^{(1)} = \int_{-1}^{1} Df_i Df_j \, dy, \qquad J_{ij}^{(1)} = \int_{-1}^{1} D^2 \bar{v} f_i f_j \, dy,$$

$$I_{ij}^{(2)} = \int_{-1}^{1} D^2 f_i D^2 f_j \, dy, \qquad J_{ij}^{(2)} = -\int_{-1}^{1} \bar{v} f_i D^2 f_j \, dy.$$

Equation (1.6.26) is derived and solved for increasing Reynolds number and a range of values of α. The occurrence of an eigenvalue with a positive imaginary part indicates an unstable flow.

Using the f_i given by eq. (1.6.23) and 20 terms in the trial solution, Platten et al. found that

$$R_{crit} = 5750 \pm 150. \qquad (1.6.28)$$

The uncertainty in the result arose from the difficulty in obtaining solutions for λ when close to the neutral stability curve with N large. Using the form of f_i given by eq. (1.6.24), Platten et al. obtained, at $\alpha = 1.02056$,

$$R_{crit} = 5772.22, \qquad (1.6.29)$$

which agrees with the result obtained by Orszag (1971) using the tau method (section 5.4.2).

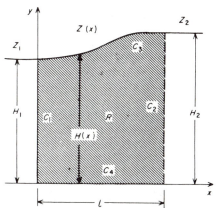

Figure 1.18. Duct geometry for acoustic transmission (after Eversman and Astley, 1981 · reprinted with permission of Academic Press)

1.6.3. Acoustic transmission in ducts

Acoustic transmission in a two-dimensional nonuniform duct has been considered by Eversman et al. (1975) for the no-flow case and by Eversman and Astley (1981) for the flow case. The nonuniformity considered has been both geometric and in relation to the acoustic properties of the duct wall. The duct geometry considered is shown in Fig. 1.18.

The nonuniform section joins two uniform sections and the acoustic impedance of the nonuniform section, $Z(x)$, is continuous with Z_1 and Z_2 (Fig. 1.18) at the upstream and downstream boundaries. The equations governing the acoustic transmission can be written

$$
U_0 \frac{\partial u}{\partial x} + \frac{1}{\rho_0} \frac{\partial p}{\partial x} + V_0 \frac{\partial u}{\partial y} + \left(ik_r + \frac{dU_0}{dx} \right) u - \frac{1}{\gamma P_0 \rho_0} \frac{dP_0}{dx} p = 0,
$$

$$
U_0 \frac{\partial v}{\partial x} + V_0 \frac{\partial v}{\partial y} + \frac{1}{\rho_0} \frac{\partial p}{\partial y} + \frac{\partial V_0}{\partial x} u + \left(ik_r + \frac{\partial V_0}{\partial y} \right) v = 0,
$$

$$
\gamma P_0 \frac{\partial u}{\partial x} + U_0 \frac{\partial p}{\partial x} + \gamma P_0 \frac{\partial v}{\partial y} + V_0 \frac{\partial p}{\partial y} + \frac{dP_0}{dx} u + \left[ik_r + \gamma \left(\frac{dU_0}{dx} + \frac{\partial V_0}{\partial y} \right) \right] p = 0,
$$

$$
\tag{1.6.30}
$$

and the boundary condition at the wall is

$$
v \cos \theta - u \sin \theta = Ap - \frac{iU_0}{k_r} \left[\frac{\partial}{\partial x} (Ap) + \tan \theta \, A \frac{\partial p}{\partial y} \right]. \tag{1.6.31}
$$

The equation (1.6.30) consists of the two momentum equations and the energy equation. The acoustic transmission is assumed to be inviscid and isentropic, so that the continuity equation is not required. u, v, and p are the acoustic perturbations on an underlying flow, characterized by U_0, V_0, and P_0, which is assumed known. In particular, U_0 and P_0 are assumed to be functions of x only. In eq. (1.6.30) k_r is the reference wave number. In eq. (1.6.31) A is the nondomensional specific acoustic admittance of the duct lining and θ is the local wall slope. The corresponding equations for the no-flow case are obtained by setting $U_0 = V_0 = 0$.

To apply the Galerkin method to eqs. (1.6.30) and (1.6.31), trial solutions of the following form are introduced:

$$
p_N(x, y) = \sum_{k=1}^{N} p_k(x) \psi_k(x, y),
$$

$$
u_N(x, y) = \sum_{k=1}^{N} u_k(x) \psi_k(x, y), \tag{1.6.32}
$$

$$
v_N(x, y) = \sum_{k=1}^{N} v_k(x) \phi_k(x, y),
$$

where

$$\psi_k(x, y) = \cos \kappa_k y, \qquad \phi_k(x, y) = \sin \kappa_k y, \qquad \kappa_k H \tan \kappa_k H = ikHA.$$
(1.6.33a)

The structure of the trial solutions is such that after application of the Galerkin method the governing equations will reduce to ordinary differential equations in x for p_k, u_k, and v_k. The trial functions ψ_k and ϕ_k account for the y dependence of the solution and also account for some x variation through the eigenvalues κ_k. In the equation for κ_k, the wall height H and the admittance A are both functions of x. The trial functions (1.6.33a) are the eigenfunctions for the no-flow problem and are also used as the trial functions for the no-flow problem.

For small values of kH more accurate results are obtained if κ_k is obtained from

$$\kappa_k H \tan \kappa_k H = ikHA \left[1 - \frac{M^2}{1 - M^2} \left(\frac{\kappa_k H}{kH} \right)^2 \right].$$
(1.6.33b)

Obtaining κ_k involves solving a nonlinear equation whether eq. (1.6.33a) or (1.6.33b) is used.

Because the trial solutions (1.6.32) do not satisfy the boundary condition (1.6.31), it is necessary to form a *boundary residual* as well as the equation residuals. Application of the Galerkin method leads to the matrix of ordinary differential equations

$$\mathbf{A} \frac{d\boldsymbol{\delta}}{dx} = \mathbf{B}\,\boldsymbol{\delta}.$$
(1.6.34)

For the flow case, $\boldsymbol{\delta}^t$ is a vector of the modal amplitudes $\{u_k, p_k, v_k\}$. The contributions to $\mathbf{A}$ and $\mathbf{B}$ are given by Eversman and Astley. The solution of eq. (1.6.34) essentially gives a transition matrix between the two ends of the nonuniform duct:

$$\boldsymbol{\delta}\big|_{x=L} = \mathbf{T}\,\boldsymbol{\delta}\big|_{x=0}.$$
(1.6.35)

This solution must be matched to the solution within the uniform ducts at $x < 0$ and $x > L$. This is done by developing an equation like eq. (1.6.34) for a uniform duct. This can be written

$$\frac{d\boldsymbol{\delta}\mathbf{u}}{dx} = \mathbf{L}\,\boldsymbol{\delta}\mathbf{u}.$$
(1.6.36)

It is assumed that $\boldsymbol{\delta}\mathbf{u} = \mathbf{q}\exp(-i\lambda_k x)$ and that λ_k can be obtained by solving the eigenvalue problem

$$\mathbf{L}\mathbf{q} = \lambda_k \mathbf{q}.$$
(1.6.37)

Subsequent solution for the modal amplitudes $\mathbf{q}^+$, $\mathbf{q}^-$, and so on permits an augmented transition matrix to be evaluated, equivalent to $\mathbf{T}$ in eq. (1.6.35).

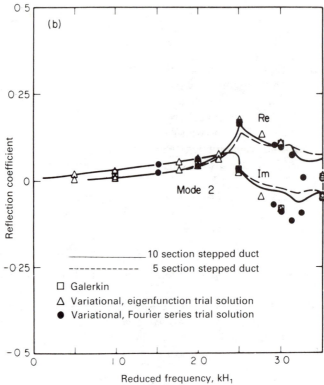

Figure 1.19. Reflection-coefficient variation in a 15° taper transition for mode 2 with mode 1 incident (after Eversman et al., 1975; reprinted with permission of Academic Press)

Subsequent manipulation leads to the establishment of overall matrices of reflection and transmission coefficients relating the amplitudes of the reflected and transmitted modes to the amplitudes of the incident modes. Thus

$$\mathbf{q}^-\big|_{x=0} = \mathbf{REF}\,\mathbf{q}^+\big|_{x=0} \quad \text{and} \quad \mathbf{q}^+\big|_{x=L} = \mathbf{TRAN}\,\mathbf{q}^+\big|_{x=0}. \qquad (1.6.38)$$

The precise form of **REF** and **TRAN** depends on whether the flow or the no-flow case is being considered.

For the no-flow case, the mode-2 reflection coefficient with mode 1 incident is shown in Fig. 1.19. The geometry here is a diverging 15° taper transition, and six trial functions were used to obtain the Galerkin results. The results are compared with stepped-duct and variational results. Agreement is good except for $kH_0 > 2.5$, where it is assumed that the stepped-duct results are inferior.

For the flow case the ratio of transmitted to incident power using the above formulation is compared with one-dimensional results of Davis and Johnson (1974) for a converging–diverging sinusoidal duct (Fig. 1.20). Agreement is

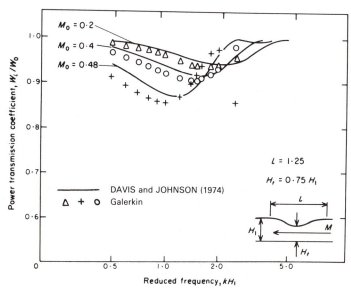

Figure 1.20. Ratio of transmitted power to incident power in a converging–diverging duct (hard wall) (after Eversman and Astley, 1981; reprinted with permission of Academic Press)

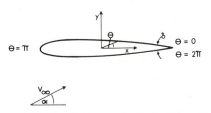

Figure 1.21. Flowfield geometry for inclined airfoil

good for low frequencies but at $kH_0 = 2.5$ and $M_0 = 0.48$ there is a sudden discrepancy associated with the cut-on of higher-order modes.

Tam (1976) has applied a traditional Galerkin method to the problem of sound transmission in curved rectangular ducts with rigid walls.

1.6.4. Flow around inclined airfoils

For the inviscid, incompressible flow around a lifting airfoil shown in Fig. 1.21 it is possible to write down a Fredholm integral equation of the second kind for the tangential velocity V at the surface,

$$g(\theta) - \frac{1}{2\pi} \int_0^{2\pi} K(\theta, \phi) g(\phi) \, d\phi = f(\theta), \qquad (1.6.39)$$

where

$$g(\theta) = (\dot{x} + \dot{y})^{1/2} \frac{V(\theta)}{V_\infty},$$

$$K(\theta, \phi) = 2\frac{\dot{y}(x - \xi) - \dot{x}(y - \eta)}{(x - \xi)^2 + (y - \eta)^2},$$

$$f(\theta) = -2(\dot{x}\cos\alpha + \dot{y}\sin\alpha), \qquad (1.6.40)$$

$$\dot{x} = \frac{dx}{d\theta}, \qquad \dot{y} = \frac{dy}{d\theta}.$$

θ is a surface coordinate defined by $x = 0.5\cos\theta$. Then $y = y(x(\theta))$ is the ordinate. (ξ, η) are dummy variables related to ϕ as (x, y) are to θ.

For airfoils with a nonzero trailing-edge angle δ, an appropriate trial solution for $g(\theta)$ (after Nugmanov, 1975) is

$$g(\theta) = \left[\sum_{j=1}^{N} A_j^* \frac{1 - \cos j\theta}{\sqrt{3\pi}} - \sum_{j=1}^{N} B_j \frac{\sin j\theta}{\sqrt{\pi}}\right]\lambda(\theta). \qquad (1.6.41)$$

The function $\lambda(\theta)$ is introduced to provide the *Kutta condition* when $\delta \neq 0$ and is defined as follows

$$\lambda(\theta) = \begin{cases} \left(\dfrac{1 - \cos\theta}{1 - \cos\theta^*}\right)^{\delta/2\pi} & \text{if } -\theta^* \leq \theta \leq \theta^*, \\ 1 \text{ otherwise.} \end{cases} \qquad (1.6.42)$$

θ^* is typically given by $\cos\theta^* = 1 - v\tau^2$, where τ is the thickness chord ratio and v is a free parameter. The coefficients A_j^* and B_j can be found iteratively by applying the Galerkin method at each step. At the first step, $\lambda^{(0)}(\theta) = 1$. Mikhlin (1964, p. 450) establishes that Galerkin solutions of the present problem will converge *in the mean* to the exact solution of eq. (1.6.39) if $K(\theta, \phi)$ and $f(\theta)$ are continuous functions of ϕ and θ and if the trial functions are orthonormal and members of a complete set.

An orthonormal trial solution based on eq. (1.6.41) is

$$g(\theta) = \left[\sum_{j=1}^{N} A_j^* \omega_j(\theta) + \sum_{j=1}^{N} B_j \sigma_j(\theta)\right]\lambda(\theta), \qquad (1.6.43)$$

where

$$\omega_1(\theta) = \frac{1 - \cos\theta}{\sqrt{3\pi}}$$

and

$$\omega_j(\theta) = M_j\left[\frac{1 - \cos j\theta}{\sqrt{3\pi}} - \frac{2}{2j - 1}\left\{\frac{1 - \cos(j - 1)\theta}{\sqrt{3\pi}} + \cdots + \frac{1 - \cos\theta}{\sqrt{3\pi}}\right\}\right]$$

with

$$M_j = \left[\frac{3(2j-1)}{2j+1}\right]^{1/2},$$

$$\sigma_j = \frac{\sin j\theta}{\sqrt{\pi}}.$$

Substitution of eq. (1.6.43) into eq. (1.6.39) and application of the Galerkin method produces the following system of algebraic equations:

$$\mathbf{DA} + \mathbf{CB} = \mathbf{E},$$
$$\mathbf{QA} + \mathbf{PB} = \mathbf{F}. \tag{1.6.44}$$

Elements of the various matrices in eq. (1.6.44) are given by

$$d_{kj} = -\frac{1}{2\pi}\int_0^{2\pi}\int_0^{2\pi} K(\theta,\phi)\omega_j(\phi)\sigma_k(\theta)\,d\phi\,d\theta,$$

$$c_{kj} = \delta_{kj} - \frac{1}{2\pi}\int_0^{2\pi}\int_0^{2\pi} K(\theta,\phi)\sigma_j(\phi)\sigma_k(\theta)\,d\phi\,d\theta,$$

$$q_{kj} = \delta_{kj} - \frac{1}{2\pi}\int_0^{2\pi}\int_0^{2\pi} K(\theta,\phi)\omega_j(\phi)\omega_k(\theta)\,d\phi\,d\theta,$$

$$p_{kj} = -\frac{1}{2\pi}\int_0^{2\pi}\int_0^{2\pi} K(\theta,\phi)\sigma_j(\phi)\omega_k(\theta)\,d\phi\,d\theta, \tag{1.6.45}$$

$$e_k = e^* - 2\sin\alpha\int_0^{2\pi} \dot{y}(\theta)\sigma_k(\theta)\,d\theta,$$

$$f_k = -2\sin\alpha\int_0^{2\pi} \dot{y}(\theta)\omega_k(\theta)\,d\theta.$$

Here $\delta_{kj} = 1$ if $k = j$, $\delta_{kj} = 0$ otherwise; and $e^* = \sqrt{\pi}\cos\alpha$ for $k = 1$, $e^* = 0$ otherwise.

If the solutions of eq. (1.6.44) are labeled $\mathbf{A}^{(0)}$ and $\mathbf{B}^{(0)}$, then

$$g^{(0)}(\theta) = \left[\sum_{j=1}^N A_j^{(0)}\omega_j(\theta) + \sum_{j=1}^N B_j^{(0)}\sigma_j(\theta)\right]\lambda(\theta). \tag{1.6.46}$$

Because of the introduction of $\lambda(\theta)$ into eq. (1.6.46), $g^{(0)}(\theta)$ will no longer satisfy eq. (1.6.39). Therefore iterative solutions are sought as

$$\mathbf{A}^{(v)} = \mathbf{A}^{(0)} + \mathbf{\Delta A}^{(1)} \cdots \mathbf{\Delta A}^{(v)},$$
$$\mathbf{B}^{(v)} = \mathbf{B}^{(0)} + \mathbf{\Delta B}^{(1)} \cdots \mathbf{\Delta B}^{(v)}, \tag{1.6.47}$$

where $\mathbf{\Delta A}^{(v)}$ and $\mathbf{\Delta B}^{(v)}$ can be obtained by solving

$$\Delta g^{(v)}(\theta) - \frac{1}{2\pi}\int_0^{2\pi} K(\theta,\phi)\Delta g^{(v)}(\phi)d\phi = f^{(v)}(\theta),$$

$$f^{(v)}(\theta) = f(\theta) - g^{(v-1)}(\theta) + \frac{1}{2\pi}\int_0^{2\pi} K(\theta,\phi)g^{(v-1)}(\phi)\Big|\,d\phi. \tag{1.6.48}$$

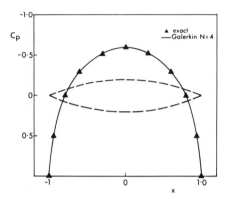

Figure 1.22. Pressure distribution for 20%-circular arc airfoils at $\alpha = 0°$

Convergence of the iterative scheme corresponds to $\|\varDelta A^{(v)}\| < \check{\varepsilon}_1$, $\|\varDelta B^{(v)}\| < \varepsilon_2$. The lift coefficient C_L is given by

$$C_L = 2 \int_0^{2\pi} g(\theta)\, d\theta = \frac{4\pi}{N} \sum_{j=1}^{N-1} g(\theta_i), \tag{1.6.49}$$

and the pressure coefficient C_p is given by

$$C_p(\theta_i) = 1 - \frac{g^2(\theta_i)}{\dot{x}_i^2 + \dot{y}_i^2}, \tag{1.6.50}$$

where $\theta_i = 2\pi i/N$, $i = 1, 2, \ldots, N - 1$.

For a 20%-circular-arc airfoil at $\alpha = 0°$, the pressure coefficient variation with x is shown in Fig. 1.22. It was found (Nugmanov, 1975) that the solution agreed with the exact one for $N \geq 4$ and that for this example, A_j and B_j were insensitive to $\lambda(\theta)$. Thus the solution could be taken as eq. (1.6.46) without the need for further iteration.

1.6.5. Microstrip disc problem

The Galerkin method is used here to obtain the capacitance of a circular microstrip disc (see Fig. 1.23), which is relevant to applications in microwave integrated circuits. The axially symmetric electrostatic potential ψ due to a charged circular disc satisfies Laplace's equation with appropriate boundary conditions. A Green's-function solution to the problem can be expressed as a Fredholm integral equation of the first kind,

$$\alpha\psi(r) = \int_0^a sG(r, s)\sigma(s)\, ds, \qquad r \leq a. \tag{1.6.51}$$

$\sigma(s)$ is the unknown charge distribution, and $G(r, s)$, the axisymmetric Green's function, can be expressed as

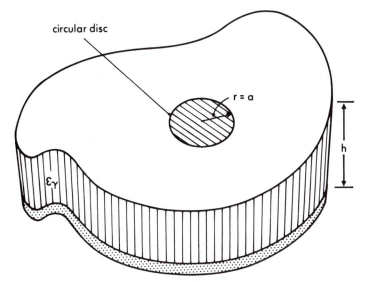

Figure 1.23. Geometry for microstrip disc problem

$$G(r, s) = k_0(r, s) + k_1(r, s), \qquad (1.6.52)$$

where $k_0(r, s)$ is the axisymmetric free-space Green's function. With suitable nondimensionalization (Coen and Gladwell, 1977), eqs. (1.6.51) and (1.6.52) can be written as

$$g(x) = \int_0^1 [k_0^*(x, y) + k_1^*(x, y)] y \eta(y) \, dy, \qquad 0 \le x \le 1. \quad (1.6.53)$$

Here x and y are nondimensional radial coordinates and η is a nondimensional charge distribution.

A trial solution for $\eta(y)$ is introduced as

$$\eta(y) = (1 - y^2)^{-1/2} \sum_{j=0}^{N} a_j P_j^*(y), \qquad (1.6.54)$$

where $P_j^*(y) = P_{2j}(\sqrt{1 - y^2})$ and P_{2j} is the Legendre polynomial. The factor $(1 - y^2)^{-1/2}$ is introduced to ensure the correct behavior at the edge of the disc. Substitution of eq. (1.6.54) and use of a weight function $x(1 - x)^{-1/2}$ $P_k^*(x)$ leads to a *generalized Galerkin* formulation. This is introduced to take advantage of the orthogonality of the Legendre functions.

The final form of the equations is

$$(4k + 1)^{-1} \lambda_k a_k + \sum_{j=0}^{N} c_{jk} a_j = \delta_{k,0}, \qquad k = 0, 1, \ldots, N, \quad (1.6.55)$$

where

$$\lambda_k = \frac{\pi}{2}\left[\frac{(2k)!}{2^{2k}(k!)^2}\right]^2,$$

$$c_{jk} = \int_0^1 P_{2k}(u)d_j(u)\,du \tag{1.6.56}$$

$$d_j(u) = \sqrt{1 - u^2}\int_0^1 k_1^*(x, t)P_{2j}(t)\,dt$$

The capacitance c is given by

$$c = 2\pi a\alpha a_0, \tag{1.6.57}$$

where $\alpha = \varepsilon_0(1 + \varepsilon_r)$.

The variation of capacitance with h/a, obtained with $N = 5$ in eq. (1.6.54), is shown in Fig. 1.24. Also shown is the comparison of the resonant frequency with the experimental results of Itoh and Mittra (1973). The resonant frequency f_{res} is related to the capacitance by

$$f_{\text{res}} = \frac{1.841v}{2\pi a\sqrt{\varepsilon_r hc}}, \tag{1.6.58}$$

where v is the speed of light in a vacuum.

The Galerkin method has been used to calculate the inductive components of microstrip discontinuities by Thomson and Gopinath (1975) and to investigate thick microstrip lines on anisotropic substrates by Alexopoulos and Uzunoglu (1978).

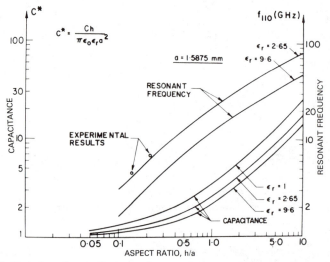

Figure 1.24. Variation of normalized capacitance and resonant frequency with aspect ratio h/a (after Coen and Gladwell, 1977; reprinted with permission of Institute of Electrical and Electronic Engineers)

1.6.6. Other applications of traditional Galerkin methods

Nield (1975) analysed the stability of a horizontal fluid layer with a non-uniform thermal gradient by introducing low-order trial solutions for the vertical velocity and perturbation temperature. The influence of different basic temperature profiles was considered. In this application one-term trial solutions were found to be adequate. Straus (1974) looked at convection in porous media and applied the Galerkin method to a velocity-related function and the perturbation temperature. With a 16-term expansion it was claimed that the Nusselt number should be accurate to 1% for Rayleigh numbers up to 400.

Belyaev et al. (1976) examined convective heat transfer in steady laminar pipe flow. For an arbitrary wall temperature distribution in the flow direction a Laplace transform was introduced to account for the longitudinal temperature variation, and a Galerkin formulation with a polynomial trial function to account for the transverse temperature variation. Results were presented for longitudinal temperature and Nusselt-number variation with up to a six-term trial solution.

Murti (1973) considered the lubrication of porous journal bearings and sought simultaneous solutions for the pressure distribution in the porous bearing and in the film. Results were obtained with typically a 12-term trial solution and used to assess load capacity and friction characteristics of porous bearings as a function of permeability.

Javeri (1978) investigated magnetohydrodynamic heat transfer in the thermal entrance region of a circular duct with a variety of fully developed velocity profiles. Convective temperature boundary conditions at the duct wall were applied. Bessel functions were assumed for the radial trial functions, and a Crank–Nicolson scheme has been used to integrate the resultant ordinary differential equations along the duct. Results for the Nusselt number obtained with 10- and 20-term trial solutions were presented at various Biot and Hartmann numbers. Quang (1974) studied the convective stability of an electrically conducting liquid subjected to Joule heating and an applied magnetic field. In this problem a layer of the liquid is bounded vertically by walls of different temperature. A trial solution accounting for the vertical variation was introduced. The problem becomes an eigenvalue problem conceptually similar to that considered in section 1.6.2. A four-term trial solution was found to be sufficient to assess the dependence of the critical Rayleigh number on the governing parameters.

Williams (1975) examined the motion of a mooring cable subsequent to the imposition of a harmonic displacement at the upper end. A traditional Galerkin method was used to obtain a function related to the dynamic tension. Buckling of stiffened rectangular plates was considered with a low-order Galerkin procedure by Wagner and Pattabiraman (1973). Buckling of circular plates due to applied normal and shear stresses was investigated by Durban (1977), using an eigenfunction expansion as the trial solution.

Durvasula (1971) has examined the flutter characteristics of clamped skew panels in supersonic flow using a 16-term beam-function expansion for the trial solution. Chopra and Durvasula (1971) have considered the vibration of trapezoidal plates with up to a six-term Fourier sine series to represent the deflected surface. Prabhu and Durvasula (1976) have analysed the postbuckling characteristics of clamped skew plates when subjected to various thermal and mechanical loads.

1.7. Closure

In this chapter traditional Galerkin methods have been examined in depth. From the simple examples provided in section 1.2 and the applications described in section 1.6 it is clear that Galerkin methods are applicable to a very broad class of problems.

In comparing traditional Galerkin methods with other methods of weighted residuals it is apparent that the Galerkin and least-squares methods produce solutions of comparable accuracy. Whereas the Galerkin method has almost no restriction on the type of problem that can be considered, the least-squares method is not suited to evolutionary or eigenvalue problems. Of the other methods of weighted residuals, the subdomain method is not as accurate as the Galerkin method, but is generally easier to implement and relates naturally to the laws of conservation of mass, momentum, and energy.

For problems that have an alternative variational formulation the Galerkin method produces equations that are identical to those produced by the Rayleigh–Ritz method. Thus the convergence properties associated with the Rayleigh–Ritz method carry over to the Galerkin method.

In applying traditional Galerkin methods, the trial (and test) functions must be chosen from a complete set to ensure convergence and from the lowest members of the complete set to ensure high accuracy. An important feature of the traditional Galerkin method, which has contributed to its widespread use, has been the ability to achieve high accuracy with few terms in the trial solution. This aspect of the method highlights the need to choose the trial functions to take advantage of prior knowledge of the expected solution. Often an eigenfunction expansion of a related (and presumably simpler) problem is used (e.g. section 1.6.3).

References

Ahlberg, J. H., Nilson, E. N., and Walsh, J. L. *The Theory of Splines and Their Applications*, Academic Press, New York (1967).

Alexopoulos, N. G., and Uzunoglu, N. K. *IEEE Trans. Mic. Th. Tech.* **MTT-26**, 455–456 (1978).

Ames, W. F. *Nonlinear Partial Differential Equations in Engineering*, Academic Press, New York (1965).

Ames, W. F. *Nonlinear Partial Differential Equations in Engineering*, Vol. II, Academic Press, New York (1972).

Anderssen, R. S., and Mitchell, A. R. *Math. Mech. Appl. Sci.*, 1, 3–15 (1979).

Anon. *Prik. Mat. Mekh.* 5, 337–341 (1941).

Bartlett, E. P., and Kendall, R. M. NASA CR-1062 (1968).

Belotserkovskii, O. M., and Chushkin, P. I. In *Basic Developments in Fluid Dynamics* (ed. M. Holt), Vol. 1, pp. 1–126 Academic Press, New York (1965).

Belyaev, N. M., Kordyuk, O. L., and Ryadno, A. A. *Inz. Fiz. Zh.* 30, 512–518 (1976).

Biezeno, C. B., and Koch, J. J. *Ingenieur* 38, 25–36 (1923).

Bickley, W. G. *Phil. Mag. (7)* 32, 50–66 (1941).

Bourke, W., McAveney, B., Puri, K., and Thurling, R. *Meth. in Comp. Phys.* 17, 267–325 (1977).

Brebbia, C. A. *The Boundary Element Method for Engineers*, Pentech Press, London (1978).

Catton, I., Ayyaswamy, P. S., and R. M. Clever *Int. J. Heat Mass Trans.* 17, 173–184 (1974).

Chandrasekhar, S. *Hydrodynamic and Hydromagnetic Stability*, Oxford U.P. (1961).

Chattot, J. J., Guiu-Roux, J., and Laminie, J. Seventh International Conference on Numerical Methods in Fluid Dynamics, Proceedings, Stanford 1980 Lecture Notes in Physics, Vol. 141, pp. 107–112, Springer-Verlag, Berlin (1981).

Chopra, I., and Durvasula, S. *J. Sound Vib.*, 19 379–392 (1971).

Citron, S. J. *J. Aero. Sci.* 27, 317–318 (1980).

Coen, S., and Gladwell, G. M. L. *IEEE Trans. Mic. Th. Tech.* **MTT-25**, 1–6 (1977).

Collatz, L. *The Numerical Treatment of Differential Equations*, Springer-Verlag, Berlin (1960).

Crandall, S. H. *Engineering Analysis*, McGraw-Hill, New York (1956).

Dahlquist, G., Björck, A., and Anderson, N. *Numerical Methods*, Prentice-Hall, Englewood Cliffs, NJ (1974).

Davis, S. S., and Johnson, M. L. *87th Meeting Acoustical Soc. of America*, Paper KK-2 (1974).

Demirdzic, I., Gosman, A. D., and Issa, R. Lecture Notes in Physics, Vol. 141, pp. 144–150, Springer-Verlag (1981).

De Vahl Davis, G. *Int. J. Heat Mass Trans.* 11, 1675–1693 (1968).

DiPrima, R. C., and Sani, R. L. *Quart. Appl. Math.* 23, 183–187 (1965).

Dorodnitsyn, A. A. *Advances in Aeronautical Sciences*, Vol. 3, Pergamon, New York (1960).

Dryden, H. L., Murnaghan, F. P., and Bateman, H. *Hydrodynamics*, p. 197, Dover, New York (1956).

Duncan, W. J. ARC R&M 1798 (1937).

Duncan, W. J. ARC R&M 1848 (1938).

Durban, D. *AIAA J.* 15, 360–365 (1977).

Durvasula, S. *AIAA J.* 7, 461–466 (1971).

Eversman, W., Cook, E. L., and Beckemeyer, R. J. *J. Sound Vib.* 38, 105–123 (1975).

Eversman, W., and Astley, R. J. *J. Sound Vib.* 74, 89–101 (1981).

Finlayson, B. A., and Scriven, L. E. *App. Mech. Rev.* 19, 735–748 (1966).

Finlayson, B. A. *Brit. Chem. Eng.* 14, 179–182 (1969).

Finlayson, B. A. *The Method of Weighted Residuals and Variational Principles*, Academic Press, New York (1972).

Fletcher, C. A. J. In *Numerical Simulation of Fluid Motion* (ed. J. Noye), pp. 537–550, North-Holland (1978).

Fletcher, C. A. J. *J. Comp. Phys.* 33, 301–312 (1979).

Fletcher, C. A. J. "Burgers' Equation: A Model for All Reasons", in *Numerical Solution of Partial Differential Equations* (ed. J. Noye), North-Holland (1982).

Fletcher, C. A. J., and Holt, M. *J. Comp. Phys.* **18**, 154–164 (1975).

Forsythe, G. E., Malcolm, M. A., and Moler, C. B. *Computer Methods for Mathematical Computations*, Prentice-Hall, Englewood Cliffs, NJ (1977).

Frazer, R. A., Jones, W. P., and Skan, S. W. ARC R&M 1799 (1937).

Fuller, E. L., Meneley, D. A., and Hetrick, D. L. *Nucl. Sci. Eng.* **40**, 206–233 (1970).

Galerkin, B. G. *Vestnik Inzhenerov, Tech.* **19**, 897–908 (1915).

Gear, C. W. *SIAM Review* **23**, 10–24 (1981).

Green, J. W. *J. Res. Nat. Bur. Std.* **51**, 127–132 (1953).

Harrington, R. F. *Field Computation by Moment Methods*, Macmillan, New York (1968).

Hamming, R. W. *Numerical Methods for Scientists and Engineers*, McGraw-Hill, New York, 2nd Edn. (1973).

Hess, J. L., and Smith, A. M. O. *J. Ship Research* **8**, 22–42 (1964).

Heywood, J. G. *Arch. Rat. Mech. Anal.* **37**, 48–60 (1970).

Holt, M. *Numerical Methods in Fluid Dynamics*, Springer-Verlag (1977).

Itoh, T., and Mittra, R. *Arch. Elek. Übertragung* **27**, 456–458 (1973).

Jain, M. K. *Appl. Sci. Res.* **A11**, 177–188 (1962).

Jameson, A., and Caughey, D. A. A Finite Volume Method for Transonic Potential Flow Calculations. AIAA Paper 77–635 (1977).

Javeri, V. *Int. J. Heat Mass Trans.* **21**, 1035–1040 (1978).

Kantorovich, L. V., and Krylov, V. I., *Approximate Methods in Higher Analysis*, Wiley, New York (1958).

Ladyzhenskaya, O. A. *The Mathematical Theory of Viscous Incompressible Flow*, Gordon and Breach, New York (1963).

Ladyzhenskaya, O. A. *Annual Review of Fluid Mechanics* **7**, 249–272 (1975).

Lee, Y., and Reynolds, W. C. Tech. Report FM-1, Dept. of Mech. Eng., Stanford Univ. (1964).

Lynn, P. P. *Int. J. Num. Meth. Eng.* **8**, 865–876 (1974).

Mikhlin, S. G. *Variational Methods in Mathematical Physics*, Pergamon, Oxford (1964).

Mikhlin, S. G. *The Numerical Performance of Variational Methods*, Noordhoff, Groningen (1971).

Milthorpe, J., and Steven, G. P. *Finite Elements in Fluids*, Vol. 3, pp. 89–110, Wiley, New York (1978).

Murphy, J. D. "Application of the Generalised Galerkin Method to the Computation of Fluid Flows", *Proceedings 1st AIAA Computational Fluid Dynamics Conference, Palm Springs*, pp. 63–68 (1973).

Murphy, J. D. *AIAA J.* **15**, 1307–1314 (1977).

Murti, P. R. K. *Wear* **26**, 95–104 (1973).

Narasimha, R., and Deshpande, S. M. *J. Fluid Mech.* **36**, 555–570 (1969).

Neuman, C. P. "Recent Developments in Discrete Weighted Residual Methods", in *Computational Methods in Nonlinear Engineering* (ed. J. T. Oden) North-Holland, Amsterdam (1974).

Nield, D. A. *J. Fluid Mechanics* **71**, 441–454 (1975).

Nugmanov, Z. Kh. *Izv. Vuz. Avia. Tekh.* **18**, 78–83 (1975).

Oden, J. T. *Finite Elements of Nonlinear Continua*, McGraw-Hill, New York (1972).

Orszag, S. A. *J. Fluid Mechanics* **50**, 689–703 (1971).

Orszag, S. A. "Numerical Simulation of Turbulent Flows", in *Handbook of Turbulence* (eds. W. Frost and T. H. Moulden), pp. 281–313, Plenum Press, New York (1977).

Orszag, S. A. *J. Comp. Phys.* **37**, 70–92 (1980).

Pallone, A. *J. Aero. Sci.* **28**, 449–456 (1961).

Panton, R. Z., and Sallee, H. B. *Computers and Fluids* **3**, 257–269 (1975).

Platten, J. K., Flandroy, P., and Vanderborck, G. *Int. J. Eng. Sci.* **12**, 995–1006 (1974).

Pomraning, G. C. *Nucl. Sci. Eng.* **24**, 291–301 (1966).

Poots, G. *Quart. J. Mech. Appl. Math.* **11**, 257–267 (1958).

Prabhu, M. S. S., and Durvasula, S. *Comp. Structures* **6**, 177–185 (1976).

Quang, V. Z. *Magnetohydrodynamics (Mg. Gidr.)* **1**, 83–88 (1974).

Rizzi, A. W., and Inouye, M. *AIAA J.* **11**, 1478–1485 (1973).

Rubbert, P. E., and Saaris, G. R. Review and Evaluation of a Three-Dimensional Lifting Potential Flow Analysis Method for Arbitrary Configurations, AIAA Paper 72-188 (1972).

Schetz, J. A., *J. Appl. Mech.* **30**, 263–268 (1963).

Shuleshko, P. *Aust. J. Appl. Sci.* **10**, 1–16 (1959).

Sigilloto, V. G. *J. Assoc. Comp. Mech.* **14**, 732–741 (1967).

Stephens, A. B. *SIAM J. Num. Anal.* **13**, 607–614 (1976).

Straus, J. M. *J. Fluid Mechanics* **64**, 51–63 (1974).

Tam, C. K. W. *J. Sound Vib.* **45**, 91–104 (1976).

Temam, R. *Numerical Analysis*, Reidel (1973).

Thomson, A. F., and Gopinath, A. *IEEE Trans. Mic. Th. Tech.* **MTT-23**, 648–655 (1975).

Thorsen, R., and Landis, F. *Int. J. Heat Mass Transfer* **8**, 189–192 (1965).

Truckenbrodt, E. *J. Aero. Sci.* **19**, 428–429 (1952).

Vichnevetsky, R. *IEEE Trans. Comp.* **C-18**, 499–512 (1969).

Villadsen, J. V., and Stewart, W. E. *Chem. Eng. Sci.* **22**, 1483–1501 (1967).

Viviand, H., and Ghazzi, W. *La Recherche Aerospatiale* **1974–5**, 5, 247–260 (1974).

Wadia, A. R., and Payne, F. R. In *Advances in Computational Methods for Partial Differential Equations* (eds. R. Vichnevetsky and R. S. Stepleman), pp. 205–219, IMACS (1979).

Wagner, H., and Pattabiraman, J. *Z. Flugwiss.* **21**, 131–140 (1973).

Williams, H. E. *J. Hydronautics* **9**, 107–118 (1975).

Yamada, H. *Rept. Res. Inst. Fluid Eng. Kyushu Univ.* **3**, 29 (1947).

Yamada, H. *Rept. Res. Inst. Fluid Eng. Kyushu Univ.*, **4**, 27–42 (1948).

CHAPTER 2

Computational Galerkin Methods

For many of the examples given in chapter 1, acceptable accuracy, and often very high accuracy, could be achieved with less than five terms in the trial solution. The advent of computers has brought both a demand for solutions of high accuracy and an interest in problems that are inherently more complex than the simple examples given in section 1.2.

In this chapter we anticipate the difficulties that would arise if a traditional Galerkin method were applied to problems that are more complex than those considered previously. The complexity might be associated with an irregular boundary, difficult boundary conditions, or complicated physical processes. Often a problem will be characterized by large gradients in a small part of the solution domain and small gradients elsewhere.

Inevitably the need to achieve solutions of acceptable accuracy manifests itself as a need to retain a large number of coefficients in the trial solution such as eq. (1.3.3). In this context a large number could be anything larger than 20, but more likely as large as 2000 or 20,000. Having identified the more intractable difficulties associated with traditional Galerkin methods, we will draw attention to those features of modern Galerkin methods that allow the difficulties to be avoided.

2.1. Limitations of the Traditional Galerkin Method

Here we reconsider the ordinary differential equation of section 1.2.1. Application of a traditional Galerkin method with test and trial functions x^j produces a system of algebraic equations for the unknown coefficients a_j that can be written

$$\mathbf{MA} = \mathbf{D}, \tag{2.1.1}$$

where a_j is an element of $\mathbf{A}$. An element of $\mathbf{M}$ is given by

$$m_{kj} = \frac{j}{j+k-1} - \frac{1}{j+k}, \tag{2.1.2}$$

and an element of $\mathbf{D}$ is given by

Table 2.1. Typical values of m_{kj} when $N = 1000$

	Column $j = 1$	Column $j = 1000$
Row $k = 999$	1.001×10^{-6}	0.50000
Row $k = 1000$	0.999×10^{-6}	0.49975
Difference	2×10^{-9}	0.00025

$$d_k = 1/k. \tag{2.1.3}$$

Clearly for large k (which implies large N),

$$m_{kj} \approx \frac{j-1}{j+k}, \tag{2.1.4}$$

and the difference in the magnitude between corresponding terms in the $(k - 1)$th and kth equations will not be very large. Table 2.1 shows that for $N = 1000$ the differences range between 2×10^{-9} and 2.5×10^{-4}. This leads us to expect that the matrix $\mathbf{M}$ will be ill conditioned for large $\mathbf{N}$.

An ill-conditioned matrix, such as $\mathbf{M}$ in eq. (2.1.1), implies that the corresponding solution $\mathbf{A}$ is very sensitive to small changes in $\mathbf{D}$ or $\mathbf{M}$. In a more complex situation the elements of $\mathbf{D}$ and $\mathbf{M}$ might be computed by numerical quadrature, in which case the small changes could be identified with roundoff error.

Following Isaacson and Keller (1966), we introduce a *condition number* μ to quantify the condition of the matrix $\mathbf{M}$. Due to roundoff error, the system of algebraic equations solved, eq. (2.1.1), is replaced by

$$(\mathbf{M} + \delta\mathbf{M})(\mathbf{A} + \delta\mathbf{A}) = \mathbf{D} + \delta\mathbf{D} \tag{2.1.5}$$

where $\mathbf{A}$ satisfies eq. (2.1.1). If A is nonsingular and if $\|\delta\mathbf{M}\| < 1/\|\mathbf{M}^{-1}\|$, then it can be shown (Isaacson and Keller, 1966, p. 38) that

$$\frac{\|\delta\mathbf{A}\|}{\|\mathbf{A}\|} \leq \frac{\mu}{1 - \mu\|\delta\mathbf{M}\|/\|\mathbf{M}\|} \left[\frac{\|\delta\mathbf{D}\|}{\|\mathbf{D}\|} + \frac{\|\delta\mathbf{M}\|}{\|\mathbf{M}\|} \right], \tag{2.1.6}$$

where μ is defined by

$$\mu = \|\mathbf{M}\| \cdot \|\mathbf{M}^{-1}\|. \tag{2.17}$$

It is convenient to calculate $\|\mathbf{M}\|$ as the maximum norm, since this is just the maximum absolute row sum

$$\|\mathbf{M}\|_\infty = \max_k \sum_{j=1}^N |m_{kj}|. \tag{2.1.8}$$

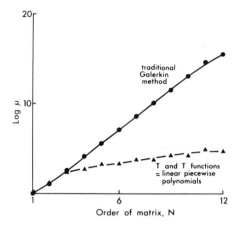

Figure 2.1. Condition-number variation

In Fig. 2.1 is shown the increase in $\log \mu$ with N for eq. (2.1.1). A useful rule of thumb (Strang, 1980) is that in solving eq. (2.1.1) by Gauss elimination, $\log \mu$ *decimal places are lost due to roundoff errors*. Clearly the effect of choosing $N = 2000$ or $20{,}000$ would be devastating.

This example raises the question: why does **M** become ill conditioned as N increases? The answer rests with the test function x^k used in eq. (1.3.4) to obtain eq. (2.1.1). In Fig. 2.2, x^k is plotted over the domain of interest, $0 \leq x \leq 1$, for various values of k. It is clear that for large enough k only the region adjacent to $x = 1$ is contributing to the evaluation of the inner product in eq. (1.3.4). Further, for large k, there is very little difference between w_{k-1} and w_k. Consequently we would expect the algebraic equations that result from evaluating eq. (1.3.4) to be almost linearly dependent. This of course shows up in Table 2.1 and in the ill conditioning of **M**.

Therefore the first problem with traditional Galerkin methods is that for large k, successive weight functions produce almost linearly dependent algebraic equations.

A second problem arises for large N, which becomes apparent after considering eq. (2.1.4). It will be noticed that all the coefficients m_{kj} are nonzero. Thus

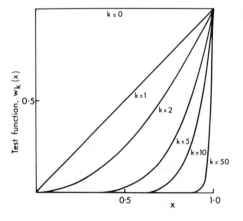

Figure 2.2. Test-function variation with k

an $O(N^2)$ process will be required to generate the matrix $\mathbf{M}$. However, an $O(N^3)$ process will be required to factorize $\mathbf{M}$ if Gaussian elimination is used as part of the solution procedure for $\mathbf{A}$. It is this process which is responsible for the large execution time if a traditional Galerkin method is used with a large N for problems that reduce to the form of eq. (2.1.1).

Where a traditional Galerkin method is used to reduce a partial differential equation to an ordinary differential equation, as in section (1.2.4), systems of equations like the following are obtained:

$$\mathbf{M\dot{A}} = -(\mathbf{BA} + \mathbf{C}). \tag{2.1.9}$$

If there is room to store lower and upper triangular matrices $\mathbf{L}$ and $\mathbf{U}$, then $\mathbf{M}$ need only be factorized once, prior to carrying out the time integration:

$$\mathbf{M} = \mathbf{LU}. \tag{2.1.10}$$

This is an $O(N^3)$ process. At each time step it is necessary to evaluate

$$\mathbf{Q_{r.h.s.}} = \mathbf{BA} + \mathbf{C},$$
$$\mathbf{Q'_{r.h.s.}} = \mathbf{L}^{-1}\mathbf{Q_{r.h.s.}}, \tag{2.1.11}$$

and to solve $\mathbf{U\dot{A}} = -\mathbf{Q'_{r.h.s.}}$. Each of these processes requires $O(N^2)$ operations. If $O(N)$ time steps are required to integrate the ordinary differential equations, then this part of the solution becomes at least as expensive as the initial factorization.

A third problem arises for large N if the underlying equation is *nonlinear*. This was the case with Burgers' equation in section 1.2.5. After application of a traditional Galerkin method, a system of ordinary differential equations for the unknown coefficients a_j is obtained, which can be written in the form of eq. (2.1.9). The various steps described above are equally applicable in this case, with one exception: Each coefficient b_{kj} depends on the current solution and is now evaluated from

$$b_{kj} = \sum_i^N a_i \left(T_j \frac{dT_i}{dx}, T_k \right). \tag{2.1.12}$$

Assuming that the inner product can be evaluated once and for all via a recursion formula, the evaluation of b_{kj} at each time step is an $O(N)$ process. Therefore the evaluation of $\mathbf{B}$ is an $O(N^3)$ process. Thus each time step now requires the same order of time as the initial factorization of M, eq. (2.1.10).

This order-of-magnitude increase in time stems directly from the quadratic nonlinearity in eq. (1.2.59). In compressible-flow problems cubic nonlinearities appear that would require an $O(N^4)$ process at each time step, if a traditional Galerkin method were used.

To store the result of evaluating the inner product shown in eq. (2.1.12) requires N^3 memory locations. For a cubic nonlinearity this would require N^4 memory locations. The storage problem can be alleviated by evaluating the inner product at each time step. However, this will further increase the execution time per time step. Thus the treatment of nonlinear terms turns out to be a very severe impediment for traditional Galerkin methods *if N is large*.

A fourth difficulty for traditional Galerkin methods relates to solving problems in a spatial domain whose boundaries do not coincide with coordinate lines. The difficulty is in constructing trial functions that automatically satisfy the boundary conditions on a distorted boundary. Of the problems considered in section 1.2, only the viscous flow in a square duct involved more than one spatial dimension, and in that example the boundaries coincided with coordinate lines. Of the examples given in section 1.6, the natural convection problem (section 1.6.1) utilized a regular geometry coinciding with coordinate lines. In the acoustic-transmission example (section 1.6.3) the problem defined on a two-dimensional domain is reduced to the solving of a system of ordinary differential equations in one spatial dimension.

In principle the difficulty could be avoided for distorted but simple geometries by mapping the distorted physical domain into a regular region. However, this would certainly increase the complexity of the governing equations. One might also expect the accuracy of the solution to be degraded in those parts of the domain where the grid is most distorted.

2.2 Solution for Nodal Unknowns

After the application of a traditional Galerkin method the unknowns in the problem are the coefficients a_j, for example as in eq. (1.3.3). These coefficients have no obvious physical interpretation. Alternatively we can define a trial solution by

$$u(x, y) = \sum_{j=1}^{N} \bar{u}_j \phi_j(x, y), \qquad (2.2.1)$$

where $\bar{u}_j$ are the *nodal* values of u. Clearly $\bar{u}_j$ have a direct physical significance. The trial functions $\phi_j(x, y)$ are necessarily *interpolatory* in nature. That is, $\phi_j = 1$ at node j, and $\phi_j = 0$ at all other nodes but not between nodes.

The form of the trial solution given by eq. (2.2.1) can be related to the form used in a traditional Galerkin method, that is,

$$u = \sum_{l=1}^{N} a_l \psi_l(x, y). \qquad (2.2.2)$$

In particular, if eq. (2.2.2) is evaluated at all the nodal points (x_j, y_j), the following matrix equation is obtained:

$$\mathbf{\Psi A} = \mathbf{U}, \qquad (2.2.3)$$

where an element of $\mathbf{\Psi}$ is $\psi_l(x_j, y_j)$, an element of $\mathbf{A}$ is a_l, and an element of $\mathbf{U}$ is $\bar{u}_j$. Thus the coefficients $\mathbf{A}$ can be obtained from the nodal unknowns $\mathbf{U}$ by

$$\mathbf{A} = \mathbf{\Psi}^{-1}\mathbf{U},$$

so that eq. (2.2.2) could be written

$$u = \sum_{l=1}^{N} \sum_{j=1}^{N} \Psi_{lj}^{-1} \bar{u}_j \psi_l(x, y) \qquad (2.2.4)$$

or

$$u = \sum_{j=1}^{N} \bar{u}_j \left\{ \sum_{l=1}^{N} \Psi_{lj}^{-1} \psi_l(x, y) \right\}. \qquad (2.2.5)$$

It follows that eq. (2.2.5) coincides with eq. (2.2.1) if

$$\phi_j(x, y) = \sum_{l=1}^{N} \Psi_{lj}^{-1} \psi_l(x, y). \qquad (2.2.6)$$

In practice the evaluation of $\mathbf{\Psi}^{-1}$ for large N would be very time consuming (i.e. an $O(N^3)$ process) and probably not very accurate. Fortunately, it is straightforward to obtain ϕ_j directly.

Obtaining the solution in terms of the nodal unknowns has a number of advantages. First, the final solution is directly useful. For example, in solving the flow external to an aerofoil (e.g. the example in section 1.6.4), the desired solution is the pressure on the surface. Often knowing the pressure at discrete intervals is sufficient. However, if the trial solution has unknown coefficients a_j, then eq. (2.2.2) has to be evaluated (an $O(N)$ process) at every point of interest.

In addition, formulation in terms of nodal unknowns allows a particular test function ϕ_k to be associated with a specific node. If the test function is of limited spatial extent, the corresponding equation, formed after application of the Galerkin method, is then associated with a limited region surrounding that node. This is very useful in solving nonlinear equations, since examination of the equation residuals then indicates the parts of the domain where the solution is far from convergence, and vice versa.

2.3. Use of Low-order Test and Trial Functions

In a traditional Galerkin method the trial functions, ϕ_j in eq. (2.2.1), span the whole domain. Here we examine the other extreme where the test and trial functions span a very small part of the domain. In particular we consider test and trial functions which are only nonzero in the immediate vicinity of the jth node (Fig. 2.3). Here the interpolating functions are linear in x between nodes $j - 1$ and j and j and $j + 1$. Thus the trial solution (2.2.1) interpolates u linearly between nodes.

Application of a Galerkin method with low-order test and trial functions

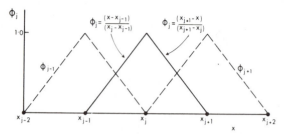

Figure 2.3. Linear interpolation function

leads to low-order integrands in evaluating the inner product (1.3.4). Where the inner products can be evaluated exactly this is of little consequence. However, in more complex problems where the inner products must be evaluated numerically, a lower-order integrand permits a lower-order quadrature formula to be used. Consequently an important computational economy is obtained when N is large.

It can be seen from Fig. 2.3 that the test functions are linearly independent and that increasing N (i.e. a mesh refinement) does not affect the linear independence. This is in clear contrast to the typical situation for a traditional Galerkin method, which was shown in Fig. 2.2. It follows that, after application of the Galerkin method, the resulting algebraic equations will also be linearly independent. Consequently it is to be expected that the condition number of the matrix **M** will grow with N, the order of the matrix, at a much lower rate than for a traditional Galerkin method.

Application of the Galerkin method with the test and trial functions shown in Fig. 2.3 to the problem of section 1.2.1 produces a matrix **M** (as in eq. (2.1.1)) whose condition-number variation with N is shown in Fig. 2.1. As expected, the growth rate of μ with N is much lower than when global polynomials are used for the test and trial functions (as in section 2.1).

The kth algebraic equation formed from applying the Galerkin method to the ordinary differential equation of section 1.2.1, with the test and trial function shown in Fig. 2.3, has the following form:

$$\frac{\bar{u}_{k+1} - \bar{u}_{k-1}}{2\,\Delta x} - \{\tfrac{1}{6}\bar{u}_{k+1} + \tfrac{2}{3}\bar{u}_k + \tfrac{1}{6}\bar{u}_{k-1}\} = 0. \qquad (2.3.1)$$

It is evident that the expression resulting from the differential term du/dx is *identical* with the centered-difference formula for du/dx.

Because the test function is zero outside the range $x_{k-1} < x < x_{k+1}$, the corresponding matrix **M**, as in eq. (2.1.1), will only have entries on and immediately adjacent to the diagonal. In fact **M** will be tridiagonal. The sparse nature of **M** can be exploited to solve systems of equations like (2.1.1) very economically (Isaacson and Keller, 1966, p. 55). This would require $5N$-4 operations, which may be compared with an $O(N^3)$ operation count for the

Figure 2.4. Element configuration:
global nodal numbering

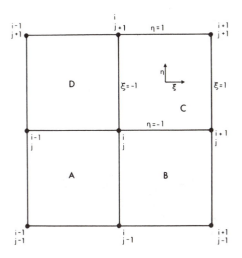

traditional Galerkin method. Even if quadratic or cubic interpolating functions replaced the linear functions shown in Fig. 2.3, the solution of eq. 2.1.1 would still be an $O(N)$ process for large N.

The local nature of the test and trial functions also makes the evaluation of nonlinear terms very economical. Thus the evaluation of $Q_{r.h.s.}$ in eq. (2.1.11) is still an $O(N)$ process when N is large, even if cubic nonlinearities appear.

This greater economy is accompanied by some loss of accuracy. As indicated in section 1.2.2, the use of local linear interpolating functions produces a solution of lower accuracy than does a solution using global test and trial functions with the same number of unknown coefficients. An example of this can also be seen in section 3.2.5.

2.4. Use of Finite Elements to Handle Complex Geometry

If a Galerkin method is to be used with local, low-order interpolating functions in a two- or three-dimensional domain, it is convenient to introduce the concept of *finite elements*. A small part of a two-dimensional domain is shown in Fig. 2.4.

The (i,j)th node is surrounded by four elements. Trial solutions can be defined independently in each element using a local coordinate system (ξ, η). Thus in element C we obtain

$$u = \sum_{j=1}^{4} \phi_j(\xi, \eta)\bar{u}_j, \tag{2.4.1}$$

where $\bar{u}_j$ are the nodal values of u (Fig. 2.5) and ϕ_j is the interpolating function (often called *shape* function) associated with the jth node. For example ϕ_1 is given by

$$\phi_1 = \tfrac{1}{4}(1 - \xi)(1 - \eta). \tag{2.4.2}$$

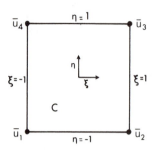

Figure 2.5. Element coordinate system

Since $\phi_1 = 1$ at node 1 and $\phi_1 = 0$ at nodes 2, 3, and 4, eq. (2.4.1) interpolates u bilinearly throughout element C. Using similar trial solutions to eq. (2.4.1) in each of the other elements produces a distribution for u which is continuous at the element boundaries. However derivatives of u *are not continuous* at element boundaries. It can be seen from Figs. 2.4 and 2.5 that $\bar{u}_1$ defined at the element level becomes $\bar{u}_{i,j}$ at the global level. Also the functions ϕ_j are associated with particular nodes. Thus the test function associated with global node (i,j) corresponds to ϕ_1 in element C, ϕ_2 in element D, and so on. In applying the Galerkin method with such test and trial functions it is clear that contributions are obtained from the four elements A, B, C, and D to the algebraic equation associated with node (i,j).

The above pattern of elements forms part of a regular grid. It is appropriate, for the use of Gaussian quadrature, if the element boundaries coincide with $\xi, \eta = \pm 1$.

If the elements, viewed as part of a global pattern, are rectangular (but not necessarily square), the transformation to define an element-based coordinate system $(-1 \le \xi, \eta \le +1)$ can be obtained easily.

By a straightforward extension of the above ideas, due to Ergatoudis et al. (1968), it is possible to define a uniform local coordinate system (ξ, η) corresponding to a distorted element in physical space. This is particularly useful in treating boundaries that do not coincide with coordinate lines (see Fig. 2.6). Element B has one boundary $(1, 2)$ that coincides with the domain boundary. A transformation between physical space (x, y) and element space (ξ, η) for element B can be defined by

$$x = \sum_{l=1}^{4} \phi_l(\xi, \eta)\bar{x}_l,$$

$$y = \sum_{l=1}^{4} \phi_l(\xi, \eta)\bar{y}_l,$$

(2.4.3)

where $\bar{x}_l, \bar{y}_l$ are the coordinates of the lth corner of element B in physical space, and $\phi_l(\xi, \eta)$ are the same interpolating functions as were used as trial functions in eq. (2.4.1). Just as eq. (2.4.1) led to continuity of u between elements, the use of eq. (2.4.3) produces a continuous mapping at element boundaries. Equation (2.4.3) is called an *isoparametric* transformation.

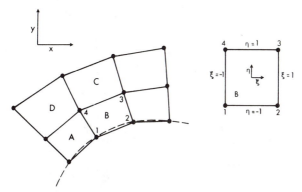

Figure 2.6. Isoparametric mapping at a boundary

The transformation is used in the evaluation of the inner product (1.3.4) to obtain the algebraic equations. In a typical situation (e.g. in section 1.2.3) the following integral is to be evaluated:

$$I = \int\int_D \frac{\partial \phi_j}{\partial x} \frac{\partial \phi_k}{\partial x} \, dx \, dy. \tag{2.4.4}$$

It is easier to evaluate the integral in (ξ, η) space. $\partial \phi_j / \partial x$, $\partial \phi_j / \partial y$ are related to $\partial \phi_j / \partial \xi$, $\partial \phi_j / \partial \eta$ by the following equations:

$$\begin{bmatrix} \dfrac{\partial \phi_j}{\partial \xi} \\[2mm] \dfrac{\partial \phi_j}{\partial \eta} \end{bmatrix} = [J] \begin{bmatrix} \dfrac{\partial \phi_j}{\partial x} \\[2mm] \dfrac{\partial \phi_j}{\partial y} \end{bmatrix}, \tag{2.4.5}$$

where the Jacobian $[J]$ is given by

$$[J] \equiv \begin{bmatrix} \dfrac{\partial x}{\partial \xi} & \dfrac{\partial y}{\partial \xi} \\[2mm] \dfrac{\partial x}{\partial \eta} & \dfrac{\partial y}{\partial \eta} \end{bmatrix}. \tag{2.4.6}$$

The elements in the Jacobian can be obtained from eq. (2.4.3). For example,

$$\frac{\partial x}{\partial \xi} = \sum_l \frac{\partial \phi_l}{\partial \xi} (\xi, \eta) \bar{x}_l. \tag{2.4.7}$$

It is apparent that the Jacobian is a function of (ξ, η) and the element corner locations $\bar{x}_l$, $\bar{y}_l$. From eq. (2.4.5) the following explicit expressions for $\partial \phi_j / \partial x$ (and $\partial \phi_k / \partial x$) are available:

$$\frac{\partial \phi_j}{\partial x} = \left(\frac{\partial y}{\partial \xi} \frac{\partial \phi_j}{\partial \eta} - \frac{\partial y}{\partial \eta} \frac{\partial \phi_j}{\partial \xi} \right) \frac{1}{\det J}. \tag{2.4.8}$$

In eq. (2.4.4), $dx\,dy = \det J\,d\xi\,d\eta$. Thus the contribution to eq. (2.4.4) from each element is

$$I_e = \int_{-1}^{1}\int_{-1}^{1}\frac{1}{\det J}\left\{\frac{\partial y}{\partial\xi}\frac{\partial\phi_j}{\partial\eta} - \frac{\partial y}{\partial\eta}\frac{\partial\phi_j}{\partial\xi}\right\}\left\{\frac{\partial y}{\partial\xi}\frac{\partial\phi_k}{\partial\eta} - \frac{\partial y}{\partial\eta}\frac{\partial\phi_k}{\partial\xi}\right\}d\xi\,d\eta. \tag{2.4.9}$$

All the terms in the integrand in eq. (2.4.9) are functions of (ξ, η), and so I_e can be evaluated. Typically this is done numerically. For the linear elements considered here the evaluation of eq. (2.4.4) would require evaluation of eq. (2.4.9) for each of the elements surrounding the ith node.

2.5. Use of Orthogonal Test and Trial Functions

Here we introduce a completely different technique to deal with the problem of test functions that approach linear dependence as N increases in eq. (1.3.4). An effective strategy is to choose the test functions so that they are *orthogonal* to all the other members of the complete set. That is, for the domain of interest we require that the test functions ϕ_k satisfy the conditions

$$(\phi_k, \phi_j)\begin{cases} = 0 & \text{if } k \neq j, \\ \neq 0 & \text{if } k = j. \end{cases} \tag{2.5.1}$$

For the Galerkin method the trial solution is defined by

$$u = \sum_{j=1}^{N} a_j\phi_j. \tag{2.5.2}$$

For a linear operator equation, $L(u) = 0$, application of the Galerkin method gives

$$\sum_{j=1}^{N} a_j(L(\phi_j), \phi_k) = 0 \tag{2.5.3}$$

and wherever the inner product (2.5.1) appears in eq. (2.5.3), only a single contribution arises. The unsteady heat-conduction problem (section 1.2.4) leads to the following matrix equation (eq. (1.2.49)):

$$\mathbf{M}\dot{\mathbf{A}} + \mathbf{B}\mathbf{A} + \mathbf{C} = 0. \tag{2.5.4}$$

If orthogonal test and trial functions are used, then the elements of $\mathbf{M}$ are zero *except on the diagonal*. Consequently the $O(N^3)$ process to factorize $\mathbf{M}$ and the $O(N^2)$ multiplications of $\mathbf{B}$ and $\mathbf{C}$ (eq. (1.2.54)) are avoided. If the problem is nonlinear, as with $\mathbf{B}$ in section 1.2.5, these $O(N^2)$ multiplications would be required at every time step, if nonorthogonal test and trial functions were used.

Since orthogonal functions are linearly independent, the resulting algebraic equations will also be linearly independent. Thus the use of orthogonal test and trial functions avoids the problem of obtaining a solution from an ill-conditioned matrix equation and also introduces the economy of avoiding a matrix factorization and subsequent matrix multiplication. However, it maintains the high accuracy of using global trial functions.

Fourier series are the most common example of using orthogonal functions. A Fourier cosine series was used in the example in section 1.2.3. For Burgers' equation (section 1.2.5) Chebyshev polynomials were used. These functions are orthogonal with respect to the weight $(1 - x^2)^{-1/2}$. Consequently they do not lead to a diagonal matrix $\mathbf{M}$ in eq. (1.2.67). However, the use of Chebyshev polynomials as test functions does produce algebraic equations that are linearly independent. Legendre polynomials are orthogonal in the interval $-1 \le x \le 1$. These are applied to Burgers' equation in section 5.2.2. It is possible to generate orthogonal functions from closely related functions to suit particular applications. This is illustrated in section 5.5.

2.6. Evaluation of Nonlinear Terms in Physical Space

The introduction of global orthogonal functions, described in section 2.5, permits the avoidance of matrix factorization. However, the problem of the nonlinear terms remains. In practice, use of (say) a Fourier series as the trial solution does produce some cancellation in evaluating quadratic nonlinear terms. However the evaluation of nonlinear terms still dominates the execution time when a global orthogonal trial solution is used.

Orszag (1969) showed how to greatly improve the economy of evaluating the nonlinear terms whilst accepting a small increase in the coding complexity. The essential idea of Orszag was to evaluate the nonlinear terms in *physical space*, where they can be handled locally, and to transform into the *parameter space* a_j once the solution has been marched to the new time step (in an unsteady problem). The transformation from physical space to parameter space is particularly efficient if advantage can be taken of the fast Fourier transforms (Cooley and Tukey, 1965). Clearly this implies using a Fourier series for the trial solution. However Orszag (1980) has pointed out that comparable transforms are available for other common orthogonal functions: Legendre polynomials, surface harmonics, Bessel functions, and so on.

2.7. Advantages of Computational Galerkin Methods

The modern counterparts of traditional Galerkin methods have evolved in two quite separate directions, although both have sought to achieve maximum computational efficiency.

First, Galerkin finite element methods have developed around the idea of using purely *local* (and hence low-order) test and trial functions. Also the test and trial functions have been almost *exclusively* polynomials. Early attempts to introduce a combined polynomial–Fourier-series solution were rather cumbersome (Chakrabarti, 1971). For piecewise linear polynomials as test and trial functions the resulting algebraic equations often coincide, for first and second derivatives in one dimension, with corresponding finite-difference expressions.

As was seen in section 2.3, by relating each test function to a different geometric location, the linear independence of the resulting algebraic equations is assured. By restricting the test function to a region close to a particular node, the resulting matrix system of algebraic equations will be *sparse*. Consequently both the factorization of a matrix like $\mathbf{M}$ in eq. (2.1.1) and the evaluation of nonlinear terms like $\mathbf{B}$ in eq. (1.2.67) can be carried out in $O(N)$ operations. By applying the trial solution in finite elements it is straightforward, through the isoparametric formulation, to cope with an irregular domain and, in particular, boundaries that do not coincide with coordinate lines.

The second direction is characterized by spectral methods which, like the traditional Galerkin method, are *global* methods, that is, the test and trial functions span the whole domain. As a result, comparable accuracy for a given number of unknowns is to be expected. However, by using orthogonal functions as test and trial functions, the test functions are by definition linearly independent, as are the resulting algebraic equations. Also, where matrix entries arise from the inner product of the test and trial functions, only entries on the diagonal will be present. Thus economy is obtained by avoiding certain matrix manipulations. The lack of economy in evaluating nonlinear terms is substantially alleviated by transforming to the physical plane and evaluating the nonlinear terms there.

Like the traditional Galerkin method, the spectral methods work best when dealing with domains whose boundaries coincide with coordinate lines. However, for distorted regions Orszag (1980) has introduced a global transformation so that the solution domain is uniform. It is not clear how serious an effect this has on the accuracy of the solution.

2.8. Closure

In this chapter we have identified properties of the traditional Galerkin method that limit its usefulness for finding solutions of high accuracy to complex problems—that is, situations where the number of unknown coefficients in the trial solution will be large. Specific ways of overcoming these deficiencies have been discussed. It has been shown that Galerkin finite-element and spectral methods possess many of these desirable features. Galerkin finite-element methods are considered in more detail in chapters 3, 4, and 7, and spectral methods in chapter 5.

References

Chakrabarti, S. *Int. J. Num. Meth. Eng.* **3**, 261–273 (1971).

Cooley, J. W., and Tukey, J. W. *Math. Comp.* **19**, 297–301 (1965).

Ergatoudis, I., Irons, B., and Zienkiewicz, O. C. *Int. J. Solids Struct.* **4**, 31–42 (1968).

Isaacson, E., and Keller, H. B. *Analysis of Numerical Methods*, Wiley, New York (1966).

Orszag, S. A. *Phys. Fluids Supplement II* **12**, 250–257 (1969).

Orszag, S. A. *J. Comp. Phys.* **37**, 93–112 (1980).

Strang, G. *Linear Algebra and Its Applications*, Academic Press, New York, 2nd Edn. (1980).

CHAPTER 3

Galerkin Finite-Element Methods

The Galerkin finite-element method has been the most popular method of weighted residuals, used with piecewise polynomials of low degree, since the early 1970s. The rise in the popularity of the Galerkin formulation and the concurrent decline in popularity of the variational finite-element formulation has coincided with the diversification of the finite-element method into areas remote from the structural birthplace of the method.

Many of these areas have included motion—for example, all the branches of fluid mechanics and convective heat transfer. However, most of the "new" areas are not easily described in terms of variational formulations.

But the variational era (say 1964–1973) did establish finite-element theory and provide it with a strong mathematical foundation. It would appear that much of the mathematical interest has remained with the subsequent development of the finite-element method. Here we have dated the variational era as extending from the appearance of the book by Mikhlin (1964) in the western literature to the appearance of the book by Strang and Fix (1973), which describes very lucidly the mathematical achievements of the period. Perhaps, from a historical perspective, the book by Strang and Fix will be seen as the epitaph of the variational era.

The later status of the finite-element method, from a mathematical perspective, is succinctly set out in Oden and Reddy (1976a) and by Mitchell and Wait (1977). An engineering perspective is given in Heubner (1975), Bathe and Wilson (1976), Zienkiewicz (1977), and Irons and Ahmad (1980). More specialized books on the finite-element method are provided by Oden (1972) on nonlinear continuum mechanics, Temam (1979b) on the Navier–Stokes equations, Pinder and Gray (1977) on groundwater and tidal flows, and Brebbia (1978) on boundary-element methods.

In this chapter we examine the finite-element method with more emphasis on those of its aspects that differentiate it from the traditional Galerkin method. This centers on the use of low-order polynomials as test and trial functions and the subsequent application of the Galerkin principle over discrete or finite elements.

After examining the more useful interpolating functions and elements, a number of examples are presented to demonstrate the mechanics and some of the ramifications of the finite-element method.

In practice many of the algebraic formulae resulting from use of lower-order trial functions reproduce formulae that are associated historically with the finite-difference method. This aspect is explored in section 3.3.

Some of the important foundation stones of finite-element theory, laid down during the variational era, are discussed in section 3.4. Later results, relating to convergence rates and error estimates, are also included. This chapter ends with some representative applications of the Galerkin finite-element method which illustrate how widespread the dissemination of the method has been.

3.1. Trial Functions and Finite Elements

We begin by considering different trial functions and elements in one dimension.

3.1.1. One-dimensional elements

The simplest situation is shown in Fig. 3.1.

The trial solution in a one-dimensional domain, $x_1 \leq x \leq x_N$, is written

$$u_a = \sum_{j=1}^{N} N_j(x)\bar{u}_{aj}, \qquad (3.1.1)$$

where the trial functions, $N_j(x)$, are linear piecewise polynomials (often called *shape* functions or *interpolating* functions in the finite element literature). $\bar{u}_{aj}$ are the values of unknowns at the nodes (i.e. the nodal unknowns), and these are to be solved for. It is apparent from Fig. 3.1a that u_a interpolates the function u linearly between nodal unknowns—that is, over each element.

The form of the shape functions N_j is shown in Fig. 3.1b. It can be seen

Figure 3.1. Finite-element interpolation using linear shape functions

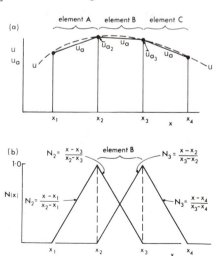

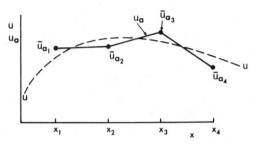

Figure 3.2. Finite-element solution using linear shape functions

that they fall linearly from a maximum value of one at a particular node to zero at the two neighboring nodes and are zero throughout the rest of the domain. So although eq. (3.1.1) is a global equation, in any particular element only two shape functions and two nodal unknowns make a nonzero contribution to eq. (3.1.1). For example, in element B only the shape functions N_2 and N_3 make a contribution to u_a in eq. (3.1.1).

It is apparent from Fig. 3.1 that although u_a is continuous throughout the domain, du_a/dx changes discontinuously at the element boundaries and higher derivatives are not defined. It is also clear that since only the nodal values coincide with u, there is some error associated with introducing u_a. This error is called the *interpolation error*.

Since the same shape functions N_j are used as test functions in the Galerkin method, the evaluation of the residual inner product (1.3.4) will only have nonzero contributions from two neighboring elements. Thus if $N_2(x)$ is used as the test function, nonzero contributions to the inner product will be obtained from elements A and B only (Fig. 3.1). Consequently in the resulting algebraic equation only nodal unknowns, $\bar{u}_{a1}$, $\bar{u}_{a2}$, and $\bar{u}_{a3}$, will appear.

If the system of algebraic equations resulting from using all the shape functions N_k as test functions are solved for the nodal unknowns, the final solution u_a might appear as in Fig. 3.2. There u represents the exact solution. One notes that the nodal values $\bar{u}_{aj}$ do not coincide with the exact solution u. The error, $u_a - u$, is called the *approximation error*. In considering the accuracy

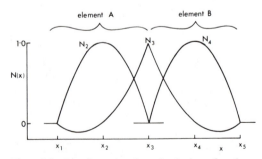

Figure 3.3. One-dimensional quadratic shape functions

of finite-element solutions we need to be aware of these two related errors: the interpolation error, which depends on the choice of trial functions, and the approximation error, which also depends on the governing equation and the method of solution. These and other errors are discussed in section 3.4.

Using the same form for the trial solution, eq. (3.1.1), we consider the situation where the $N_j(x)$ are quadratic piecewise polynomials. These are shown in Fig. 3.3. They can be constructed directly from Lagrangian interpolation functions. Thus N_3 is given as follows:

$$\text{in element } A, \qquad N_3 = \frac{(x - x_1)(x - x_2)}{(x_3 - x_1)(x_3 - x_2)}, \qquad (3.1.2)$$

$$\text{in element } B, \qquad N_3 = \frac{(x - x_4)(x - x_5)}{(x_3 - x_4)(x_3 - x_5)}, \qquad (3.1.3)$$

$$\text{for } x < x_1, \qquad N_3 = 0,$$

$$\text{for } x > x_5, \qquad N_3 = 0.$$

As discussed in section 2.2, it would be possible to replace eq. (3.1.1) with

$$u_a = c_1 + c_2 x + c_3 x^2 \qquad (3.1.4)$$

over each element and to require that u_a be equal to $\bar{u}_{a1}, \bar{u}_{a2}, \bar{u}_{a3}$ (for element A in Fig. 3.3), thereby generating sufficient equations to solve for c_1, c_2 and c_3. This could also be done at the shape-function level. Thus if N_3 is written in the form

$$N_3 = b_0 + b_1 x + b_2 x^2, \qquad (3.1.5)$$

eq. (3.1.2) can be constructed by requiring that $N_3 = 1$ at x_3 and $N_3 = 0$ at $x = x_1, x_2$.

The approximate solution u_a is interpolated quadratically within each element by eq. (3.1.1), and contributions arise from three shape functions. For example, over element B (Fig. 3.3), the shape functions N_3, N_4, and N_5 make nonzero contributions to eq. (3.1.1). Although up to second derivatives are defined within each element, only u_a is continuous at element boundaries.

Application of the Galerkin method with N_3 as the test function would generate nonzero contributions from elements A and B only (Fig. 3.3), and in the ensuing algebraic equation there would appear the nodal unknowns $\bar{u}_{a1}$ to $\bar{u}_{a5}$. However, if N_2 is used as the test function, nonzero contributions arise only from element A, and consequently in the algebraic equation only nodal unknowns $\bar{u}_{a1}$ to $\bar{u}_{a3}$ would appear.

Clearly the form of eq. (3.1.1) sets no limit on the degree of the polynomial used in the shape function. Cubic shape functions are shown in Fig. 3.4. The shape function N_4 is as follows:

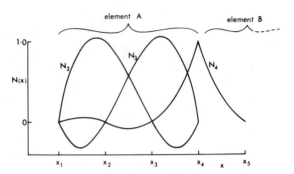

Figure 3.4. One-dimensional cubic shape functions

in element A, $N_4 = \dfrac{(x - x_1)(x - x_2)(x - x_3)}{(x_4 - x_1)(x_4 - x_2)(x_4 - x_3)}$,   (3.1.6)

in element B, $N_4 = \dfrac{(x - x_5)(x - x_6)(x - x_7)}{(x_4 - x_5)(x_4 - x_6)(x_4 - x_7)}$,   (3.1.7)

for $x \leq x_1$, $N_4 = 0$,

for $x \geq x_7$, $N_4 = 0$.

The function u_a, eq. (3.1.1), is interpolated cubically in each element with up to the third derivative nonzero. However, as with the linear and quadratic shape functions, only the function value u_a is continuous between elements.

Use of N_4 (Fig. 3.4) as the test function in the Galerkin method will produce an algebraic equation with nodal unknowns $\bar{u}_{a1}$ to $\bar{u}_{a7}$ appearing. Use of N_2 or N_3 (Fig. 3.4) with the Galerkin method will produce an algebraic equation in which only nodal values $\bar{u}_{a1}$ to $\bar{u}_{a4}$ appear.

The three one-dimensional elements considered so far are referred to as C^0 elements because they guarantee continuity of up to the *zeroth-order derivative* of u_a at element boundaries. For most of the problems that arise in fluid mechanics and heat transfer, C^0 continuity is sufficient. However, for some problems in the bending of plates it is ncessary to require continuity of the first derivative.

Hermite polynomials can be introduced directly for the shape functions, since they ensure continuity of the function and first derivative at element boundaries. The application of Hermite polynomials with the finite-element method is described by Heubner (1975) and by Strang and Fix (1973). If continuity of higher derivatives is required, B-splines (Schultz, 1973) may be used.

However, more demanding continuity requirements necessitate higher-order interpolation within each element and algebraic equations with many nodal unknowns coupled together. This leads, typically, to a less economical but more accurate solution.

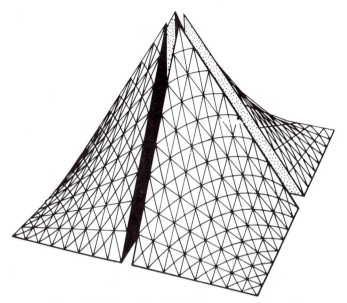

Figure 3.5. Bilinear shape function on a rectangular grid

3.1.2. Rectangular elements in two and three dimensions

The use of Lagrange interpolation functions as shape functions in one dimension extends quite naturally to two and three dimensions. The variation of a "linear" shape function over four adjacent rectangular elements is shown in Fig. 3.5. However, it is convenient, in more than one dimension, to introduce element coordinates (ξ, η) as in Fig. 3.6. The variation of u_a in a two-dimensional domain can be represented by

$$u_a = \sum_i \sum_j N_{ij}(x, y)\bar{u}_{aij}, \tag{3.1.8}$$

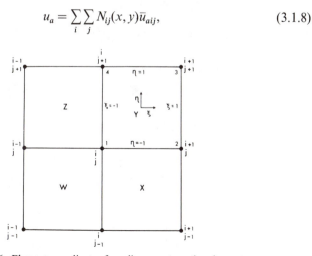

Figure 3.6. Element coordinates for a linear rectangular element

that is, eq. (3.1.1) extended to two dimensions. The notation in eq. (3.1.8) is *global* in nature. However, within any element, say element Y in Fig. 3.6, eq. (3.1.8) could be written

$$u_a = \sum_{l=1}^{4} N_l(\xi, \eta) \bar{u}_{al}. \tag{3.1.9}$$

The notation and the coordinate system is now *element based*. The ease with which equations can be evaluated at the element level and the result assembled into global expressions is one of the strengths of the finite-element method.

For the "linear" shape functions shown in Fig. 3.5, N_l takes the following form (in element Y):

$$N_1 = 0.25(1 - \xi)(1 - \eta),$$

$$N_2 = 0.25(1 + \xi)(1 - \eta),$$

$$N_3 = 0.25(1 + \xi)(1 + \eta),$$

$$N_4 = 0.25(1 - \xi)(1 + \eta),$$

$$\tag{3.1.10a}$$

or

$$N_l = 0.25(1 + \xi_l \xi)(1 + \eta_l \eta). \tag{3.1.10b}$$

For lines of constant ξ or η the shape functions revert to the one-dimensional form. In global coordinates eq. (3.1.10) are equivalent to

$$N_l = c_0 + c_1 x + c_2 y + c_3 xy \tag{3.1.11}$$

with different values of c_0, c_1, c_2, and c_3 for each element. For interpolation in any element, contributions come from the nodal unknowns at the four corners ($\bar{u}_{ai}$) and the associated shape functions.

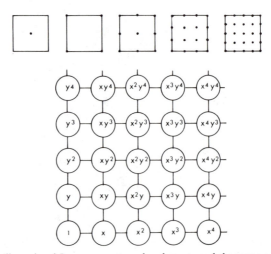

Figure 3.7. Two-dimensional Lagrange rectangular elements and the terms contributing to the global trial functions

The shape function associated with the global node (i,j) is zero outside of elements W, X, Y, Z. Using this shape function as the test function in the Galerkin method requires evaluation of the inner product (1.3.4) over the four elements W, X, Y, Z only. The resulting algebraic equation will have nonzero contributions associated with the nine nodes shown in Fig. 3.6.

A "quadratic" shape function on a rectangular element can be derived from the product of the corresponding one-dimensional Lagrange interpolation functions. The nodal placement and the terms contributing to the element interpolation functions are shown in Fig. 3.7. Also shown are other members of the two-dimensional Lagrange family. The lowest-order member corresponds to a constant value of u_a in each element, so that u_a changes discontinuously at element boundaries for this element.

The "quadratic" shape function can be written

$$N_i = \prod_{r \neq i} \frac{\xi - \xi_r}{\xi_i - \xi_r} \frac{\eta - \eta_r}{\eta_i - \eta_r}. \tag{3.1.12}$$

Because of the nature of the element coordinate system it is perhaps clearer if eq. (3.1.12) is written as follows:

corner nodes: $\quad N_i = 0.25\, \xi_i\xi(1 + \xi_i\xi)\eta_i\eta(1 + \eta_i\eta),$

midside nodes: $\begin{cases} \xi_i = 0, & N_i = 0.5(1 - \xi^2)\eta_i\eta(1 + \eta_i\eta), \\ \eta_i = 0, & N_i = 0.5(1 - \eta^2)\xi_i\xi(1 + \xi_i\xi), \end{cases} \tag{3.1.13}$

internal node: $\quad N_i = (1 - \xi^2)(1 - \eta^2).$

The form of these shape functions can be visualized by considering constant values of ξ or η; then the shape function is given by the corresponding one-dimensional configuration, Fig. 3.3.

The use of "quadratic" Lagrange shape functions as test functions generates a contribution from four elements for a corner node, two elements for a midside node, and one element for an internal node. The corresponding number of contributing nodes in the ensuing algebraic equation is shown in Table 3.1 below.

The nodal locations for a "cubic" shape function on a rectangular element are shown in Fig. 3.7. The algebraic formulae for the shape function may be obtained from the products of one-dimensional formulae like eq. (3.16). The actual formulae, in element coordinates, are given by Pinder and Gray (1977). The number of contributing nodes in the algebraic equations after application of the Galerkin method with a "cubic" test function is shown in Table 3.1.

A related family of rectangular elements can be obtained by taking the Lagrange shape functions and modifying them to eliminate the internal nodes (Ergatoudis et al., 1968). These are called elements of the *serendipity* family and are shown in Fig. 3.8. The name was given to this family because they were discovered by chance. The "linear" element coincides with the "linear"

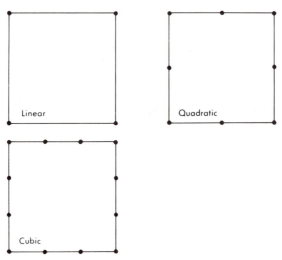

Figure 3.8. Two-dimensional serendipity rectangular elements

Lagrange element. For the "quadratic" serendipity element the shape functions are

corner nodes: $\quad N_i = 0.25(1 + \xi_i\xi)(1 + \eta_i\eta)(\xi_i\xi + \eta_i\eta - 1),$

midside nodes: $\xi_i = 0, \quad N_i = 0.5(1 - \xi^2)(1 + \eta_i\eta),$ $\qquad\qquad$ (3.1.14)

$\qquad\qquad \eta_i = 0, \quad N_i = 0.5(1 + \xi_i\xi)(1 - \eta^2).$

The shape-function formulae for "cubic" serendipity elements are given by Zienkiewicz (1977). For the linear, quadratic, and cubic serendipity shape functions used as test functions the number of contributing nodes in the algebraic equation arising from applying the Galerkin method is shown in Table 3.1.

Both the Lagrange and the serendipity family of elements extend naturally to three dimensions. The terms represented in the interpolating function and the nodal locations for members of the Lagrange family are shown in Fig. 3.9. The shape functions are obtained from a product of the one-dimensional Lagrange interpolating functions. Comparable data for the serendipity family are given by Zienkiewicz (1977). For members of both families used as test functions the number of contributing nodes in the algebraic equations, obtained after applying the Galerkin method, are shown in Table 3.1.

An examination of Table 3.1 indicates that the number of contributing nodes in an algebraic equation increases rapidly with the increase in the degree of the shape functions particularly in two and three dimensions. This implies a substantial increase in execution time, which would require a proportionately substantial increase in accuracy to justify using, say, cubic

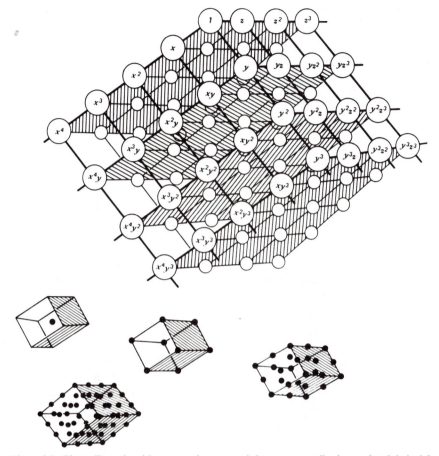

Figure 3.9. Three-dimensional Lagrange elements and the terms contributing to the global trial functions (after Oden and Reddy, 1976a; reprinted with permission of John Wiley and Sons, Inc.)

elements rather than linear elements. Alternatively the solution resulting from fewer cubic elements would have to be more accurate than the solution resulting from linear elements but requiring the same execution time.

For a typical linear elliptic problem the main contribution to the execution time is the solution of the matrix equation

$$\mathbf{Kq} = \mathbf{R}, \qquad (3.1.15)$$

where $\mathbf{K}$ is the stiffness matrix and $\mathbf{q}$ is the solution vector. One of the advantages of the finite-element method is that it leads to a sparse stiffness matrix $\mathbf{K}$. Perhaps less than 5% of the entries in a row of $\mathbf{K}$ (corresponding to the algebraic equation at one node) would be nonzero. To satisfy this criterion using cubic

Table 3.1. Nonzero entries in one row of **K** (internal global node)

Dim.	Order of shape function	Lagrange fin. el.[a]				Serendipity fin. el.[a]			Fin. diff.
		C	MS$_4$	MS$_2$	I	C	MS$_4$	MS$_2$	
1	Linear	3				3			3
	Quadratic	5		3	3	5		3	5
	Cubic	7		4		7		4	7
2	Linear	9				9			5
	Quadratic	25		15	9	21		13	9
	Cubic	49		28	16	33		20	13
3	Linear	27				27			7
	Quadratic	125	75	45	27	117	71	43	13
	Cubic	343	196	112	64	279	164	96	19

[a] C = corner, MS$_4$ = midside (four elements), MS$_2$ = midside (two elements), I = internal.

Lagrange elements in three dimensions implies solving a problem with more than 6000 nodal unknowns.

Also shown in Table 3.1 is the number of nonzero terms in each row of **K** resulting from a finite-difference formulation of the governing equation. The number of terms in one dimension has been chosen to match the number of terms corresponding to corner nodes in the finite-element method. This might be expected to lead to roughly comparable accuracy on a coarse grid.

For finite-difference equations, formed at internal global nodes, all rows of **K** have the same number of nonzero terms. To compare with this, an average number of nonzero terms per row has been evaluated for Lagrange finite elements. This number has been computed as the weighted root mean square of the number of nonzero terms in each row. This gives a rough measure of the number of operations and hence execution time. More precise operation counts depend on the particular strategy used to factorize eq. (3.1.15) (Jennings, 1977).

In computing the values shown in Table 3.2 it has been assumed that the number of internal global nodes is so large that economies associated with known function values on the boundary are insignificant. The data in Table 3.2 indicate that using higher-order schemes, particularly in two or three dimensions, produces a stiffness matrix that is far sparser for a finite-difference formulation than for a finite-element formulation. A crude comparison of the execution times can be obtained by squaring the numbers given in Table 3.2. It is clear that the finite-element solution must be significantly more accurate than the finite-difference solution, in two and three dimensions, if greater computational efficiency is to be achieved.

Although the convergence rate for different methods and interpolation functions can be predicted with reasonable precision (section 3.4), the accuracy for a particular mesh size depends on the smoothness of the exact solution

Table 3.2. Average number of nonzero entries in one row of **K**

Dimension	Order of shape function	Lagrange fin. el.: av. no. terms	Fin. diff.: no. terms
1	Linear	3	3
	Quadratic	4	5
	Cubic	5	7
2	Linear	9	5
	Quadratic	19	9
	Cubic	30	13
3	Linear	27	7
	Quadratic	80	13
	Cubic	165	19

and the problem being solved. Questions of computational efficiency are considered further in chapter 6.

All the elements considered so far are suitable candidates for use with the isoparametric formulation in section 2.4. All the elements discussed above provide C^0 interelement continuity. Hermite polynomials can be used in two and three dimensions to provide C^1 continuity, but there are rather severe restrictions on the permitted geometry of the elements (Strang and Fix, 1973, p. 89).

3.1.3. Triangular elements

Before the discovery of the isoparametric mapping, straight-sided triangular elements were popular because they could approximate an irregular boundary. For many problems very simple formulae are available for the integrations over triangular elements to evaluate the inner products (1.3.4) associated with generating the algebraic equations (Zienkiewicz, 1977, p. 168). Some typical two dimensional triangular elements and the contributing terms to the trial functions are shown in Fig. 3.10.

The shape functions are most efficiently defined in terms of element coordinates. For triangular elements *area coordinates* are the natural element coordinates. Thus, with reference to Fig. 3.11, we can define

$$L_1 = \frac{a_1 + b_1 x + c_1 y}{2A_T},$$

$$L_2 = \frac{a_2 + b_2 x + c_2 y}{2A_T}, \qquad (3.1.16)$$

$$L_3 = 1 - L_1 - L_2,$$

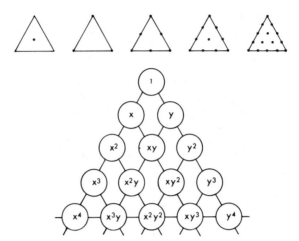

Figure 3.10. Two-dimensional triangular elements and the terms contributing to the global trial functions

where A_T is the area of the triangle and

$$a_1 = x_2 y_3 - x_3 y_2,$$
$$b_1 = y_2 - y_3, \qquad\qquad (3.1.17)$$
$$c_1 = x_3 - x_2,$$

and similarly for a_2, b_2, c_2, and so on.

For a linear triangular element the shape functions are just the area functions themselves:

$$N_1 = L_1, \qquad N_2 = L_2, \qquad N_3 = L_3. \qquad (3.1.18)$$

A consideration of eqs. (3.1.16) to (3.1.17) will indicate that $N_1 = 1$ at node 1

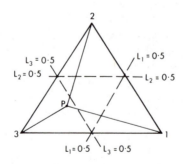

Co-ordinates of P : L_1 = (area P23)/(area 123)
L_2 = (area P31)/(area 123)
L_3 = (area P12)/(area 123)

Figure 3.11. Area coordinates in triangular elements

and $N_1 = 0$ at nodes 2 and 3 as required. For quadratic triangular elements we have

corner nodes:
$$N_1 = L_1(2L_1 - 1),$$
$$N_2 = L_2(2L_2 - 1), \qquad (3.1.19)$$
$$N_3 = L_3(2L_3 - 1),$$

midside nodes:
$$N_4 = 4L_1L_2,$$
$$N_5 = 4L_2L_3, \qquad (3.1.20)$$
$$N_6 = 4L_3L_1.$$

For cubic triangular elements,

corner nodes: $\quad N_1 = 0.5L_1(3L_1 - 1)(3L_1 - 2), \quad$ etc.,

midside nodes: $\quad N_4 = 4.5L_1L_2(3L_1 - 1), \quad$ etc., $\qquad (3.1.21)$

internal nodes: $\quad N_{10} = 27L_1L_2L_3, \quad$ etc.

The extension of the triangular elements to three dimensions produces *tetrahedral* elements (Zienkiewicz, 1977, p. 172) and *volume coordinates*. A major difficulty in three dimensions is coding the nodal points of the elements into a global configuration.

One could construct tables similar to Tables 3.1 and 3.2 for triangular and tetrahedral elements. Similar conclusions concerning the use of higher-order elements in more than one dimension would be expected.

The isoparametric formulation could be used with any of the triangular or tetrahedral elements considered above. However, the use of rectangular elements with an isoparametric formulation is much more common.

A number of the elements described above are used in the examples considered in the next section.

3.2. Examples

In this section we consider specific examples to illustrate the features of the finite-element method described in chapter 2 and the effect of using the different shape functions and elements described in section 3.1.

The first example is a one-dimensional second-order ordinary differential equation with boundary conditions at both ends of the interval. This example will exhibit the mechanics of generating and solving the system of algebraic equations which results from applying the Galerkin finite-element method.

The second example considers viscous flow in a two-dimensional channel. This example demonstrates the use of linear and quadratic rectangular elements. The third example is concerned with inviscid, incompressible flow past a circular cylinder. This example introduces the complication of solving two equations per node point. Use is made of the isoparametric transformation with both triangular and rectangular elements.

The last two examples are governed by parabolic partial differential equations. For these examples, application of the Galerkin finite-element method generates systems of ordinary differential equations. The fourth example is the unsteady heat-conduction problem considered in section 1.2.4. The fifth is based on a Burgers'-equation problem. This introduces the complication of solving a nonlinear equation.

3.2.1. A simplified Sturm–Liouville equation

The general Sturm–Liouville equation is

$$\frac{d}{dx}\left(p(x)\frac{dy}{dx}\right) + q(x)y = f \tag{3.2.1}$$

with boundary conditions at $x = 0$ and 1. Here we will solve the Sturm–Liouville equation with $p = 1$, $q = 1$ and subject to the boundary conditions

$$y(0) = 0, \qquad \frac{dy}{dx}(1) = 0. \tag{3.2.2}$$

If f is chosen to be

$$f = -\sum_{l=1}^{M} a_l \sin(l - 0.5)\pi x, \tag{3.2.3}$$

then the exact solution is

$$y = -\sum_{l=1}^{M} \frac{a_l}{1 - ((l - 0.5)\pi)^2} \sin\left[(l - 0.5)\pi x\right]. \tag{3.2.4}$$

A solution to eq. (3.2.1) is sought using the Galerkin finite-element method and a trial solution based on linear elements

$$y_a = \sum_{j=1}^{N} \bar{y}_{a_j} N_j(x), \tag{3.2.5}$$

where the shape function $N_j(x)$ is defined in an element coordinate system $x(\xi)$ as follows:

$$\begin{aligned} \text{in element } j, \qquad & N_j(\xi) = 0.5(1 + \xi), \\ \text{in element } j + 1, \quad & N_j(\xi) = 0.5(1 - \xi). \end{aligned} \tag{3.2.6}$$

In element j (Fig. 3.12) the nodal coordinate ξ is defined by

$$\xi = 2\frac{x - 0.5(x_{j-1} + x_j)}{x_j - x_{j-1}}. \tag{3.2.7}$$

The equation residual is formed by substituting eq. (3.2.5) into eq. (3.2.1), with $p = q = 1$. This gives

$$R = \frac{d^2 y_a}{dx^2} + y_a - f. \tag{3.2.8}$$

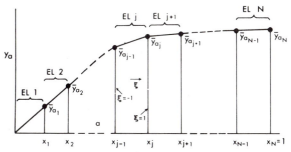

Figure 3.12. Element and nodal-point configurations for linear elements in one dimension

The Galerkin method forms the algebraic equations by evaluating the inner product $(R, N_k) = 0$:

$$\int_0^1 R N_k \, dx = 0 \quad \text{for } k = 1, \ldots, N. \tag{3.2.9}$$

Substituting eq. (3.2.8) into eq. (3.2.9) gives

$$\left(\frac{d^2 y_a}{dx^2}, N_k\right) + (y_a, N_k) - (f, N_k) = 0. \tag{3.2.10}$$

The finite-element method requires that the elements used should provide interelement continuity of derivatives of degree one less than the maximum that appears in the weak form of the governing equation, that is, eq. (3.2.10). Therefore it is necessary to integrate the first term by parts:

$$\left(\frac{d^2 y_a}{dx^2}, N_k\right) \equiv \int_0^1 \frac{d^2 y_a}{dx^2} N_k \, dx = \left[\frac{dy_a}{dx} N_k\right]_0^1 - \int_0^1 \frac{dy_a}{dx} \frac{dN_k}{dx} \, dx.$$

For $k = 1$, the node adjacent to $x = 0$ (Fig. 3.12), we have $N_1 = 0$ at $x = 0$. Therefore there is no contribution to $[(dy_a/dx)N_k]$ at $x = 0$. At $x = 1$ the boundary condition requires that $dy_a/dx = 0$. For $k = 2, \ldots, N - 1$, we have $N_k = 0$ at $x = 0$ and 1. Therefore for all the equations obtained from eq. (3.2.10) we can set

$$\left(\frac{d^2 y_a}{dx^2}, N_k\right) = -\left(\frac{dy_a}{dx}, \frac{dN_k}{dx}\right). \tag{3.2.11}$$

Substituting eq. (3.2.11) into eq. (3.2.10) permits C^0 elements to be used, since no derivatives higher than the first appear.

Substituting eq. (3.2.5) into eq. (3.2.10) gives

$$\sum_{j=1}^N \bar{y}_{aj} \left[-\left(\frac{dN_j}{dx}, \frac{dN_k}{dk}\right) + (N_j, N_k)\right] = (f, N_k), \quad k = 1, \ldots, N, \tag{3.2.12}$$

or as a matrix equation,

$$\mathbf{B} \mathbf{Y}_a = \mathbf{G}. \tag{3.2.13}$$

Because of the structural origins of the finite-element method, $\mathbf{B}$ is referred to as the *stiffness* matrix. A typical element b_{kj} of $\mathbf{B}$ is given by

$$b_{kj} = -\left(\frac{dN_j}{dx}, \frac{dN_k}{dk}\right) + (N_j, N_k). \tag{3.2.14}$$

Since $N_k = 0$ for $x < x_{k-1}$ and $x > x_{k+1}$, only three values of b_{kj} will be non-zero. These are

$$b_{k\,k-1} = + \int_{x_{k-1}}^{x_k} \left(-\frac{dN_k}{dx}\frac{dN_{k-1}}{dx} + N_k N_{k-1}\right) dx,$$

$$b_{kk} = \int_{x_{k-1}}^{x_{k+1}} \left(-\left(\frac{dN_k}{dx}\right)^2 + N_k^2\right) dx, \tag{3.2.15}$$

$$b_{k\,k+1} = \int_{x_k}^{x_{k+1}} \left(-\frac{dN_k}{dx}\frac{dN_{k+1}}{dx} + N_k N_{k+1}\right) dx.$$

It can be seen that $b_{k\,k-1}$ requires an integration over element j only (Fig. 3.12), b_{kk} requires an integration over elements j and $j + 1$, and $b_{k\,k-1}$ requires an integration over element $j + 1$ only. The integrals are most easily evaluated in element coordinates. Thus

$$b_{k\,k-1} = -\frac{2}{\Delta x_j}\int_{-1}^{1}\frac{dN_k}{d\xi}\frac{dN_{k-1}}{d\xi}d\xi + \frac{\Delta x_j}{2}\int_{-1}^{1} N_k N_{k-1} d\xi$$

$$= \frac{1}{\Delta x_j} + \frac{\Delta x_j}{6}, \tag{3.2.16}$$

where $\Delta x_j = x_j - x_{j-1}$. In a similar manner

$$b_{kk} = \frac{-1}{\Delta x_j} + \frac{\Delta x_j}{3} - \frac{1}{\Delta x_{j+1}} + \frac{\Delta x_{j+1}}{3} \tag{3.2.17}$$

and

$$b_{k\,k+1} = \frac{1}{\Delta x_{j+1}} + \frac{\Delta x_{j+1}}{6}. \tag{3.2.18}$$

For the equation formed using N_1 as a test function we do not need to evaluate $b_{k\,k-1}$, since $\bar{y}_a(o) = 0$. However, if $\bar{y}_a(o) = y_g$, then it would be necessary to evaluate $b_{1,0}y_g$ and substract it from g_1 on the right-hand side of eq. (3.2.13).

When N_N is used as the test function, contributions are obtained only from element N. Then $b_{N\,N-1}$ is as given by eq. (3.2.16), but b_{NN} is given by

$$b_{N\,N} = \frac{-1}{\Delta x_N} + \frac{\Delta x_N}{3} \tag{3.2.19}$$

and $b_{N\,N+1} = 0$.

It is possible to evaluate the right-hand side of eq. (3.2.12) exactly. However, a technique that is very useful for more complicated situations is to assume that f can be interpolated in the same manner as y_a. Thus let

$$f = \sum_{j=1}^{N} \bar{f}_j N_j(x), \tag{3.2.20}$$

where $\bar{f}_j$ are the nodal values of f, given by eq. (3.2.3):

$$\bar{f}_j = -\sum_{l=1}^{M} a_l \sin{(l - 0.5)}\pi x_j. \tag{3.2.21}$$

Consequently the right-hand side of eq. (3.2.12) becomes

$$(f, N_k) = \sum_{j=1}^{N} \bar{f}_j(N_j, N_k). \tag{3.2.22}$$

Evaluation of the inner product with linear elements gives the following expression for g_k, an element of $\mathbf{G}$, at internal nodes:

$$g_k = \frac{\Delta x_k}{6}\bar{f}_{k-1} + \frac{\Delta x_k + \Delta x_{k+1}}{3}\bar{f}_k + \frac{\Delta x_{k+1}}{6}\bar{f}_{k+1}. \tag{3.2.23}$$

At node N, g_N is

$$g_N = \frac{\Delta x_N}{6}\bar{f}_{N-1} + \frac{\Delta x_N}{3}\bar{f}_N. \tag{3.2.24}$$

It is apparent from eqs. (3.2.16) to (3.2.18) that $\mathbf{B}$ in eq. (3.2.13) has a *tri-diagonal* form. This can be taken advantage of in solving eq. (3.2.13). Using the Thomas algorithm (Isaacson and Keller, 1966, p. 55) the solution only requires $5N - 4$ operations (only counting multiplications and divisions). Solutions have been obtained with the values

$$a_1 = 1.0, \quad a_2 = -0.5, \quad a_3 = 0.3, \quad a_4 = -0.2, \quad a_5 = 0.1$$

in eq. (3.2.3). The nodal errors for 11 equally spaced points in the interval $0 \le x \le 1$ are shown in Fig. 3.13. It is apparent that the error is largest adjacent

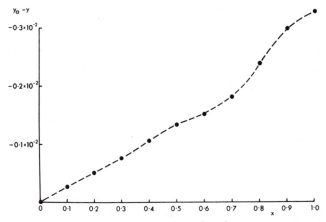

Figure 3.13. Error distribution for the solution of $d^2y/dx^2 + y = f$ using linear elements

Table 3.3. Error in rms norm for Sturm–
Liouville problem, $p = 1$, $q = 1$

N	h	$\|y_a - y\|_{rms}$
6	0.2	7.830×10^{-3}
11	0.1	1.860×10^{-3}
21	0.05	0.447×10^{-3}
41	0.025	0.109×10^{-3}
81	0.0125	0.027×10^{-3}

to the boundary where the Neumann (derivative) boundary condition is applied. For a range of uniform grid sizes the error in the rms norm is shown Table 3.2. Here the rms norm is defined by $\|y_a - y\|_{rms} = \{\Sigma_{i=1}^{N} (y_{a_i} - y_i)^2\}^{1/2} / N^{1/2}$. The results shown in Table 3.3 demonstrate a reduction in error that varies like h^2. This is to be expected for linear elements.

3.2.2. Viscous flow in a channel

This is the same problem that was considered in section 1.2.3. Viscous flow through a square channel (Fig. 1.6) is governed by the following nondimensional equation:

$$\frac{\partial^2 w}{\partial x^2} + \frac{\partial^2 w}{dy^2} + 1 = 0, \tag{3.2.25}$$

with boundary conditions

$$w = 0 \quad \text{on } x = \pm 1 \text{ and } y = \pm 1.$$

A trial solution is introduced by

$$w_a = \sum_{j=1}^{N} \bar{w}_{aj} N_j(x, y), \tag{3.2.26}$$

where $\bar{w}_{aj}$ are the nodal values of w_a, and $N_j(x, y)$ are bilinear rectangular shape functions, which in element coordinates (ξ, η) (Fig. 3.6) are

$$N_j = 0.25(1 + \xi_j \xi)(1 + \eta_j \eta), \tag{3.2.27}$$

where (ξ_j, η_j) are the four corners $(\xi_j = \pm 1, \eta_j = \pm 1)$. Substitution of eq. (3.2.26) into eq. (3.2.25) and application of the Galerkin method produces the system of algebraic equations

$$\left(\frac{\partial^2 w_a}{\partial x^2}, N_k\right) + \left(\frac{\partial^2 w_a}{\partial y^2}, N_k\right) = (-1, N_k). \tag{3.2.28}$$

To be able to use C^0 elements it is necessary to apply Green's theorem (the two-dimensional analogue of integration by parts) to the two inner products

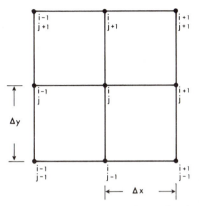

Figure 3.14. Global mesh notation for viscous flow in a channel

on the left-hand side of eq. (3.2.28). For equations formed with nodes adjacent to the boundary as test functions, a line integral along the domain boundary arises after application of Green's theorem. However, this line integral makes no contribution because the corresponding N_k is zero on the boundary. Thus eq. (3.2.28) becomes

$$- \sum_{j=1}^{N} \left(\int_{-1}^{1} \int_{-1}^{1} \frac{\partial N_j}{\partial x} \frac{\partial N_k}{\partial x} + \frac{\partial N_j}{\partial y} \frac{\partial N_k}{\partial y} dx \, dy \right) \bar{w}_{aj}$$
$$= - \int_{-1}^{1} \int_{-1}^{1} 1 N_k \, dx \, dy, \qquad k = 1, \ldots, N. \tag{3.2.29}$$

If the boundary conditions were such that equations had to be formed on the boundary, then in general a line-integral contribution to the left-hand side of eq. (3.2.29) would be expected.

Evaluation of the integrals in eq. (3.2.29) is best done in element coordinates, as with the example in section 3.2.1. On a rectangular grid $(\Delta x, \Delta y)$ the equation formed with a test function associated with the grid point (i, j) (Fig. 3.14) has the following form:

$$\frac{(\bar{w}_a)_{i-1,j+1} - (2\bar{w}_a)_{i,j+1} + (\bar{w}_a)_{i+1,j+1}}{6 \Delta x^2} + \frac{2[(\bar{w}_a)_{i-1,j} - (2\bar{w}_a)_{i,j} + (\bar{w}_a)_{i+1,j}]}{3 \Delta x^2}$$

$$+ \frac{(\bar{w}_a)_{i-1,j-1} - (2\bar{w}_a)_{i,j-1} + (\bar{w}_a)_{i+1,j-1}}{6 \Delta x^2} + \frac{(\bar{w}_a)_{i-1,j-1} - (2\bar{w}_a)_{i-1,j} + (\bar{w}_a)_{i-1,j+1}}{6 \Delta y^2}$$

$$+ \frac{2[(\bar{w}_a)_{i,j-1} - (2\bar{w}_a)_{i,j} + (\bar{w}_a)_{i,j+1}]}{3 \Delta y^2} + \frac{(\bar{w}_a)_{i+1,j-1} - (2\bar{w}_a)_{i+1,j} + (\bar{w}_a)_{i+1,j+1}}{6 \Delta y^2}$$

$$= -1. \tag{3.2.30}$$

Table 3.4. Solutions for viscous flow in a square channel

N	Linear shape functions		Quadratic (Lagrange) shape functions	
	w_{CL}	$\dot{q}$	w_{CL}	$\dot{q}$
3	0.3107	0.5116	0.2949	0.5584
5	0.3014	0.5394	0.2947	0.5612
7	0.2984	0.5494	0.2946	0.5619
9	0.2970	0.5540	0.2947	0.5621
11	0.2963	0.5565	0.2947	0.5622
25	0.2950	0.5610		
45	0.2947	0.5618		
Exact	0.2947	0.5623	0.2947	0.5623

The first three terms on the left-hand side are associated with $\partial^2 w_a/\partial x^2$ in eq. (3.2.28), and the last three terms with $\partial^2 w_a/\partial y^2$. An examination of these terms indicates a resemblance to centered-difference formulae distributed over adjacent grid lines with weights $\frac{1}{6}$, $\frac{2}{3}$, and $\frac{1}{6}$. This aspect of the finite-element method will be pursued in section 3.3.

A system of equations like (3.2.30) could be written down in the form of eq. (3.2.13). However, unless one dimension is very much narrower than the other, it is not possible to number the nodes so that **B** is banded. That is, all of the nonzero elements should be close to the diagonal.

Fortunately, **B** formed from eq. (3.2.30) is *diagonally dominant* (Jennings, 1977), that is,

$$|b_{kk}| \geq \sum_{j \neq k} |b_{kj}| \quad \text{and} \quad b_{kk} \text{ all have the same sign.}$$

This implies that an iterative solution of eq. (3.2.30) is possible. It can be constructed by first defining an *algebraic-equation residual* R_a. This should not be confused with the partial-differential-equation residual R. For an arbitrary choice of $(\bar{w}_a)_{i,j}$, R_a will not be zero. The solution of eq. (3.2.30) will make $R_a = 0$. However, R will still be nonzero except in the limit as $\Delta x, \Delta y \to 0$. The algebraic-equation residual R_a can be written (if $\Delta x = \Delta y$)

$$R_a = \tfrac{1}{3}[(\bar{w}_a)_{i-1,j+1} + (\bar{w}_a)_{i,j+1} + (\bar{w}_a)_{i+1,j+1} + (\bar{w}_a)_{i-1,j} - (8\bar{w}_a)_{i,j}$$
$$+ (\bar{w}_a)_{i+1,j} + (\bar{w}_a)_{i-1,j-1} + (\bar{w}_a)_{i,j-1} + (\bar{w}_a)_{i+1,j-1} + 3\,\Delta x^2]. \tag{3.2.31}$$

Then the correction $(\Delta\bar{w}_a)_{i,j}$ can be defined by

$$(\Delta\bar{w}_a)_{i,j} = -b_{kk}^{-1} R_a = \tfrac{3}{8} R_a,$$

and an iterative sequence can be defined by

$$(\bar{w}_a^{(v+1)})_{i,j} = (\bar{w}_a^{(v)})_{i,j} + \lambda (\Delta\bar{w}_a^{(v+1)})_{i,j}, \tag{3.2.32}$$

where λ is a *relaxation factor* introduced to speed convergence.

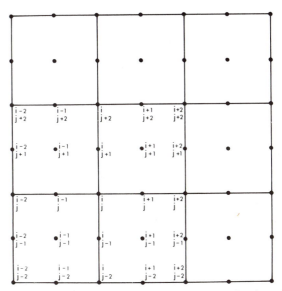

Figure 3.15. Distribution of quadratic Lagrange elements for channel flow

Solutions of eq. (3.2.30), using the iterative technique (3.2.32), are indicated in Table 3.4 for various degrees of mesh refinement. w_{CL} is the centerline velocity, and $\dot{q}$ is the nondimensional mass flow, given by

$$\dot{q} = \int_{-1}^{1} \int_{-1}^{1} w_a \, dx \, dy. \tag{3.2.33}$$

A comparison with the results shown in Table 1.6 indicates that the results obtained with "linear" finite elements are considerably less accurate than the results obtained with the traditional Galerkin method with the same number of unknowns, N^2. The relatively high accuracy achieved with one "quadratic" term in eq. (1.2.40) (see Table 1.6) suggests that the use of "quadratic" shape functions would be considerably more accurate than the use of "linear" shape functions. A typical arrangement of "quadratic" Lagrange elements is shown in Fig. 3.15.

The shape functions are given by eq. (3.1.13). After application of the Galerkin method, algebraic equations equivalent to eq. (3.2.30) are obtained. The contributing nodes depend on which node the test function is associated with. Thus for a corner node (i, j in Fig. 3.15) contributions to the left-hand side of eq. (3.2.30) come from the 25 nodes labeled, whereas for a test function associated with node $i + 1, j + 1$ contributions come from the nine immediately adjacent nodes.

The algebraic equations associated with each node can be solved using the iterative scheme (3.2.32). Results for various N are shown in Table 3.4, and it is clear that the solution for the same N is considerably more accurate than that using linear elements.

3.2.3. Inviscid, incompressible flow

This example is of interest because it is of practical importance in predicting the flow around aircraft wings and fuselages (Rubbert and Saaris, 1972). Here we consider flow around a two-dimensional circular cylinder for which an exact solution is readily available. The equations governing this type of flow will be expressed in Cartesian coordinates. However, to suit the body geometry the computational domain will be described in polar coordinates. This provides an opportunity to demonstrate the isoparametric transformation in action.

The governing equations for this problem can be expressed in terms of the velocity components u and v as

$$\frac{\partial u}{\partial x} + \frac{\partial v}{\partial y} = 0, \tag{3.2.34}$$

$$\frac{\partial u}{\partial y} - \frac{\partial v}{\partial x} = 0. \tag{3.2.35}$$

Eq. (3.2.34) is a statement of conservation of mass, and eq. (3.2.35) states that the flow is irrotational. Solutions to eq. (3.2.34) and (3.2.35) will be sought using the Galerkin finite-element method in the quarter plane shown in Fig. 3.16. Because of symmetry, only a quarter plane need be considered.

Appropriate boundary conditions are

$$\frac{\partial u}{\partial y} = v = 0 \quad \text{on } AB,$$

$$v_n = 0 \quad \text{on } BC, \tag{3.2.36}$$

$$\frac{\partial u}{\partial x} = v = 0 \quad \text{on } DC,$$

$$u = u_{\text{FS}} \text{ and } v = v_{\text{FS}} \quad \text{on } DA.$$

u_{FS} and v_{FS} must be specified. If DA is far enough from the body, then

$$u_{\text{FS}} \approx u_\infty \quad \text{and} \quad v_{\text{FS}} \approx 0. \tag{3.2.37}$$

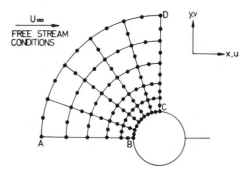

Figure 3.16. Element distribution for flow past a circular cylinder

Trial solutions are introduced for u and v as

$$u = \sum_j N_j \bar{u}_j,$$

$$v = \sum_j N_j \bar{v}_j. \tag{3.2.38}$$

$\bar{u}_j$ and $\bar{v}_j$ are the nodal values of u and v. For this problem "linear" and "quadratic" shape functions with rectangular and triangular elements have been considered. Therefore for the moment the particular choice for the shape function will not be specified.

Substitution of eqs. (3.2.38) into eqs. (3.2.34) and (3.2.35), and application of the Galerkin method, produces the algebraic equations

$$\sum_j a_{kj} \bar{u}_j + \sum_j b_{kj} \bar{v}_j = 0, \qquad k = 1, \ldots, N,$$

$$\sum_j b_{kj} \bar{u}_j - \sum_j a_{kj} \bar{v}_j = 0, \qquad k = 1, \ldots, N. \tag{3.2.39}$$

The form of the coefficients a_{kj}, b_{kj} depends on whether Green's theorem is applied or not. Since only first derivatives appear in the governing equations and since finite elements with C^0 interelement continuity are to be used, application of Green's theorem is not required for convergence. However, it does alter the application of the boundary conditions.

After application of Green's theorem the expressions for a_{kj} and b_{kj} are

$$a_{kj} = \iint \frac{\partial N_k}{\partial x} N_j \, dx \, dy - \int_s N_k N_j l_x \, ds, \tag{3.2.40}$$

$$b_{kj} = \iint \frac{\partial N_k}{\partial y} N_j \, dx \, dy - \int N_k N_j l_y \, ds, \tag{3.2.41}$$

where l_x and l_y are the direction cosines. The line integrals contribute only if the test function N_k is associated with a boundary node. However, with an appropriate choice of l_x, l_y or a particular boundary condition, the line integral may be zero.

The integrals in eqs. (3.2.40) and (3.2.41) can be evaluated most conveniently via the isoparametric transformation. As indicated in section 2.4, the integrations are carried out over elements in (ξ, η) space rather than over the irregular elements in (x, y) space (Fig. 3.16). A contribution from one element to a_{kj} and b_{kj}, when k corresponds to an internal node, can be written

$$(a_{kj})_e = \sum_{l=1}^{Le} c_{jkl} \bar{y}_l \tag{3.2.42}$$

and

$$(b_{kj})_e = -\sum_{l=1}^{Le} c_{jkl} \bar{x}_l \tag{3.2.43}$$

where $(\bar{x}_l, \bar{y}_l)$ are the coordinates of the nodes of the element in physical space and Le is the total number of nodes in the element. The coefficients c_{jkl} are given by

$$c_{jkl} = \int\int N_j \left\{ \frac{\partial N_k}{\partial \eta} \frac{\partial N_l}{\partial \xi} - \frac{\partial N_k}{\partial \xi} \frac{\partial N_l}{\partial \eta} \right\} d\xi \, d\eta. \qquad (3.2.44)$$

The form of eqs. (3.2.42) to (3.2.44) is interesting because the location of the element in physical space enters only through the algebraic expressions in eqs. (3.2.42) and (3.2.43). The computationally expensive part is the evaluation of c_{jkl} from eq. (3.2.44); but this need only be done once. Thus for this example the application of the isoparametric transformation is very economical.

This economy stems directly from the appearance of a *single* first derivative in the integrand (eq. 3.2.40). The determinant of the Jacobian associated with transforming the derivative into an equivalent expression in (ξ, η) space cancels with the derivative associated with transforming the integral.

The more general situation can be appreciated by applying the Galerkin finite-element method to $\partial^2 u/\partial x^2$. After application of Green's theorem to permit C^0 elements to be used, the isoparametric transformation gives eq. (2.4.9), which contains det J in the denominator. This requires an integration *over every element* in the domain. If the evaluation of the isoparametric integrals is very time consuming, this may justify introducing first derivatives as unknowns and including two extra equations per node, so that the use of equations like (3.2.42) to (3.2.44) becomes available.

The use of an isoparametric formulation, and in particular the form of eqs. (3.2.42) and (3.2.43), permits the grid to be refined in regions where rapid variation is expected. Thus for the present example elements have been clustered close to the cylinder surface.

From the nature of the boundary conditions (3.2.36) and (3.2.37), only one Galerkin equation is required on AB, BC, and CD, and no equation is required on AD. The condition of no flow normal to the body surface provides the second equation on BC. Thus

$$v_n = u \cos\theta - v \sin\theta = 0. \qquad (3.2.45)$$

In applying one Galerkin equation on AB, BC, and CD a choice exists between eq. (3.2.34) and eq. (3.2.35). For the results shown in Tables 3.5 and 3.6, eq. (3.2.34) has been used on these boundaries.

If the body shown in Fig. 3.16 were of a more practical shape (e.g. an airfoil), the main parameter of interest would be the pressure at the body surface. But this is directly related to the velocity at the surface. Consequently the rms difference between the finite-element solution for the tangential velocity component at the body surface, q_T, and the corresponding exact solution, q_e, has been used to compare the accuracy achieved with different elements and element densities. The rms difference σ is defined as follows

Table 3.5. Comparison of elements for inviscid incompressible flow with a coarse grid

Element type	Shape function[a]	Number of unknowns	Number of elements	CPU time (sec)	σ
Rectangular	Linear	199	100	5.1	0.043
Rectangular	Quadratic (S)	149	25	4.5	0.049
Rectangular	Quadratic (L)	199	25	7.5	0.047
Triangular	Linear	199	200	5.1	0.061
Triangular	Quadratic	199	50	7.1	0.120

[a] S = serendipity, L = Lagrange.

Table 3.6. Comparison of elements for inviscid, incompressible flow with a moderate grid

Element type	Shape function[a]	Number of unknowns	Number of elements	CPU time (sec)	σ
Rectangular	Linear	399	200	18	0.034
Rectangular	Quadratic (S)	299	50	15	0.023
Rectangular	Quadratic (L)	399	50	25	0.023
Triangular	Linear	399	400	17	0.034
Triangular	Quadratic	399	100	23	0.018

[a] S = serendipity, L = Lagrange.

$$\sigma = \left[\sum_{i=1}^{N_s} \frac{(q_T - q_e)^2}{N_s} \right]^{1/2} \tag{3.2.46}$$

where N_s is the number of grid points on the body surface.

The system of equations (3.2.39) supplemented by the use of eq. (3.2.45) on the body is solved using a sparse Gauss elimination procedure (Fletcher, 1976). The results for different grid refinements and elements are summarized in Tables 3.5 and 3.6.

The CPU times quoted in Tables 3.5 and 3.6 are for an IBM 370-168. For coarse grids it is clear that higher-order shape functions do not improve the accuracy. The data in Table 3.5 indicate that rectangular elements produce more accurate solutions than triangular elements for comparable execution times. The data also indicate that the use of serendipity elements is more economical than the use of Lagrange elements with comparable accuracy. The results shown in Table 3.6 have been obtained with double the number of elements for Table 3.5. The influence of grid refinement on the accuracy achieved from quadratic shape functions is much larger than the influence on the accuracy from linear shape functions. At this level of mesh refinement quadratic shape functions are more efficient than linear ones. Also it is apparent that triangular elements are as accurate as rectangular ones.

It can be seen from Tables 3.5 and 3.6 that quadratic shape functions require more execution time than linear ones for the same number of unknowns. This is mainly due to the larger number of nonzero elements that occur in the stiffness matrix associated with eqs. (3.2.39). A more extensive discussion of different elements for this problem is given by Fletcher (1977).

3.2.4. Unsteady heat conduction

We wish to determine the temporal temperature variation along a bar (Fig. 1.7, section 1.2.4). Here we use the problem to illustrate the semidiscrete Galerkin method. That is, a finite-element representation is introduced only for the spatial temperature variation. The decay of temperature with time is then governed by a system of ordinary differential equations. With a suitable nondimensionalization (see section 1.2.4), the governing equation becomes

$$\frac{\partial \theta}{\partial t} - \frac{\partial^2 \theta}{\partial x^2} = 0 \tag{3.2.47}$$

with initial and boundary conditions

$$\theta(x, 0) = \sin \pi x + x$$

$$\theta(0, t) = 0 \quad \text{and} \quad \theta(1, t) = 1.$$

A trial solution for θ is introduced as

$$\theta_a = \sum_j N_j(x)\bar{\theta}_j(t). \tag{3.2.48}$$

We have dropped the a from $\bar{\theta}_{aj}$ for notational convenience. In contrast to the trial solution assumed in applying the traditional Galerkin method to this problem (eq. (1.2.47)), no attempt is made here to introduce an auxiliary function θ_0 to account for the initial and boundary conditions. The weak form of eq. (3.2.47) can be written

$$\left(\frac{\partial \theta}{\partial t}, N_k\right) - \left(\frac{\partial^2 \theta}{\partial x^2}, N_k\right) = 0. \tag{3.2.49}$$

Integration by parts and introduction of eq. (3.2.48) gives

$$\sum_j (N_j, N_k)\frac{d\bar{\theta}_j}{dt} + \sum_j \left(\frac{dN_j}{dx}, \frac{dN_k}{dx}\right)\bar{\theta}_j = 0. \tag{3.2.50}$$

Since equations associated with boundary nodes are not required, no additional contribution arises from the integration by parts.

A linear shape function will be introduced for N_j (and N_k). Evaluation of eq. (3.2.50) on a uniform grid gives

$$\frac{1}{6}\frac{d\bar{\theta}_{j-1}}{dt} + \frac{2}{3}\frac{d\bar{\theta}_j}{dt} + \frac{1}{6}\frac{d\bar{\theta}_{j+1}}{dt} = \frac{1}{\Delta x^2}(\bar{\theta}_{j-1} - 2\bar{\theta}_j + \bar{\theta}_{j+1}). \tag{3.2.51}$$

The right-hand side is identical with a second-order finite-difference expression for $\partial^2 \theta / \partial x^2$.

To create an efficient algorithm from eq. (3.2.51), terms like $d\bar{\theta}/dt$ are replaced by $(\bar{\theta}^{n+1} - \bar{\theta}^n)/\Delta t$ and the right-hand side is evaluated as a weighted average of the values at the nth and $(n + 1)$th time levels. This gives

$$\tfrac{1}{6} \Delta\bar{\theta}_{j-1}^{n+1} + \tfrac{2}{3} \Delta\bar{\theta}_j^{n+1} + \tfrac{1}{6} \Delta\bar{\theta}_{j+1}^{n+1}$$

$$= \frac{\Delta t}{\Delta x^2} [\lambda(\bar{\theta}_{j-1}^{n+1} - 2\bar{\theta}_j^{n+1} + \bar{\theta}_{j+1}^{n+1}) + (1 - \lambda)(\bar{\theta}_{j-1}^n - 2\bar{\theta}_j^n + \bar{\theta}_{j+1}^n)]. \tag{3.2.52}$$

When the parameter $\lambda = 1$, the right-hand side is evaluated at time level $n + 1$; when $\lambda = 0$, it is evaluated at time level n.

The form of the left-hand side of eq. (3.2.52) makes an implicit solution mandatory except for the particular value $\lambda = \Delta x^2/6 \Delta t$. However, eq. (3.2.52) can be manipulated into the following tridiagonal form:

$$A_j^n \Delta\bar{\theta}_{j-1}^{n+1} + B_j^n \Delta\bar{\theta}_j^{n+1} + C_j^n \Delta\bar{\theta}_{j+1}^{n+1} = D_j^n, \tag{3.2.53}$$

where

$$\Delta\bar{\theta}^{n+1} = \bar{\theta}^{n+1} - \bar{\theta}^n,$$

$$A_j^n = \frac{1}{6} - \frac{\lambda \Delta t}{\Delta x^2},$$

$$B_j^n = \frac{2}{3} + \frac{\lambda \Delta t}{\Delta x^2},$$

$$C_j^n = \frac{1}{6} - \frac{\lambda \Delta t}{\Delta x^2},$$

$$D_j^n = \Delta t \frac{\bar{\theta}_{j-1}^n - 2\bar{\theta}_j^n + \bar{\theta}_{j+1}^n}{\Delta x^2}.$$

The problem has an exact solution

$$\theta_{ex} = (\sin \pi x) \exp(-\pi^2 t) + x. \tag{3.2.54}$$

In the limit that the steady state is reached, $D_j^n = 0$ and $\Delta\bar{\theta}^{n+1} = 0$. Since the boundary conditions at $x = 0$ and 1 are not functions of time, $\Delta\bar{\theta}_1^{n+1} = \Delta\bar{\theta}_N^{n+1} = 0$.

Solutions to eq. (3.2.53) have been obtained with $\lambda = 0.5$. This gives the *Crank–Nicolson* scheme, which is second-order accurate in time. A time-step $\Delta t = 0.001$ has been used. At time intervals $\Delta t = 0.05$, the finite-element solution has been compared with the corresponding exact solution. Table 3.7 gives a comparison of the error in the solution for various mesh refinements. The rms error σ shown in Table 3.7 has been evaluated over the nodal points in the spatial domain at each time interval. The error σ is related to the error in the discrete L_2 norm by $\|\theta_a - \theta_{ex}\|_{2,d} = N^{-1/2}\sigma$. The time step is sufficiently small that the rms errors, shown in Table 3.7, are associated with the spatial approximation only. The results indicate a rapid improvement in accuracy

Table 3.7. Comparison of solutions for unsteady heat conduction: linear elements

t	σ			θ_{ex}
	$N = 11$	$N = 21$	$N = 51$	$(x = 0.5)$
0	0	0	0	1.5000
0.05	0.1847×10^{-2}	0.0464×10^{-2}	0.0086×10^{-2}	1.1105
0.10	0.2251×10^{-2}	0.0568×10^{-2}	0.0106×10^{-2}	0.8727
0.15	0.2055×10^{-2}	0.0517×10^{-2}	0.0094×10^{-2}	0.7275
0.20	0.1668×10^{-2}	0.0420×10^{-2}	0.0076×10^{-2}	0.6389
0.25	0.1271×10^{-2}	0.0320×10^{-2}	0.0058×10^{-2}	0.5848
0.30	0.0929×10^{-2}	0.0235×10^{-2}	0.0043×10^{-2}	0.5518
0.35	0.0661×10^{-2}	0.0168×10^{-2}	0.0031×10^{-2}	0.5316
0.40	0.0461×10^{-2}	0.0118×10^{-2}	0.0022×10^{-2}	0.5193
0.45	0.0317×10^{-2}	0.0082×10^{-2}	0.0016×10^{-2}	0.5118
0.50	0.0216×10^{-2}	0.0057×10^{-2}	0.0012×10^{-2}	0.5072

with grid refinement and close agreement with the exact solution for the finest grid ($N = 51$). For $\Delta x = 0.02$ the temporal development of the midpoint temperature, $\theta(x = 0.5)$, is shown in Fig. 3.17.

3.2.5. Burgers' equation

We reconsider here the problem described in section 1.2.5 to illustrate the traditional Galerkin method. Burgers' equation is

$$\frac{\partial u}{\partial t} + u\frac{\partial u}{\partial x} - \frac{1}{Re}\frac{\partial^2 u}{\partial x^2} = 0. \tag{3.2.55}$$

This equation is similar to the unsteady heat-conduction problem considered in the previous section. However, the appearance of the nonlinear convective

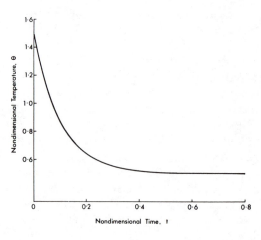

Figure 3.17. Variation of temperature with time for the node at $x = 0.5$

term in eq. (3.2.55) creates large gradients in the flow if Re is large. It also complicates the computational solution. Eq. (3.2.55) will be solved in conjunction with the boundary conditions

$$u(x_L, t) = 1, \qquad u(x_R, t) = 0, \tag{3.2.56}$$

and initial conditions

$$u(x, 0) = \begin{cases} 1 & \text{if } x_L \le x \le 0, \\ 0 & \text{if } 0 < x \le x_R. \end{cases} \tag{3.2.57}$$

x_L and x_R are chosen to be sufficiently large that eq. (3.2.56) is true at the time and for the value of the Re at which a comparison is made with the exact solution (1.2.61). Typically $1 \le |x_L|, x_R \le 2$.

The equations (3.2.55) to (3.2.57) describe the viscous propagation of a shock wave.

A trial solution, similar to that used in section 3.2.4, is introduced as

$$u_a = \sum_j N_j(x) \bar{u}_j(t). \tag{3.2.58}$$

Application of the Galerkin method produces a system of ordinary differential equations that can be written

$$M\dot{U} + (B + C)U = 0. \tag{3.2.59}$$

An element of U is just the nodal unknown $\bar{u}_j$. An element of the *mass matrix* M is

$$m_{kj} = (N_k, N_j). \tag{3.2.60}$$

An element of C is

$$c_{kj} = \frac{1}{Re}\left(\frac{dN_k}{dx}, \frac{dN_j}{dx}\right), \tag{3.2.61}$$

and an element of B is

$$b_{kj} = \sum_i \bar{u}_i \left(N_k, N_i \frac{dN_j}{dx}\right). \tag{3.2.62}$$

It can be seen that the coefficients b_{kj} depend on the solution. The structure of the ordinary differential equations (3.2.59) is particularly simple if linear elements are used with a uniform grid. Then eq. (3.2.59) takes the form

$$\frac{1}{6}\frac{d\bar{u}_{k-1}}{dt} + \frac{2}{3}\frac{d\bar{u}_k}{dt} + \frac{1}{6}\frac{d\bar{u}_{k+1}}{dt} + \frac{\bar{u}_{k-1} + \bar{u}_k + \bar{u}_{k+1}}{3}\frac{\bar{u}_{k+1} - \bar{u}_{k-1}}{2\,\Delta x} \tag{3.2.63}$$

$$-\frac{1}{Re}\frac{\bar{u}_{k-1} - 2\bar{u}_k + \bar{u}_{k+1}}{\Delta x^2} = 0.$$

The nonlinear term may be compared with

$$u\frac{du}{dx} \Rightarrow \bar{u}_k \frac{\bar{u}_{k+1} - \bar{u}_{k-1}}{2\,\Delta x} \tag{3.2.64}$$

that would arise through introduction of a centered finite-difference approximation of du/dx. The Galerkin formulation smooths the nonlinear contribution over three neighboring grid points when used with linear elements.

An alternative treatment of the nonlinear term is possible (Swartz and Wendroff, 1969). Eq. (3.2.55) can be written

$$\frac{\partial u}{\partial t} + \frac{1}{2}\frac{\partial (u^2)}{\partial x} - \frac{1}{Re}\frac{\partial^2 u}{\partial x^2} = 0. \tag{3.2.65}$$

If, in addition to eq. (3.2.58), the following trial solution is introduced for u^2:

$$u_a^2 = \sum_j N_j(x)\bar{u}_j^2(t), \tag{3.2.66}$$

application of the Galerkin method produces the following system of ordinary differential equations instead of eq. (3.2.59):

$$\mathbf{M\dot{U}} + \mathbf{BU}^2 + \mathbf{CU} = 0, \tag{3.2.67}$$

where an element of $\mathbf{U}^2$ is $\bar{u}_j^2$ and an element of $\mathbf{B}$ is given by

$$b_{kj} = \frac{1}{2}\left(N_k, \frac{dN_j}{dx}\right). \tag{3.2.68}$$

For linear elements on a uniform grid, eq. (3.2.63) is replaced by

$$\frac{1}{6}\frac{d\bar{u}_{k-1}}{dt} + \frac{2}{3}\frac{d\bar{u}_k}{dt} + \frac{1}{6}\frac{d\bar{u}_{k+1}}{dt} + \frac{\bar{u}_{k+1}^2 - \bar{u}_{k-1}^2}{4\,\Delta x} - \frac{1}{Re}\frac{\bar{u}_{k-1} - 2\bar{u}_k + \bar{u}_{k+1}}{\Delta x^2} = 0. \tag{3.2.69}$$

The advantage of the form of the nonlinear representation shown in eq. (3.2.69) is that far less algebraic manipulation is required than with eq. (3.2.63) if higher-order shape functions, more than one dimension, or cubic nonlinearities occur (see section 4.1.2).

The equations (3.2.59) have been integrated from $t = 0$ to $t = 0.5$ typically. The numerical integration has been carried out with a fourth-order Runge–Kutta scheme and a sufficiently small time step that any inaccuracies in the finite-element solution at $t = 0.5$ are due to the spatial discretization alone. At the final time, the solution is compared with the exact solution. Some typical results for the spatial distribution of the error are shown in Fig. 3.18. The finite element solutions are based on the form (3.2.63). A coarse mesh corresponds to $N = 11$ in Table 3.8, and a fine mesh corresponds to $N = 21$. It is clear that the linear finite-element solution is more accurate than a conventional three-point finite-difference scheme. The results shown in Table 3.8 are not all integrated to the same time, so that a comparison of the error in the rms norm is the best way of judging the accuracy of the alternative methods. Clearly grid refinement and a higher-order shape function produce a significant improvement in the accuracy.

A solution ($N = 21$) based on eq. (3.2.69) is also shown in Table 3.8. For this particular mesh refinement the solution is less accurate than that obtained

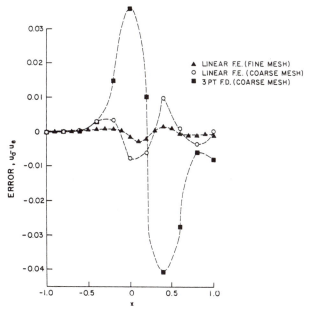

Figure 3.18. Spatial error distribution for Burgers' equation, Re = 10, $t = 0.5$

Table 3.8. Solution of Burgers' equation by Galerkin finite-element method

x	3-pt. F.D. $N = 11$ $t = 0.81$	Lin. F.E. $N = 11$ $t = 0.47$	Lin. F.E. $N = 21$ $t = 0.50$	Lin. F.E. Eq. (3.2.69) $N = 21$ $t = 0.50$	Quad. F.E. $N = 21$ $t = 0.50$	Exact $t = 0.50$
-1.0	1.0000	1.0000	1.0000	1.0000	1.0000	1.0000
-0.9			1.0000	1.0000	1.0000	1.0000
-0.8	1.0000	1.0000	1.0000	1.0000	1.0000	1.0000
-0.7			1.0000	1.0000	0.9999	0.9999
-0.6	1.0000	1.0003	0.9998	0.9998	0.9995	0.9996
-0.5			0.9991	0.9990	0.9987	0.9987
-0.4	1.0000	0.9989	0.9967	0.9966	0.9959	0.9960
-0.3			0.9907	0.9898	0.9893	0.9891
-0.2	1.0000	0.9746	0.9734	0.9728	0.9724	0.9725
-0.1			0.9367	0.9358	0.9370	0.9366
0	0.9731	0.8514	0.8657	0.8650	0.8671	0.8672
0.1			0.7482	0.7488	0.7501	0.7509
0.2	0.8021	0.5608	0.5864	0.5900	0.5882	0.5884
0.3			0.4065	0.4130	0.4053	0.4062
0.4	0.4696	0.2339	0.2463	0.2537	0.2440	0.2448
0.5			0.1307	0.1368	0.1293	0.1301
0.6	0.1948	0.0540	0.0613	0.0652	0.0611	0.0619
0.7			0.0256	0.0277	0.0263	0.0268
0.8	0.0622	0.0049	0.0094	0.0105	0.0103	0.0106
0.9			0.0029	0.0037	0.0036	0.0039
1.0	0.0082	0.0005	0.0000	0.0013	0.0011	0.0013
$\|u_a - u_e\|_{\mathrm{rms}}$	0.0215	0.0050	0.0011	0.0022	0.0005	

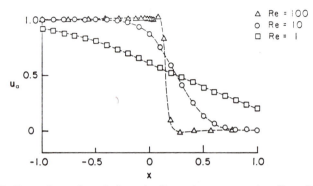

Figure 3.19. Burgers' equation solution using linear elements, at various Reynolds numbers

with a conventional finite-element method. The greater accuracy arises from the smoothing action of $(\bar{u}_{k-1} + \bar{u}_k + \bar{u}_{k+1})/3$ in eq. (3.2.63). This is important where the solution is changing rapidly with x. Further results using the *conservation form*, as eq. (3.2.65) is called, and different time integration schemes are given by Fletcher (1982b).

For the data with $N = 11$ a comparison can be made with the solution obtained with the traditional Galerkin method (Table 1.8). It is apparent that the traditional Galerkin method is more accurate. This is due to the use of high-order trial functions spanning the whole domain.

For linear elements and $N = 21$, solutions at various Reynolds numbers are presented in Fig. 3.19 and Table 3.9. The results at Re = 100 have been obtained on a variable grid constructed to cluster more grid points in the location where the solution is changing most rapidly. If the ratio of element widths on either side of the kth node is given (for linear elements) by

$$r = \frac{x_{k+1} - x_k}{x_k - x_{k-1}},\qquad (3.2.70)$$

then the equation equivalent to eq. (3.2.63) is

$$\frac{1}{6}\frac{d\bar{u}_{k-1}}{dt} + \frac{1+r}{3}\frac{d\bar{u}_k}{dt} + \frac{r}{6}\frac{d\bar{u}_{k+1}}{dt} + \frac{\bar{u}_{k-1} + \bar{u}_k + \bar{u}_{k+1}}{3}\frac{\bar{u}_{k+1} - \bar{u}_{k-1}}{2\,\Delta x}$$
$$-\frac{1}{\text{Re}}\left\{\frac{\bar{u}_{k-1} - (1 + 1/r)\bar{u}_k + (1/r)\bar{u}_{k+1}}{\Delta x^2}\right\} = 0. \qquad (3.2.71)$$

An examination of the results in Table 3.9 indicates that the accuracy decreases with increasing Reynolds number. However, Fig. 3.19 indicates that this is directly associated with the steepening of the shock wave at higher Reynolds number.

A computer program, BURG4, that solves the propagating shock problem is described and listed in appendix 2.

Table 3.9. Solution of Burgers' equation for various Reynolds numbers using linear finite elements ($N = 21$)

Re = 1.0, $t = 0.49$			Re = 10, $t = 0.50$			Re = 100, $t = 0.32$		
x	Approx.	Exact	x	Approx.	Exact	x	Approx.	Exact
−1.0	0.9179	0.9174	−1.0	1.0000	1.0000	−1.000	1.0000	1.0000
−0.9	0.8998	0.8996	−0.9	1.0000	1.0000	−0.762	1.0000	1.0000
−0.8	0.8792	0.8789	−0.8	1.0000	1.0000	−0.567	1.0000	1.0000
−0.7	0.8556	0.8554	−0.7	1.0000	0.9999	−0.412	1.0000	1.0000
−0.6	0.8289	0.8288	−0.6	0.9998	0.9996	−0.290	1.0000	1.0000
−0.5	0.7991	0.7991	−0.5	0.9991	0.9987	−0.197	1.0000	1.0000
−0.4	0.7664	0.7664	−0.4	0.9967	0.9960	−0.128	1.0000	1.0000
−0.3	0.7307	0.7308	−0.3	0.9901	0.9891	−0.079	1.0000	1.0000
−0.2	0.6923	0.6926	−0.2	0.9734	0.9725	−0.045	1.0000	1.0000
−0.1	0.6517	0.6520	−0.1	0.9367	0.9366	−0.020	1.0000	1.0000
0	0.6091	0.6095	0	0.8657	0.8672	0	1.0000	0.9998
0.1	0.5651	0.5656	0.1	0.7482	0.7509	0.020	1.0003	0.9994
0.2	0.5202	0.5207	0.2	0.5864	0.5884	0.045	0.9924	0.9976
0.3	0.4750	0.4757	0.3	0.4065	0.4062	0.079	1.0195	0.9846
0.4	0.4302	0.4309	0.4	0.2463	0.2448	0.128	0.8190	0.8281
0.5	0.3862	0.3871	0.5	0.1307	0.1301	0.197	0.0942	0.1236
0.6	0.3438	0.3447	0.6	0.0613	0.0619	0.290	−0.0212	0.0009
0.7	0.3032	0.3043	0.7	0.0256	0.0268	0.412	0.0045	0.
0.8	0.2651	0.2662	0.8	0.0095	0.0106	0.567	−0.0010	0.
0.9	0.2299	0.2309	0.9	0.0029	0.0039	0.762	0.0002	0.
1.0	0.1969	0.1984	1.0	0.0000	0.0013	1.000	0.	0.
$\|u_a - u_e\|_{rms}$	0.0007			0.0011			0.0119	

3.3. Connection with Finite-Difference Formulae

The use of low-order shape functions on a regular grid often produces algebraic equations that are identical with those arising from substituting finite-difference expressions for the derivatives. The use of one-dimensional linear shape functions generated the following finite-difference expressions in eq. (3.2.63):

$$\frac{du}{dx} \Rightarrow \frac{u_{k+1} - u_{k-1}}{2\,\Delta x} \qquad (3.3.1)$$

and

$$\frac{d^2u}{dx^2} \Rightarrow \frac{u_{k-1} - 2u_k + u_{k+1}}{\Delta x^2}. \qquad (3.3.2)$$

Here we have dropped the use of the overbar to denote nodal values. Swartz and Wendroff (1969) used a Galerkin method with linear trial functions to generate Crank–Nicolson versions of eq. (3.2.63) and eq. (3.2.69). However, they preferred to refer to such equations as *generalized finite-difference schemes*. The Crank–Nicolson version of eq. (3.2.63) is obtained by replacing the time derivative terms with a first-order finite-difference expression and evaluating the spatial terms at the average of the nth and ($n + 1$)th time levels (equivalent to setting $\lambda = 0.5$ in eq. (3.2.52)).

To solve two-dimensional, rotational, inviscid, and incompressible flows, Arakawa (1960) obtained a nine-point finite-difference scheme for the convection of vorticity that conserves many physically conserved parameters, such as the kinetic energy (see section 6.5.3). Jespersen (1974) demonstrated that Arakawa's scheme is a finite-element scheme based on linear trial functions in rectangular elements.

For linear and quadratic one-dimensional elements (section 3.1.1) the algebraic expressions for first and second derivatives are shown in Table 3.10, along with three-point and five-point finite-difference formulae. It can be seen that the three-point formulae are the same whether obtained from a finite-difference or a finite-element approach. However, the five-point formulae are not. The finite-element five-point formulae have *truncation errors* of $O(\Delta x^2)$, whereas the finite-difference five-point formulae have truncation errors of $O(\Delta x^4)$. However, the corresponding *solution error* for the quadratic finite-element formulation is $O(\Delta x^3)$.

The treatment of nonderivatives (or terms that do not require the trial functions to be differentiated) is completely different in the two methods even for three-point formulae. Thus if we construct a vector of time-derivative terms in eq. (3.2.63) as

$$\dot{\mathbf{u}}^t = \{\dot{u}_{k-1}, \dot{u}_k, \dot{u}_{k+1}\},\tag{3.3.3}$$

the finite element method (with linear elements) effectively replaces $\dot{\mathbf{u}}$ with

$$\dot{\mathbf{u}} \Rightarrow \{\tfrac{1}{6}, \tfrac{2}{3}, \tfrac{1}{6}\}\dot{\mathbf{u}},\tag{3.3.4}$$

whereas the conventional finite-difference method replaces $\dot{\mathbf{u}}$ with

$$\dot{\mathbf{u}} \Rightarrow \{0, 1, 0\}\dot{\mathbf{u}}.\tag{3.3.5}$$

It is the spreading out of the time derivative that produces more accurate solutions. The combination of eq. (3.3.4) and the three-point formula for $\partial u/\partial x$ creates a scheme that has a truncation error of $O(\Delta x^4)$. However this also implies the added complexity of an implicit solution procedure.

The concept of a distributed time derivative has been combined with a finite-difference representation in the *compact implicit* methods (Adam, 1977; Ciment et al., 1978). The application of compact implicit methods to Burgers' equation is reviewed by Fletcher (1982a).

The treatment of the nonlinear term, $u\,\partial u/\partial x$, in eqs. (3.2.63) and (3.2.69) can be unified following Khosla and Rubin (1979), who used the following finite-difference representation of eq. (3.2.55):

Figure 3.20. An arrangement of linear triangles to simulate the five-point Laplacian difference formula

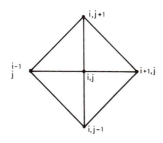

$$\frac{u_k^{n+1} - u_k^n}{\Delta t} + \left(\frac{u_{k-1}^{n+1} + \gamma u_k^{n+1} + u_{k+1}^{n+1}}{2 + \gamma}\right)\left(\frac{u_{k+1}^{n+1} - u_{k-1}^{n+1}}{2\,\Delta x}\right)$$
$$- \frac{1}{Re}\frac{u_{k-1}^{n+1} - 2u_k^{n+1} + u_{k+1}^{n+1}}{\Delta x^2} = 0. \tag{3.3.6}$$

Khosla and Rubin wished to suppress nonlinear instabilities when solving Burgers' equation at large values of Re on a coarse grid. Different values of γ correspond to different degrees of smoothing or filtering. $\gamma = \infty$ corresponds to no filtering (a conventional finite-difference approach), $\gamma = 1$ corresponds to eq. (3.2.63), and $\gamma = 0$ corresponds to eq. (3.2.69). The optimum choice of γ is such that

$$Re\,\Delta x = 2\frac{2 + \gamma}{1 + \gamma}. \tag{3.3.7}$$

However, for cell Reynolds numbers Re Δx greater than 2 and large Courant numbers $u\,\Delta t/\Delta x$, two solutions may occur.

In two dimensions the equivalence between finite-element formulae and finite-difference formulae is not so obvious. However, the five-point difference formula (with $\Delta x = \Delta y = h$) for Laplace's equation in two dimensions,

$$\frac{u_{i-1,j} + u_{i+1,j} + u_{i,j-1} + u_{i,j+1} - 4u_{i,j}}{h^2} = 0, \tag{3.3.8}$$

can be obtained with linear triangular elements arranged as in Fig. 3.20. However, in general, this is not a preferred element distribution (Strang and Fix, 1973, p. 153).

If uniform linear rectangular elements are applied to x derivatives, the same formulae as in Table 3.10 are produced, but with weights in the y direction, $(\frac{1}{6}, \frac{2}{3}, \frac{1}{6})$, as in eq. (3.3.4). Thus $\partial u/\partial x$ produces the following algebraic formula (see Fig. 3.14 for notation):

$$\frac{1}{2\Delta x}\{(\tfrac{1}{6}u_{i+1,j+1} + \tfrac{2}{3}u_{i+1,j} + \tfrac{1}{6}u_{i+1,j-1}) - (\tfrac{1}{6}u_{i-1,j+1} + \tfrac{2}{3}u_{i-1,j} + \tfrac{1}{6}u_{i-1,j-1})\}.$$
$$\tag{3.3.9}$$

The relationship to the one-dimensional formulae can be made more explicit by writing eq. (3.3.9) as

Table 3.10. One-dimensional algebraic formulae for derivatives on a uniform grid

	$\partial u/\partial x$	$\partial^2 u/\partial x^2$
Linear finite element	$\dfrac{u_{k+1} - u_{k-1}}{2\,\Delta x}$	$\dfrac{u_{k-1} - 2u_k + u_{k+1}}{\Delta x^2}$
Quadratic f.e. (corner node)	$\dfrac{u_{k-2} - 4u_{k-1} + 4u_{k+1} - u_{k+2}}{4\,\Delta x}$	$\dfrac{-u_{k-2} + 8u_{k-1} - 14u_k + 8u_{k+1} - u_{k+2}}{4\,\Delta x^2}$
Quadratic f.e. (midside node)	$\dfrac{u_{k+1} - u_{k-1}}{2\,\Delta x}$	$\dfrac{u_{k-1} - 2u_k + u_{k+1}}{\Delta x^2}$
3-pt. finite diff.	$\dfrac{u_{k+1} - u_{k-1}}{2\,\Delta x}$	$\dfrac{u_{k-1} - 2u_k + u_{k+1}}{\Delta x^2}$
5-pt. finite diff.	$\dfrac{u_{k-2} - 8u_{k-1} + 8u_{k+1} - u_{k+2}}{12\,\Delta x}$	$\dfrac{-u_{k-2} + 16u_{k+1} - 30u_k + 16u_{k+1} - u_{k+2}}{12\,\Delta x^2}$

$$\left.\frac{\partial u}{\partial x}\right|_{i,j} \rightarrow \frac{1}{\Delta x}m_y \otimes \delta_x u_{i,j}, \qquad (3.3.10)$$

where δ_x and m_y are one-dimensional operators. For example,

$$\delta_x u_{i,j} = 0.5(u_{i+1,j} - u_{i-1,j}), \quad \text{i.e.} \quad \delta_x \equiv 0.5\{-1,0,1\},$$

and

$$m_y u_{i,j} = \tfrac{1}{6}u_{i,j-1} + \tfrac{2}{3}u_{i,j} + \tfrac{1}{6}u_{i,j+1}, \quad \text{i.e.} \quad m_y^t \equiv \tfrac{1}{6}\{1,4,1\}.$$

In eq. (3.3.10) $\otimes$ denotes a *tensor product*, that is, each component of m_y multiplies all components of δ_x. For linear rectangular elements on a uniform two-dimensional grid the second derivative

$$\left.\frac{\partial^2 u}{\partial y^2}\right|_{i,j} \rightarrow \frac{1}{\Delta y^2}m_x \otimes \delta_y^2 u_{i,j},$$

where

$$\delta_y^2 u_{i,j} \equiv u_{i,j+1} - 2u_{i,j} + u_{i,j-1}.$$

Thus for a uniform grid the two-dimensional algebraic expressions only differ from the one-dimensional expressions in the appearance of the mass operators m_x, m_y.

For first derivatives the particular coefficients in the mass operators ensures that the algebraic formulae have a truncation error of $O(\Delta x^4)$. However the second derivatives have a truncation error of $O(\Delta x^2)$.

For quadratic Lagrange elements the same formulae as appeared in Table 3.10 are appropriate, with coefficients in the mass operators (corner nodes) given by

$$m_x = m_y^t \equiv \tfrac{2}{3}\{-\tfrac{1}{10}, \tfrac{1}{5}, \tfrac{4}{5}, \tfrac{1}{5}, -\tfrac{1}{10}\}.$$

For midside and internal nodes the coefficients are

$$m_x = m_y^t \equiv \tfrac{4}{3}\{\tfrac{1}{10}, \tfrac{4}{5}, \tfrac{1}{10}\}.$$

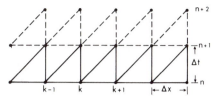

Figure 3.21. Triangular space–time elements

Gray and Pinder (1976) have considered linear and quadratic rectangular elements on a regular grid. For quadratic shape functions, the truncation error of the derivative representation is fourth-order accurate if five points are available in the direction of the derivative, and second-order accurate if three points are available. However, the representation of first derivatives with linear rectangular elements is exceptional, since fourth-order accuracy is obtained with only three points in the derivative direction.

Cushman (1979) has shown that, by constructing finite elements in (x, t) space, a number of the finite-difference formulae for the one-dimensional wave equation

$$\frac{\partial u}{\partial t} + c\frac{\partial u}{\partial x} = 0 \qquad (3.3.11)$$

can be obtained from a conventional Galerkin formulation. Thus the first-order difference scheme

$$\frac{u_k^{n+1} - u_k^n}{\Delta t} = c\frac{u_k^n - u_{k-1}^n}{\Delta x} \qquad (3.3.12)$$

can be obtained by introducing linear triangular elements as shown in Fig. 3.21. Each element (in the k direction) is considered in turn to obtain u_k^{n+1} at the $(n + 1)$th time level, assuming that all data at the nth level are known.

The two-step Lax–Wendroff scheme requires the introduction of additional points, $u_{k+1/2}^{n+1/2}$. Thus

step 1: $\qquad u_{k+1/2}^{n+1/2} = 0.5(u_k^n + u_{k+1}^n) - 0.5\frac{c\Delta t}{\Delta x}(u_{k+1}^n - u_k^n),$

$$\qquad (3.3.13)$$

step 2: $\qquad u_k^{n+1} = u_k^n - \frac{c\Delta t}{\Delta x}(u_{k+1/2}^{n+1/2} - u_{k-1/2}^{n+1/2}).$

For step 1 the required linear triangular element is shown in Fig. 3.22a. Only one triangular element contributes to the solution at node $\binom{n+\frac{1}{2}}{k+\frac{1}{2}}$. For step 2 two triangular elements are arranged back to back to make a diamond shape. Clearly the two elements serve to connect the solution at node $\binom{n+1}{k}$ with nodal values at

$$\binom{n+\frac{1}{2}}{k+\frac{1}{2}}, \qquad \binom{n+\frac{1}{2}}{k-\frac{1}{2}}, \quad \text{and} \quad \binom{n}{k}.$$

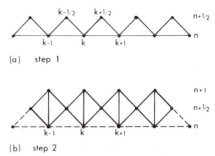

Figure 3.22. Space–time triangular elements to obtain Lax–Wedroff formula

Cushman (1981) has generalized the above technique to generate a family of Lax–Wendroff schemes.

It can be seen from the material presented in this section that a strong connection with the finite-difference method only occurs with linear elements in one dimension. The finite-element method has the property of distributing the algebraic representation for a derivative over a region. Obviously this is more noticeable in two and three dimensions. The distributed nature of the derivative representation leads to a greater solution accuracy but also a greater solution cost, since more nonzero algebraic terms are introduced.

This poses the interesting question: does the greater accuracy associated with the distributed nature of the finite-element method more than compensate for the extra cost? In short, is the finite-element method computationally more efficient than the finite-difference method? We defer consideration of computational efficiency until chapter 6.

3.4. Theoretical Properties

In this section we set down some of the important results, particularly concerning convergence and error estimation, that have been discovered during the theoretical development of the finite-element method. Theoretical aspects of the method have received close attention since the late 1960s, so that now the mathematical basis of the method is well established.

Two aspects of the method permit substantial mathematical development. First, the trial solution is constructed to make use of *polynomial interpolation* between the nodal values; and polynomial interpolation is a well-established discipline in its own right (Birkhoff and de Boor, 1965). Second, for problems that have an equivalent variational formulation, the theoretical properties of the Ritz method (Mikhlin, 1964) carry over to the Galerkin finite-element method.

3.4.1. Convergence

The Galerkin method has the desirable feature of reducing the highest derivative that appears in the weak form of the equation,

$$(L(u), \phi_k) = 0, \tag{3.4.1}$$

by shifting derivatives onto the test functions ϕ_k. Thus if the highest derivative appearing in $L(u)$ is of order $2m$, application of Green's theorem m times causes the highest derivative in eq. (3.4.1) to be of order m.

As the elements become smaller and smaller, a derivative of order m in eq. (3.4.1) will assume a constant value over an element.

For convergence to the exact solution it is necessary that the trial functions be capable of representing a constant value of the highest derivative, m, exactly. Assuming that polynomial trial functions are used, this requires a polynomial that is complete to order m. Thus if second derivatives appear in the weak form (3.4.1), then at least quadratic shape functions are required for convergence.

Second, there is a required that elements be *conforming*. That is, if m is the highest derivative appearing in eq. (3.4.1), the trial function and its first $(m - 1)$ derivatives should be continuous across element boundaries. This is a requirement that the trial functions be in the function space H^m. The H^m norm is defined by

$$\|u_a\|^2_{H^m, R} = \sum_{|\alpha| \le m} \int_R \left| \frac{\partial^{|\alpha|} u_a(x_1, x_2, \ldots, x_n)}{\partial x_1^{\alpha_1} \cdots \partial x_n^{\alpha_n}} \right|^2 dx. \tag{3.4.2}$$

As an example, if in eq. (3.4.2) $m = 2$ and $n = 2$, the following would be obtained:

$$\|u_a\|_{H^2, R} = \left[\int\int_R \left[|u_a|^2 + \left| \frac{\partial u_a}{\partial x} \right|^2 + \left| \frac{\partial u_a}{\partial y} \right|^2 \right. \right.$$
$$\left. \left. + \left| \frac{\partial^2 u_a}{\partial x^2} \right|^2 + \left| \frac{\partial^2 u_a}{\partial y^2} \right|^2 + \left| \frac{\partial^2 u_a}{\partial x \partial y} \right|^2 \right] dx\, dy \right]^{1/2}.$$

Conforming elements ensure that the H^m norm of u_a, the trial solution, is finite.

We recall that the L_2 norm of u_a is defined by

$$\|u_a\|_2 = \left[\int_R u_a^2 \, dx \right]^{1/2}, \tag{3.4.3}$$

so that the L_2 norm and the H° norm coincide.

Elements that ensure continuity of the $m - 1$ derivatives of the trial solution are said to be $C^{(m-1)}$ elements. All the elements described in section 3.1 are C^0 elements. Thus only the trial solution is continuous across element boundaries. For many problems, particularly in the fluid-dynamics and heat-transfer area, the highest derivative appearing in the governing partial differential equation(s) is two. After application of Green's theorem to the weak form, $m = 1$, so that elements with C^0 continuity are sufficient.

For problems in which higher derivatives appear it is generally possible to replace, say, a single equation containing high-order derivatives with a system of equations containing lower-order derivatives. This is achieved by introducing *auxiliary variables* for the derivatives. This technique also improves the condition number of the matrix system of equations to be solved (Strang and Fix, 1973, p. 121).

It was pointed out above that conforming elements ensure the interelement continuity required for convergence. However, in practice convergence may still occur even when nonconforming elements are used. To test the convergence of nonconforming elements, Irons has developed the *patch test*. This is described in Irons and Razzaque (1972).

Irons proposed considering a patch of elements with boundary conditions adjusted so that the exact solution is just $u = p_m$, where p_m is a polynomial of degree m. Here m is the highest derivative appearing in the weak form (3.4.1). To pass the patch test the finite-element solution must also coincide with P_m. On a regular grid the patch test becomes equivalent to a Taylor expansion of the algebraic equation to establish *consistency* with the governing partial differential equation.

This provides a link with the convergence principle that is invoked for finite difference methods, namely, that

$$\text{consistency} + \text{stability} = \text{convergence}.$$

Strictly the above "equation" is valid only under the conditions of the Lax equivalence theorem (Richtmyer and Morton, 1967):

Given a properly posed initial value problem and a finite-difference approximation to it that satisfies the consistency condition, stability is the necessary and sufficient condition for convergence.

However, the general principle is still a very useful guide for conditions where convergence is not guaranteed. For instance, in dealing with nonlinear boundary-value problems it is a common practice to locally freeze enough of the nonlinear terms to construct an equivalent linear equation so that a linear stability analysis can be applied. The same general principle carries over to the finite-element method (Oden and Reddy, 1976a). Oden and Reddy (1976b) used the principle to establish the convergence of a mixed Galerkin formulation.

On a regular grid consistency can be checked for the algebraic equations derived from the finite-element formulation in exactly the same way as for the finite-difference equations. That is, a Taylor expansion is made about a reference point. If the algebraic equations are consistent, the original partial differential equation is recovered, plus higher-order terms (in Δx^s, Δy^t for a two-dimensional problem) which constitute the truncation error. This is essentially the approach of Gray and Pinder (1976).

There is a difficulty associated with elements of higher order than linear. The algebraic equations vary in their truncation error depending on whether they are associated with corner, midside or internal nodes (section 3.3). In addition, points in an element may exist where much higher solution accuracy is achieved. Such points might be certain nodal points or Gauss (numerical integration) points. Thus for the finite-element method the algebraic-equation expansion is useful for establishing consistency but is less useful in obtaining a measure of the solution accuracy.

The approach of starting with the algebraic equations to establish consistency and stability is particularly useful for the semidiscrete Galerkin formulations used in the examples in sections 3.2.4 and 3.2.5. Then the consistency of the final algebraic equation can be considered without taking into account the fact that part of the algebraic equation arises from a finite-element representation and part from a finite-difference representation.

For steady problems that have a variational formulation, the equivalent Galerkin finite-element method produces a positive definite stiffness matrix, and stability of the solution of the algebraic equations is guaranteed (Strang and Fix, 1973, p. 124).

For problems that have a variational formulation it can be shown (Mitchell and Wait, 1977) that the Ritz approximation is best in the energy norm (1.5.7). That is,

$$\|u - u_R\|_A = \inf_{\tilde{u} \in S_h} \|u - \tilde{u}\|_A, \tag{3.4.4}$$

where u_R is the Ritz solution to the variational problem and S_h is the finite-dimensional subspace containing the trial functions. The expression inf stands for infimum or greatest lower bound.

Mitchell and Wait indicate that the Galerkin method will be near best in some Sobolev norm (3.4.2). That is,

$$\|u - u_G\|_{H^r, R} \le C \inf_{\tilde{u} \in S_h} \|u - \tilde{u}\|_{H^r, R} \tag{3.4.5}$$

for some r, $C > 0$. For the second-order problem,

$$\frac{\partial}{\partial x}\left(p_1(x, y)\frac{\partial u}{\partial x}\right) + \frac{\partial}{\partial y}\left(p_2(x, y)\frac{\partial u}{\partial y}\right) + f(x, y) = 0, \tag{3.4.6}$$

where $p_1, p_2 > 0$ and $u = 0$ on ∂R, eq. (3.4.5) becomes

$$\|u - u_G\|_{H^1, R} \le C \inf_{\tilde{u} \in \mathring{S}_h} \|u - \tilde{u}\|_{H^1, R}. \tag{3.4.7}$$

3.4.2. Error estimates

For a given problem the error between the solution read from the computer output and the true or exact solution of the governing partial differential equation can have contributions from many sources.

We will interpret the *approximation error* as being the difference between the Galerkin solution and the exact solution. As the number of nodes increases, we expect this error to decrease. The approximation error will also depend on the method used. For example, for a steady problem we might expect it to be larger if the collocation method is used instead of the Galerkin method.

A major contribution to the approximation error is the *interpolation error* that arises from the nature of the trial functions used within each element. The approximation (and thus the interpolation) error is the main source of error that arises in the finite element method. However, additional errors can arise from using numerical quadrature to evaluate the inner products, from interpolating the boundary conditions, from the failure of the finite-element boundaries to match the actual boundaries, and from roundoff in the computer.

For the operator equation

$$A(u) = f, \tag{3.4.8}$$

which satisfies the *ellipticity* condition

$$(\sigma \|u\|_{H^m,R}^2 \le (A(u), u) \le K\|u\|_{H^m,R}^2, \tag{3.4.9}$$

a connection between the approximation error and the interpolation error can be readily obtained (Oden and Reddy, 1976a, p. 328). The approximation error e can be expressed as

$$\|e\|_{H^m,R} = \|u - u_a\|_{H^m,R} \le \left(1 + \frac{K}{\sigma}\right)\|u - u_I\|_{H^m,R}, \tag{3.4.10}$$

where $\|u - u_I\|_{H^m,R}$ is the H^m norm (i.e. eq. (3.4.2)) of the interpolation error, and the weak form of eq. (3.4.8) contains derivatives up to order m.

If the exact solution u is sufficiently smooth, the error (both approximation and interpolation) in the finite-element solution u_a is given by Strang and Fix (1973, p. 166) as

$$\|u - u_a\|_{H^s,R} \le Ch^{k+1-s}\|u\|_{H^{k+1},R} \qquad \text{if } s \ge 2m - k - 1, \quad (3.4.11)$$

$$\|u - u_a\|_{H^s,R} \le Ch^{2(k+1-m)}\|u\|_{H^{k+1},R} \quad \text{if } s \le 2m - k - 1, \quad (3.4.12)$$

where C is a positive constant, and k is the order of the interpolating function. Strang and Fix note that eq. (3.4.11) is more likely to be limiting; it comes from the interpolation error. Eq. (3.4.12) represents an upper limit. In eqs. (3.4.11) and (3.4.12) h is a characteristic element dimension, such as the maximum width.

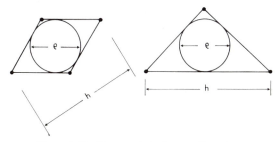

Figure 3.23. Characteristic element dimensions

If u is not as smooth as the interpolation (i.e., $u \in H^r(R)$, $r \leq k + 1$), then eq. (3.4.11) is replaced by

$$\|u - u_a\|_{H^s, R} \leq Ch^{r-s}\|u\|_{H^r, R}. \tag{3.4.13}$$

Oden and Reddy (1976a, p. 343) extend the above estimates to include the effect of boundary conditions.

The effect of element distortion can be appreciated by defining two lengths, h and ρ in Fig. 3.23, to characterize the element. Then eq. (3.4.11) is replaced by (Oden and Reddy, 1976a, p. 279)

$$\|u - u_a\|_{H^s, R} \leq C\frac{h^{k+1}}{\rho^s}|u|_{H^{k+1}, R}, \tag{3.4.14}$$

where $|u|_{H^{k+1}, R}$ is the seminorm, defined by

$$|u|_{H^\alpha, R} = \left(\int_R |D^\alpha u|^2 \, dx \right)^{1/2}. \tag{3.4.15}$$

Gaussian quadrature is widely used in finite-element practice. Mitchell and Wait (1977) note that the quadrature scheme is optimal if the quadrature error is of the same or higher order than the error (3.4.13). An estimate for the quadrature error is given by Oden and Reddy (1976a, p. 357) as

$$\|u_a - u_a^q\|_{H^m, R} \leq C_1 h^\mu \tag{3.4.16}$$

where

$\mu = q + 1 - p + m$,
q is the order of the quadrature scheme,
m is highest-order derivative appearing in the weak form, and
p is the largest exponent appearing in the trial function.

For convergence $q \geq m$. Thus if numerical quadrature is expensive, q can be adjusted to give convergence but will lead to a suboptimal error estimate (see eq. (3.4.18)).

The roundoff error typically varies like h^{-2m} (Strang and Fix, 1973, p. 110) and is not normally a problem on modern computers when solving low-order equations, say $m \leq 2$. Roundoff error affects the numerical solution via the condition number of the stiffness matrix (section 2.1). If a given accuracy of solution is required, roundoff error is reduced by using higher-order elements and a coarser grid.

If the isoparametric formulation is used to ensure that the finite-element grid boundary matches the true boundary, the errors introduced are no worse than the errors given by eq. (3.4.13) provided the elements do not become distorted in physical space. For a second-order partial differential equation, this implies the element edges should be displaced from straight lines by no more than $O(h^2)$ (Strang and Fix, 1973, p. 163).

3.4.3. Optimal error estimates and superconvergence

Much theoretical work has been undertaken to obtain the highest possible accuracy when applying the Galerkin finite-element method to different categories of equations and associated boundary conditions. This leads to so-called optimal error estimates.

Following Oden and Reddy (1976a, p. 337) a general finite-element subspace $S_h^{k,m}(R)$ is considered in order to obtain such estimates. The parameter h is a measure of the element size and is such that $0 < h < 1$. The parameter k is the highest-order complete polynomial that can be represented exactly. The parameter m is a measure of the smoothness of the space. Thus

$$S_h^{k,m}(R) \subset H^m(R), \qquad k + 1 > m \geq 0. \tag{3.4.17}$$

If the exact solution u of the governing partial differential equation is such that $u \in H^r(R)$ and if $0 \leq s \leq \min(r, m)$, then a Galerkin solution u_a chosen from $S_h^{k,m}$ is said to have an *optimal error* in the H^s norm if

$$\|u - u_a\|_{H^s, R} \leq Ch^\sigma \|u\|_{H^r, R}, \tag{3.4.18}$$

where

$$\sigma = \min(k + 1 - s, r - s).$$

Thus the error estimates given by eqs. (3.4.11) and (3.4.13) are optimal.

We now consider typical parabolic and hyperbolic partial differential equations to see if optimal error estimates have been obtained. We will only mention problems with homogeneous Dirichlet boundary conditions. The influence of more general boundary conditions on the global error estimates are described by Oden and Reddy (1976a) and Fairweather (1978).

Strang and Fix (1973, p. 245) discuss linear parabolic equations of the form

$$\frac{\partial u}{\partial t} + L(u) = f, \tag{3.4.19}$$

where $L(u)$ is an elliptic operator of order $2m$ that possesses some equivalent variational form. It is shown that the error between the Galerkin and the exact solution in the L_2 norm (3.4.3) is of order h^{k+1}, which is the same as for the elliptic equation

$$L(u) = f \tag{3.4.20}$$

—that is, using eq. (3.4.11). Comparable error estimates for eq. (3.4.19) have been given by Showalter (1975).

Error estimates for parabolic equations of the following quasilinear form are given by Fairweather (1978):

$$\frac{\partial u}{\partial t} = \nabla \cdot (a(x, u)\nabla u) + f(x, t, u), \tag{3.4.21}$$

with initial conditions

$$u(x, 0) = u_0(x)$$

and boundary conditions

$$u(x, t) = 0 \quad \text{on } \partial R, \text{ the boundary of } R.$$

It is assumed that $a(x, u)$ is bounded by positive constants c_1 and c_2:

$$c_1 \leq a(x, u) \leq c_2.$$

For the weak form of eq. (3.4.21), if $u(x, t) \in H^r(R)$ and if the initial data $u_0(x)$ are sufficiently smooth, it is possible to obtain an optimal error estimate in L_2 of the form

$$\|u - u_a\|_2 < Ch^r, \tag{3.4.22}$$

where all second derivatives of a in eq. (3.4.21) are bounded in the domain R.

The result (3.4.22) is applicable to the semidiscrete Galerkin representation of eq. (3.4.21). However, it is common to introduce a finite-difference representation for $\partial u/\partial t$. Wheeler (1973) considered a number of discrete-time Galerkin approximations of second-order nonlinear parabolic partial differential equations. For Hermite polynomials in a rectangular region, optimal L_2 error estimates, like eq. (3.4.22), were obtained.

For sufficiently smooth parabolic partial differential equations the Galerkin finite-element method leads to solutions of comparable accuracy to those obtained for elliptic partial differential equations. However, such estimates rely on the initial data being smooth and also assume that the projection w_0 of u_0 into the finite-dimensional subspace containing u_a is sufficiently smooth. Thus eq. (3.4.22) relies on the following error estimate for the initial data:

$$\|u_0 - w_0\|_2 \leq Ch^r. \tag{3.4.23}$$

For hyperbolic partial differential equations error estimates are given by Strang and Fix (1973) for the model first-order and second-order equations

$$\begin{aligned}
\frac{\partial u}{\partial t} &= \frac{\partial u}{\partial x}, \\
\frac{\partial^2 u}{\partial t^2} &= c^2 \frac{\partial^2 u}{\partial x^2}.
\end{aligned} \tag{3.4.24}$$

For both situations the L_2 error of the Galerkin approximation is typically $O(h^k)$. However for linear test and trial functions, $O(h^2)$ is obtained (Dupont, 1973). Similar estimates (i.e. $O(h^k)$ in the L_2 norm) have also been obtained by Wellford and Oden (1975) for first-order hyperbolic equations that arise in linear convection problems. In general, the accuracy of the Galerkin finite-element method applied to hyperbolic equations is one order lower than that for elliptic or parabolic equations. Thus the standard Galerkin finite-element method does not provide optimal error estimates when applied to hyperbolic equations.

Various techniques to obtain the optimal convergence in the L_2 norm for hyperbolic equations have been developed and are described in Fairweather (1978, chapter 5). However, these techniques are considerably more complicated than the conventional Galerkin method or involve replacing the original equation with an equivalent higher-order equation. The usefulness of such techniques for the complicated hyperbolic equations that arise in, say, weather forecasting may be limited.

The error estimates given above have been global in nature. However, it is well known (Douglas and Dupont, 1974) that the accuracy *at specific points* may exceed the global accuracy estimates. Such points might be the corner nodes of high-order elements or the Gauss points associated with numerical integration.

For a weak formulation of a two-point boundary-value problem in one dimension,

$$\frac{d}{dx}\left(a(x)\frac{du}{dx}\right) - b(x)u - f = 0, \qquad (3.4.25)$$

where $a \in C^s$ and $b \in C^{s-1}$,

Douglas and Dupont (1974) obtained the error estimate

$$|(u - u_a)(x_i)| \leq Ch^{s+t}\|u\|_{H^{t+1},R} \qquad (3.4.26)$$

for $0 \leq t \leq s$, where u_a is the Galerkin solution and x_i is the point of interest. In addition if $t = s$ and $\hat{u}_a$ is a polynomial of degree $2s - 1$ that interpolates $u_a(x_i)$, then the following *superconvergence* estimate is obtained:

$$\|u - \hat{u}_a\|_{L^\infty} \leq Ch^{2s}. \qquad (3.4.27)$$

For a Poisson equation on a unit square Douglas and Dupont (1974) found that the L_∞ error in the Galerkin solution at the nodes is $O(h^{k+2})$. For this problem an error of $O(h^{k+1})$ would be optimal. Similar results for general elliptic equations have been obtained for the interpolated solution $\hat{u}_a$ by Bramble and Schatz (1977) and for derivatives of $\hat{u}_a$ by Thomée (1977).

Superconvergence results have also been obtained for parabolic equations (Fairweather, 1978, p. 178) and hyperbolic equations (Fairweather, 1978, p. 207; Dougalis and Serbin, 1980).

3.4.4. Numerical convergence results

Some numerical experiments have also been undertaken to establish convergence rates and accuracies of the Galerkin method via direct measurement. Culham and Varga (1971) considered a nonlinear parabolic equation of the form

$$\beta(p)\frac{\partial p}{\partial t} - \frac{\partial}{\partial x}\left(a(p)\frac{\partial p}{\partial x}\right) = \lambda S(x, t), \tag{3.4.28}$$

and for linear elements in x obtained a convergence rate in the L_2 norm of h^α, where $\alpha = 1.93$ (compare $\alpha_{th} = 2$). Lagrangian cubic and Hermitian cubic elements gave convergence rates $\alpha = 3.80$ and 3.58 respectively (compare $\alpha_{th} = 4$). Baker and Soliman (1979) found that for the hyperbolic convection equation (u given)

$$\frac{\partial q}{\partial t} + u\frac{\partial q}{\partial x} = 0, \tag{3.4.29}$$

$\alpha = 2$ for linear elements in x, as expected theoretically. For the model parabolic equation

$$\frac{\partial q}{\partial t} - \frac{\partial}{\partial x}\left(k\frac{\partial q}{\partial x}\right) = 0, \tag{3.4.30}$$

the theoretically expected results were obtained for linear, quadratic, and cubic elements. However, for the Hermitian cubic element, $\alpha = 2$ (compare $\alpha_{th} = 4$) in the L_2 norm.

Fletcher (1982b) has considered the convergence properties of the Galerkin finite-element method applied to the modified Burgers' equation

$$\frac{\partial u}{\partial t} + (u - \alpha)\frac{\partial u}{\partial x} - \frac{1}{Re}\frac{\partial^2 u}{\partial x^2} = 0 \tag{3.4.31}$$

with linear, quadratic, and cubic Lagrangian elements in x. By using a value of $\alpha = 0.5$ and the boundary conditions given by eq. (3.2.56), the convection of the shocklike disturbance to the right does not occur (section 1.2.5).

The time step has been kept sufficiently small that errors in the solution after a finite time are due to the spatial finite-element representation only. Steady-state solutions have been obtained by integrating to large t. By plotting the error in the L_2 norm namely

$$\|e\|_2 = \|u_a - u_e\|_2 = \left[\int_{x_L}^{x_R}(u_a - u_e)^2\,dx\right]^{1/2}, \tag{3.4.32}$$

against the step size Δx, a spatial convergence rate can be obtained in the form

$$\|e\|_2 = k\Delta x^\alpha. \tag{3.4.33}$$

Table 3.11. Measured spatial convergence rates

Finite-element method	Spatial convergence rate α		Finite-difference method	Spatial convergence rate α
	Conventional	Conservation		
Linear	2.00	2.04	Three-point	2.00
Quadratic	2.98	2.90	Five-point	3.87
Cubic	3.90	3.94	Seven-point	5.58

Table 3.12. Theoretical spatial convergence rates

Finite-element method	Spatial convergence rate α in Δx^{α}	Finite-difference method	Spatial convergence rate α in Δx^{α}
Linear	2	Three-point	2
Quadratic	3	Five-point	4
Cubic	4	Seven-point	6

For Re = 10 and a range of step sizes $0.002 < \Delta x < 0.2$, average values of α are shown in Table 3.11. The corresponding theoretical values of α (from section 3.4.2) are shown in Table 3.12.

The conventional finite-element results are based on eq. (3.4.31). For linear elements this corresponds to eq. (3.2.63) with the addition of the α term. Also shown in Table 3.11 are convergence rates for the conservation finite-element formulation, eq. (3.2.65) plus the α term. It is apparent that both formulations produce convergence rates in good agreement with the theoretical predictions (Table 3.12).

Spatial convergence rates have also been obtained for three-point, five-point, and seven-point finite-difference representations of eq. (3.4.31). The errors are measured in the rms norm

$$\|u_a - u_e\|_{\mathrm{rms}} = \frac{\left[\sum_j (u_{aj} - u_{ej})^2\right]^{1/2}}{N^{1/2}}. \tag{3.4.34}$$

As with the finite-element results, the finite-difference convergence rates (Table 3.11) substantially achieve the theoretical convergence rates (based on truncation error) shown in Table 3.12. It was found (Fletcher, 1982b) that the finite-difference convergence rates improved with grid refinement.

Spatial convergence rates have also been determined for small values of time—that is, a parabolic situation. However, it was found that the initial data had a significant influence (Fletcher, 1982b). If the discontinuous initial conditions (3.2.57) were used, the convergence rate for the cubic finite-element and five- and seven-point finite-difference schemes was reduced to second order.

However if the exact solution at $t = 0.01$ was used for the initial conditions, convergence rates comparable to those shown in Table 3.11 were obtained for all methods. Similar convergence rates, for continuous initial conditions, were also obtained with Re = 1 and 100.

3.5. Applications

Galerkin finite-element methods now encompass so many areas of interest that an attempt to even catalogue, let alone describe, all the applications would fill more than one book (Whiteman, 1975; Norrie and De Vries, 1976). Here we restrict our attention to the areas of convective heat transfer and fluid flow. From this large category we have selected a few specific examples to illustrate the way Galerkin finite-element methods are combined with other techniques to produce practical algorithms.

The first four examples deal with problems with substantial dissipation where the system of governing equations is either elliptic or parabolic in character. This is where Galerkin finite-element methods should work best, according to the variational connections and the theoretical results of section 3.4. The last two examples are geophysical in origin, and here the dissipative processes are less significant, so that the governing equations are effectively hyperbolic in character. This puts more emphasis on the treatment of the time-stepping algorithm. As will be seen, Galerkin finite-element methods are very effective in treating such problems.

3.5.1. Convective heat transfer

Many interesting problems involve a fluid thermal interaction. Such problems can be handled by including the energy equation with the incompressible Navier–Stokes equations. Following Gartling (1977), these equations can be written as follows:

momentum:

$$\rho u_j \frac{\partial u_i}{\partial x_j} - \rho g_i + \rho g_i \beta (T - T_{ref}) - \frac{\partial \tau_{ij}}{\partial x_j} = 0, \qquad (3.5.1)$$

where

$$\tau_{ij} + p\delta_{ij} - \mu \left(\frac{\partial u_i}{\partial x_j} - \frac{\partial u_j}{\partial x_i} \right) = 0; \qquad (3.5.2)$$

continuity:

$$\frac{\partial u_i}{\partial x_i} = 0;$$ (3.5.3)

energy:

$$\rho C_p u_j \frac{\partial T}{\partial x_j} + \frac{\partial q_i}{\partial x_i} - Q = 0,$$ (3.5.4)

where

$$q_i = -k_{ij} \frac{\partial T}{\partial x_j}.$$ (3.5.5)

In the above equations u_i is the velocity component in the x_i direction, g is the gravitational constant, β is the coefficient of volume expansion, τ_{ij} is the stress tensor, C_p is the specific heat, q_i is the heat flux vector, k_{ij} is the thermal-conductivity tensor, and Q is the volumetric heat source.

Boundary conditions for eqs. (3.5.1) to (3.5.3) require specification of either the velocity or surface stress, that is

$$u_i = f_i \quad \text{on } \partial R_v$$ (3.5.6a)

or

$$t_i = \tau_{ij} n_j \quad \text{on } \partial R_t.$$ (3.5.6b)

The appropriate boundary conditions for the energy equation (3.5.4) and (3.5.5) are on either the temperature or the heat flux:

$$T = T_w \quad \text{on } \partial R_T,$$
$$q_i n_i = g \quad \text{on } \partial R_q.$$ (3.5.7)

The conventional Galerkin finite-element method proceeds by introducing trial solutions for the velocity components, temperature, and pressure:

$$u_i(x_i) = \sum_j N_j^v(x_i) \bar{u}_{ij},$$
$$P(x_i) = \sum_j N_j^P(x_i) \bar{P}_j,$$ (3.5.8)
$$T(x_i) = \sum_j N_j^T(x_i) \bar{T}_j.$$

Substitution of eqs. (3.5.8) into the momentum equations (3.5.1) and (3.5.2), the continuity equation (3.5.3), and the energy equations (3.5.4) and (3.5.5) produces residuals R_M, R_C, and R_E respectively. Algebraic equations are generated by evaluating the following inner products:

$$(R_M, N^v) = 0,$$
$$(R_C, N^P) = 0,$$ (3.5.9)
$$(R_E, N^T) = 0.$$

The details are given by Gartling (1977). The resulting matrix of equations can be written

$$C(u)v + Kv = F(T) \tag{3.5.10}$$

and

$$D(u)T + LT = G, \tag{3.5.11}$$

where

$$u^t = \{u_1^t, u_2^t\},$$
$$v^t = \{u_1^t, u_2^t, p^t\}. \tag{3.5.12}$$

Eq. (3.5.10) is derived from the momentum and continuity equations, and eq. (3.5.11) is derived from the energy equation. The nonlinear terms $C(u)v$ and $D(u)T$ arise from the convection operators on the left-hand sides of eqs. (3.5.1) and (3.5.4) respectively.

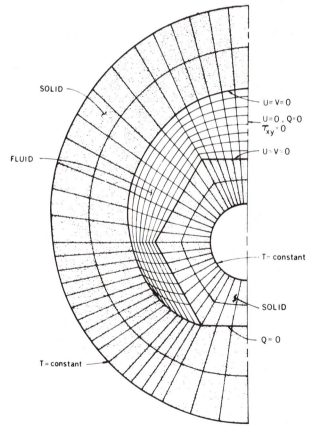

Figure 3.24. Finite-element grid for hexagonal–cylindrical enclosure (after Gartling, 1977; reprinted with permission of North-Holland Publishing Co.)

Gartling (1977) reports using eight-node isoparametric serendipity rectangular elements and six-node isoparametric triangular elements with numerical integration. In eqs. (3.5.8) the velocity components and temperature are interpolated quadratically over each element and pressure is interpolated linearly.

For strongly coupled problems an iterative procedure is used to solve eqs. (3.5.10) and (3.5.11). This can be written

$$\mathbf{D}(\mathbf{u}^n)\mathbf{T}^{n+1} + \mathbf{L}\mathbf{T}^{n+1} = \mathbf{G} \qquad (3.5.13)$$

and

$$\mathbf{C}(\mathbf{u}^n)\mathbf{v}^{n+1} + \mathbf{K}\mathbf{v}^{n+1} = \mathbf{F}(\mathbf{T}^{n+1}). \qquad (3.5.14)$$

The order shown is used for buoyancy-driven flows; for forced-convection problems the order is reversed. At each stage of the iteration the matrix factorization is carried out using a modified frontal method (Irons, 1970).

A typical problem associated with the convective transport of energy in a spent nuclear shipping cask is represented in Fig. 3.24. The solid hexagonal fuel assembly is assumed to have a constant core temperature. For a Grashof number GR $(= g\beta\Delta T R^3 \rho^2/\mu^2)$ of 9×10^4, typical streamline and isotherm plots are shown in Fig. 3.25. Other examples using the same computer code are given by Gartling (1980). A similar finite-element treatment of non-Newtonian heat-transfer problems is given by Phuoc and Tanner (1980).

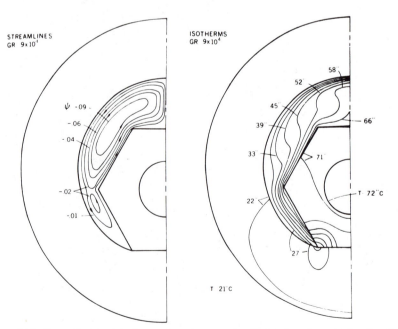

Figure 3.25. Streamlines and isotherms for a hexagonal–cylindrical enclosure (after Gartling, 1977; reprinted with permission of North-Holland Publ. Co.)

3.5.2. Viscous incompressible flow

This topic, particularly the driven-cavity problem (Tuann and Olson, 1978), has received considerable attention in the finite-element literature. Here we describe *penalty methods* that have been introduced to avoid explicit consideration of the incompressibility constraint

$$\frac{\partial u_i}{\partial x_i} = 0. \tag{3.5.15}$$

The steady Navier–Stokes equations that govern incompressible viscous flow can be written

$$-\frac{1}{\text{Re}}\frac{\partial^2 u_i}{\partial x_j \partial x_j} + u_j\frac{\partial u_i}{\partial x_j} + \frac{\partial p}{\partial x_i} = f_i \quad \text{in } R, \tag{3.5.16}$$

$$\frac{\partial u_i}{\partial x_i} = 0 \quad \text{in } R, \tag{3.5.17}$$

and

$$u_i = g_i \quad \text{on } \partial R, \tag{3.5.18}$$

where $u_i = \{u_1, u_2\}$ is the velocity, p is the pressure, and g_i is known. If a general test function w_i satisfies

$$\frac{\partial w_i}{\partial x_i} = 0 \quad \text{in } R,$$
$$w_i = 0 \quad \text{on } \partial R, \tag{3.5.19}$$

then a *weak* formulation of eqs. (3.5.15) to (3.5.18) can be written (Temam, 1979a)

$$\frac{1}{\text{Re}}a(u_i, w_i) + b(u_j, u_i, w_i) = (f_i, w_i), \tag{3.5.20}$$

where

$$a(u_i, w_i) = \int_R \frac{\partial u_i}{\partial x_i}\frac{\partial w_i}{\partial x_i}\,d\mathbf{x}, \tag{3.5.21}$$

$$b(u_j, v_i, w_i) = \int_R \left[\left(u_j\frac{\partial}{\partial x_j}\right)v_i\right]w_i\,d\mathbf{x}. \tag{3.5.22}$$

Because of the form of eq. (3.5.19), p does not appear explicitly in eq. (3.5.20) (Temam, 1979b). By considering the analogy of slightly compressible flow it can be shown rigorously (Temam, 1979b) that the pressure p is given by

$$p = -\lambda\frac{\partial u_i}{\partial x_i}, \tag{3.5.23}$$

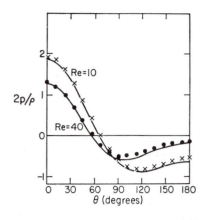

Figure 3.26. Pressure distribution for flow around a sphere: ——analytic results (Dennis and Walker, 1971); •, × FEM results (after Hughes et al., 1979; reprinted with permission of Academic Press)

where λ is a parameter that is made arbitrarily large. A Galerkin formulation is introduced by using the same trial functions in the expansions u_{hi} and w_{hi} which replace u_i and w_i. To avoid having to impose the discrete form of eq. (3.5.19), the following algorithm is introduced:

$$\frac{1}{Re}a_h(u_{hi}^{m+1}, w_{hi}) + b_h(u_{hj}^{m+1}, u_{hi}^{m+1}, w_{hi}) + \lambda\left(\frac{\partial u_{hi}^{m+1}}{\partial x_i}, \frac{\partial w_{hi}}{\partial x_i}\right) - \left(p_h^m, \frac{\partial w_{hi}}{\partial x_i}\right) = (f, w_{hi}) \tag{3.5.24}$$

and

$$p_h^{m+1} = p_h^m - \rho\frac{\partial u_{hi}^{m+1}}{\partial x_i}. \tag{3.5.25}$$

Teman and Thomasset (1976) have used this algorithm to study the driven-cavity problem using linear triangular elements.

Hughes et al. (1978) have also considered a penalty formulation in the (Galerkin) form

$$\frac{2}{Re}a_h(u_{hi}, w_{hi}) + b_h(u_{hj}, u_{hi}, w_{hi}) + \lambda\left(\frac{\partial u_{hi}}{\partial x_i}, \frac{\partial w_{hi}}{\partial x_i}\right) = (f, w_{hi}). \tag{3.5.26}$$

The pressure does not appear explicitly in eq. (3.5.26), but is recovered at the end of the calculation using a discrete analogue of eq. (3.5.23). Hughes et al. (1979) have used bilinear isoparametric rectangular elements to represent the velocities in eq. (3.5.26). Consequently the pressure solution obtained from eq. (3.5.23) is discontinuous at element boundaries, and a least-squares smoothing procedure is found necessary to extract useful pressure information.

Where the pressure on the boundary is required (e.g. in the classical aerofoil problem), special extrapolation procedures are required. A typical pressure solution for the flow over a sphere is shown in Fig. 3.26 (after Hughes et al., 1979). The comparative data are from analytic results due to Dennis and Walker (1971).

It is found necessary, in constructing the matrix equations from eq. (3.5.26), to numerically integrate the λ term using a reduced integration formula, in

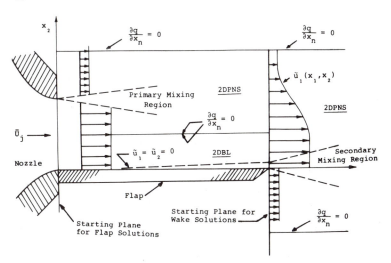

Figure 3.27. Idealized jet-flap flowfield geometry

order to avoid a singular matrix. The theoretical implications of penalty methods and reduced integration are discussed by Oden (1980).

3.5.3 Jet-flap flows

This problem is relevant to the use of two-dimensional jets exhausting over the upper surface of aircraft wings to increase the lift without requiring undue structural complexity (flap retraction mechanisms etc.)

Computationally this problem introduces the technique of parabolizing the Navier–Stokes equations (Davis and Rubin, 1980), namely, the deletion of the second-derivative terms in one particular coordinate direction on an order-of-magnitude basis. The solution can then be obtained by marching in that direction. For the present problem this is the downstream direction.

Second, the flow is turbulent and to account properly for the noise generated by the jet it is necessary to supplement the Navier–Stokes equations with transport equations for the turbulent kinetic energy k and the dissipation function ε (Baker and Manhardt, 1978).

An idealized solution domain is shown in Fig. 3.27. The solution domain can be conveniently split into two parts. From the origin at the exit of the slot nozzle to the end of the flap, the problem is solved by marching downstream, simultaneously considering the region adjacent to the flap (2DBL) and the region far from the flap (2DPNS). 2DBL stands for two-dimensional boundary-layer equations, and 2DPNS stands for two-dimensional parabolized Navier–Stokes equations. Downstream of the flap trailing edge (wake solution), the 2DPNS equations are solved across the whole region (in the direction x_2) as one marches in the direction x_1.

For this problem the governing equations can be written (see Baker and Manhardt for the detailed derivation)

$$L_c(\rho) = \frac{\partial}{\partial x_i}(\rho u_i) = 0, \qquad (3.5.27)$$

$$L_m(u_i) = \frac{\partial}{\partial x_j}(\rho u_i u_j) - \frac{\partial}{\partial x_l}\left[\mu^e\left(\frac{\partial u_i}{\partial x_l} + \frac{\partial u_l}{\partial x_i}\right)\right] + \frac{\partial p}{\partial x_i} = 0, \qquad (3.5.28)$$

$$L_k(k) = \frac{\partial}{\partial x_j}(\rho u_j k) - \frac{\partial}{\partial x_l}\left[C_k \mu^e \frac{\partial k}{\partial x_l}\right] - \mu^e \frac{\partial u_1}{\partial x_l}\frac{\partial u_1}{\partial x_l} + \rho\varepsilon = 0, \qquad (3.5.29)$$

$$L_\varepsilon(\varepsilon) = \frac{\partial}{\partial x_j}(\rho u_j \varepsilon) - \frac{\partial}{\partial x_l}\left[C_\varepsilon \mu^e \frac{\partial \varepsilon}{\partial x_l}\right]$$

$$- C_\varepsilon^1 \frac{\varepsilon}{k}\mu^e\frac{\partial u_1}{\partial x_l}\frac{\partial u_1}{\partial x_l} + C_\varepsilon^2 \rho\frac{\varepsilon^2}{k} = 0, \tag{3.5.30}$$

where

$$\mu^e = \frac{\mu}{\mathrm{Re}} + \rho v_T,$$

$$v_T = \frac{C_v k^2}{\varepsilon}. \tag{3.5.31}$$

For three-dimensional flow $1 \leq i,j \leq 3$ and $2 \leq l \leq 3$. Since $l \neq 1$, the solution can be obtained by marching in the x_1 direction.

The Galerkin finite-element procedure is applied by introducing trial solutions for ρ, u_i, k, and ε. For example,

$$\rho = \sum_j N_j^\rho(x_2, x_3)\bar{\rho}_j(x_1). \tag{3.5.32}$$

This form is typical of that for time-dependent problems (e.g. section 3.2.4). In the present problem the coordinate x_1 has a timelike role. Thus the Galerkin finite-element method is only applied over the x_2, x_3 domain.

Substitution of equations like eq. (3.5.32) into eqs. (3.5.27) to (3.5.30) produces residuals $L_c(\bar{\rho}_j, \bar{u}_{ij})$ and so on. Evaluation of the inner products like

$$(L_c, N_k) = 0 \tag{3.5.33}$$

produces a system of ordinary differential equations. A finite-difference representation is introduced for the x_1 derivatives, allowing a marching algorithm to be constructed. The 2DBL formulation is obtained from the above 3DPNS formulation by discarding eq. (3.5.28) when $j = 2$ and only letting $l = 2$ in the other equations.

Baker and Manhardt present data for the turbulent properties of the downstream jet development. A typical longitudinal velocity profile downstream of the jet flap is shown in Fig. 3.28. Excellent agreement with the experimental data of Schreker and Maus (1974) is indicated.

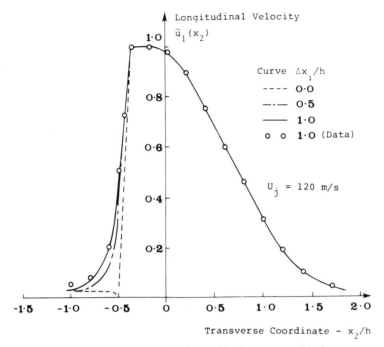

Figure 3.28. Longitudinal velocity profiles downstream of jet flap

3.5.4. Acoustic transmission in ducts

Here we describe the application of a Galerkin finite-element method, after
Astley and Eversman (1981), to the problem of acoustic transmission in a
nonuniform two-dimensional duct. Application of the traditional Galerkin
method to this problem was described in section 1.6.3. The duct geometry is
given in Fig. 1.18, and the governing equations and related boundary conditions for the perturbation velocity components u, v and pressure p are given
by eqs. (1.6.30) and (1.6.31).

The purpose of solving for the perturbation flowfield in the nonuniform
duct is to obtain a transition matrix connecting the vector of modal coeffi-
cients, $\mathbf{b}$, downstream ($x > l$) with the vector of modal coefficients, $\mathbf{a}$, upstream
($x < 0$) of the nonuniform section (Fig. 1.18). The upstream and downstream
sections are uniform, so that an eigenvalue problem is solved in those regions.

The unknowns u, v, p have been represented by quadratic trial functions in
two-dimensional, isoparametric serendipity elements. Typical arrangements
of elements are shown in Fig. 3.29. If the nodal unknowns are treated as a
vector $\boldsymbol{\delta}$, it is convenient to represent the global trial solution $\mathbf{f} = \{u, v, p\}$ by

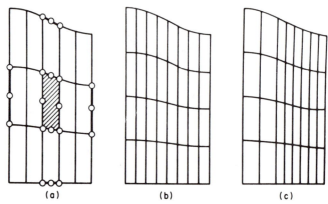

Figure 3.29. Typical element distributions for cosine converging duct (after Astley and Eversman, 1981; reprinted with permission of Academic Press)

$$\mathbf{f} = [\mathbf{N}_1 \; \mathbf{N}_i \; \mathbf{N}_2] \begin{bmatrix} \boldsymbol{\delta}_1 \\ \boldsymbol{\delta}_i \\ \boldsymbol{\delta}_2 \end{bmatrix}, \tag{3.5.34}$$

where $\boldsymbol{\delta}_1$ is the solution at $x = 0$ and $\boldsymbol{\delta}_2$ is the solution at $x = l$. But these solutions are required to match the upstream and downstream solutions. Thus

$$\boldsymbol{\delta}_1 = \mathbf{D}_1 \mathbf{a} \quad \text{and} \quad \boldsymbol{\delta}_2 = \mathbf{D}_2 \mathbf{b}, \tag{3.5.35}$$

where $\mathbf{D}_1$ and $\mathbf{D}_2$ are known from the uniform-duct solution. Thus eq. (3.5.34) can be written

$$\mathbf{f} = \mathbf{N}\mathscr{D}\boldsymbol{\Delta}, \tag{3.5.36}$$

where

$$\boldsymbol{\Delta}^T = \{\mathbf{a}\boldsymbol{\delta}_i\mathbf{b}\},$$

$$\mathscr{D} = \begin{bmatrix} \mathbf{D}_i & & \\ & \mathbf{I} & \\ & & \mathbf{D}_2 \end{bmatrix}, \tag{3.5.37}$$

$$\mathbf{N} = [\mathbf{N}_1 \; \mathbf{N}_i \; \mathbf{N}_2].$$

In applying the Galerkin method it is found that more accurate results are obtained if the boundary conditions on C_3 and C_4 (Fig. 1.18) are formed into boundary residuals which are introduced to partially cancel related terms in the equation residual formed at the boundary. The result of applying the Galerkin method can be written in matrix form as

$$\mathbf{K}_G\boldsymbol{\Delta} = 0. \tag{3.5.38}$$

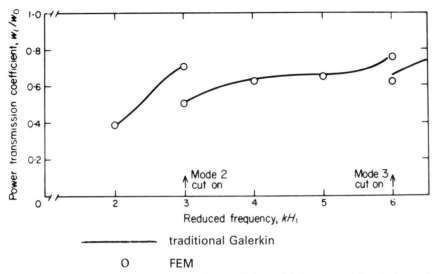

Figure 3.30. Variation of power transmission coefficient with frequency (after Astley and Eversman, 1981; reprinted with permission of Academic Press)

If it is assumed that $\mathbf{a}$ is given, eq. (3.5.38) can be partitioned into

$$\mathbf{K}_{22}\boldsymbol{\delta}_i + \mathbf{K}_{23}\mathbf{b} = -\mathbf{K}_{21}\mathbf{a},$$
$$\mathbf{K}_{32}\boldsymbol{\delta}_i + \mathbf{K}_{33}\mathbf{b} = -\mathbf{K}_{31}\mathbf{a},$$
$$(3.5.39)$$

where $\mathbf{K}_{22}$ and the rest are appropriate partitions of $\mathbf{K}_G$. From the solution of eqs. (3.5.39) the appropriate transition matrix $\mathbf{T}$ can be obtained. Thus

$$\mathbf{b} = \mathbf{Ta}. \qquad (3.5.40)$$

Manipulation of $\mathbf{T}$ permits matrices of reflection and transmission coefficients (like eq. (1.6.38)) to be obtained.

The ratio of transmitted to incident power is compared with the traditional Galerkin solution (section 1.6.3) for a low-Mach-number flow in Fig. 3.30. Agreement is very close. For higher-Mach-number flows the agreement is not as good because of a lack of grid refinement (Fig. 3.29). Solutions to this problem were also obtained using a least-squares finite-element formulation, but these solutions were less accurate (Astley and Eversman, 1981). For a related one-dimensional problem, Astley and Eversman (1982) report more accurate results with a Hermitian cubic element, and conjecture an improvement would also occur in the two-dimensional problem discussed here.

3.5.5 Tidal flows

Tidal flows and surge flows due to sea-bed eruptions are conveniently modeled using the depth-averaged *shallow-water equations*. These are given by Taylor and Davis (1975) as

C: $\qquad \dfrac{\partial \zeta}{\partial t} + \dfrac{\partial}{\partial x}\Big[(H + \zeta)u\Big] + \dfrac{\partial}{\partial y}\Big[(H + \zeta)v\Big] = 0,$ $\qquad$ (3.5.41)

XM: $\quad \dfrac{\partial u}{\partial t} + u\dfrac{\partial u}{\partial x} + v\dfrac{\partial u}{\partial y} + g\dfrac{\partial}{\partial x}(\zeta - \zeta_0)$

$$= \Omega v - \frac{gqu}{C^2(H + \zeta)} + \frac{w_x}{H + \zeta},$$ (3.5.42)

YM: $\quad \dfrac{\partial v}{\partial t} + u\dfrac{\partial v}{\partial x} + v\dfrac{\partial v}{\partial y} + g\dfrac{\partial}{\partial y}(\zeta - \zeta_0)$

$$= -\Omega u - \frac{gqv}{C^2(H + \zeta)} + \frac{w_y}{H + \zeta},$$ (3.5.43)

where

$$u = \frac{1}{H + \zeta}\int_{-H}^{\zeta} U\,dz$$ (3.5.44)

and

$$v = \frac{1}{H + \zeta}\int_{-H}^{\zeta} V\,dz.$$ (3.5.45)

Here H is the mean sea-level height above the sea bottom, ζ is the wave height above mean sea level, and ζ_0 is the pressure height. Ω is the Coriolis parameter, w_x, w_y are wind stresses at the water surface, and C is the Chézy friction coefficient. In eqs. (3.5.42) and (3.5.43) $q = (U^2 + V^2)^{1/2}$. Depending on the problem, the boundary conditions are the specification of ζ, u, and v and the requirement that the normal velocity be zero, that is, that there be no flow through a solid boundary.

Application of the Galerkin method in space to eqs. (3.5.41) to (3.5.43) proceeds by introducing trial solutions for each of the dependent variables, for example,

$$\zeta = \sum_j N_j(x, y)\bar{\zeta}_j(t),$$ (3.5.46)

and requiring that the inner product of the equation residuals, R_C, R_{XM}, and R_{YM} with $N_k(x, y)$ be zero, for example,

$$(R_C, N_k) = 0.$$ (3.5.47)

The result is a system of ordinary equations that can be conveniently written

$$\mathbf{M}\dot{p}_i = -F_i(\bar{\zeta}, \bar{u}, \bar{v}),$$ (3.5.48)

where

$\dot{p}_i = \dot{\zeta}$, $\dot{u}$, or $\dot{v}$ as $i = $ C, XM, or YM,

$\mathbf{M}$ is the mass matrix, and

F_i contains all the terms arising from the spatial derivatives.

A major difficulty is to choose an appropriate discretization for $\dot{p}_i$. The system of equations (3.5.41) to (3.5.43) is hyperbolic in character, and the damping terms are weak, so that waves tend to be propagated (physically) with little loss of amplitude. It is important that the spatial and temporal discretizations do not introduce significant higher-order spurious terms to modify the amplitude and speed of the wave propagation. These aspects are discussed by Pinder and Gray (1977).

Taylor and Davis (1975) considered three alternative temporal discretizations:

    (i) an Adams–Moulton multistep predictor–corrector scheme,
    (ii) a finite-difference trapezoidal scheme,
    (iii) a finite-element (in time) scheme.

A four-time-level Adams–Moulton predictor–corrector was used. This requires four levels of storage for the P_i and the repeated use of the corrector formula at each time step until convergence is obtained. It was found that small time steps were required to achieve a satisfactory accuracy.

If a linear interpolation function in time is used with the Galerkin method for method (iii), then methods (ii) and (iii) can be put into the same framework. Namely,

$$\mathbf{M}p_i^{n+1} = \mathbf{M}p_i^n - \Delta t((1 - \lambda)F_i^n + \lambda F_i^{n+1}). \qquad (3.5.49)$$

Method (ii) corresponds to $\lambda = \frac{1}{2}$, and method (iii) to $\lambda = \frac{2}{3}$. Method (iii) was used earlier to study the same problem by Grotkop (1973). However Taylor

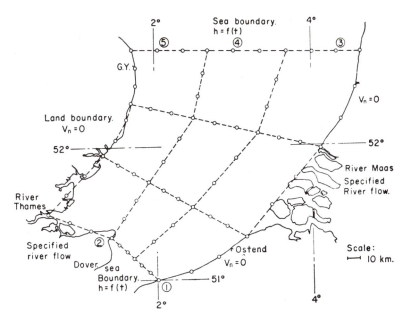

Figure 3.31. Element distribution and boundary conditions, North Sea (after Taylor and Davis, 1975; reprinted with permission of Pergamon Press)

Interpolated currents
at nine points within
each element are
also shown.

Scale:
— 1 m / sec

Figure 3.32. North Sea currents at 1600 hr (after Taylor and Davis, 1975; reprinted with permission of Pergamon Press)

and Davis found that this method did not handle the wave amplitude correctly. This aspect is discussed at some length by Pinder and Gray (1977). Taylor and Davis found method (ii) to be satisfactory.

Taylor and Davis were forced to extrapolate to evaluate F_i^{n+1} and to refine the value iteratively at each time step. This had the advantage that the mass matrix $\mathbf{M}$, which is not a function of p, need only be factorized once. If F_i^{k+1} had been combined with $\mathbf{M}$ to form an augmented mass matrix $\mathbf{M}^a$, method (ii) would be the Crank–Nicolson scheme, but this would require a factorization of $\mathbf{M}^a$ at each time step. Since F_i depends nonlinearly on p_i, an iteration at each time step is still required. The alternating-direction implicit scheme described in section 4.1 gives an economical alternative treatment of the repeated factorization of $\mathbf{M}^a$.

Taylor and Davis also considered a cubic finite element in time, but found it computationally too expensive. They used method (ii) with cubic, isoparametric serendipity elements to study tidal flows in the southern part of the North Sea. Figures 3.31 and 3.32 show the element configuration and a typical distribution of currents respectively. Good agreement with experimental data was reported.

Kawahara et al. (1978) have considered the propagation of tsunami waves. This is associated with an eruption of the sea bed that generates a transverse surface wave that propagates towards the coast—causing, typically, substantial property damage. Kawahara et al. used essentially eqs. (3.5.41) to (3.5.43) with the right-hand sides set equal to zero. In contrast to Taylor and Davis, linear triangular elements were used, and the mass matrix $\mathbf{M}$ in eq. (3.5.48) was diagonalized (*lumped*), thus avoiding the relatively expensive matrix factorization at every time step.

The time variation was obtained by introducing a two-step Lax–Wendroff scheme that, with reference to eq. (3.5.48), can be written

$$M_d P_i^{n+1/2} = M_d P_i^n - \frac{\Delta t}{2} F_i^n,$$

$$M_d P_i^{n+1} = M_d P_i^n - \Delta t \, F_i^{n+1/2}. \tag{3.5.50}$$

The scheme is second-order accurate in time and requires no iteration at each time level. The scheme was used with a time step sufficiently small to maintain accuracy; the time step was less than that required to satisfy the stability restriction for this problem, namely,

$$\Delta t \leq C \, \Delta x / \sqrt{gH}. \tag{3.5.51}$$

The characteristic properties (wave speed, amplitude, etc.) of the computed tsunami wave at various locations on the Japanese coastline were found to be in good agreement with those measured during an actual occurrence.

A review of more recent applications of the Galerkin finite-element method to tidal computations is given by Taylor (1980).

3.5.6. Weather forecasting

The finite-element method has been applied to meteorological forecasting by Staniforth and Daley (1979) and by Cullen and Hall (1979). More recent developments are given by Staniforth (1982). Staniforth and Daley used linear "brick" elements on a three-dimensional grid covering the Northern Hemisphere. Because they were interested in obtaining a high-resolution forecast over North America, a graded mesh, as shown in Fig. 3.33, was used. A mixed velocity, vorticity, stream-function, pressure, and temperature system of equations has been solved. A *semiimplicit leapfrog* scheme was used to integrate in time. In the notation of eq. (3.5.48) this would be

$$\mathbf{M} \frac{P_i^{n+1} - P_i^{n-1}}{2 \, \Delta t} = -(0.5 F_{ia}^{n-1} + 0.5 F_{ia}^{n+1} + F_{ib}^n) \tag{3.5.52}$$

or

$$\left(\mathbf{M} + \Delta t \frac{\partial F_{ia}^{n-1}}{\partial P_i} \right) (P_i^{n+1} - P_i^{n-1}) = -2 \, \Delta t \, F_{ia}^{n-1} - 2 \, \Delta t \, F_{ib}^n. \tag{3.5.53}$$

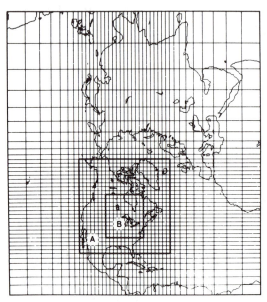

Figure 3.33. Variable-resolution grid for North America (after Staniforth and Daley, 1979; reprinted with permission of the American Meterological Society)

A factorization of the left-hand side is required at every time step. The authors indicate that the finite-element model is suitable for detailed 24-hr forecasts over North America. Comparisons with an operational spectral model indicate that the finite-element method is of comparable accuracy and efficiency.

Cullen and Hall combined a finite-element grid of linear triangles on a latitude–longitude spherical surface with a five-level-finite difference representation in the vertical direction. They used both explicit and semiimplicit leapfrog time-differencing. A special two-stage Galerkin procedure was used for evaluating terms of the form $u\,dv/dx$. This technique is described by Cullen and Morton (1980) for a uniform one-dimensional grid. If Z is an intermediate variable for $\partial v/\partial x$ and W is an intermediate variable for $u\,\partial v/\partial x$, a two-stage Galerkin representation is

$$\tfrac{1}{6}Z_{i-1} + \tfrac{2}{3}Z_i + \tfrac{1}{6}Z_{i+1} = \frac{V_{i+1} - V_{i-1}}{2\,\Delta x} \qquad (3.5.54)$$

and

$$\tfrac{1}{6}W_{i-1} + \tfrac{2}{3}W_i + \tfrac{1}{6}W_{i+1}$$
$$= \tfrac{1}{12}(u_{i-1}Z_{i-1} + u_{i-1}Z_i + u_iZ_{i-1} + u_iZ_{i+1} + u_{i+1}Z_i + u_{i+1}Z_{i+1}) + \tfrac{1}{2}u_iZ_i.$$
$$(3.5.55)$$

Cullen and Morton show that this scheme is more accurate than a conventional Galerkin method.

For the meteorological problem Cullen and Hall found it necessary to introduce smoothing to integrate what are essentially hyperbolic equations. Comparisons were made with a spectral model and with a three-dimensional finite-difference model. For specific cases the finite-element model is competitive, but it produced better results when used with a stream function ψ and velocity potential ϕ, rather than the velocity components u and v.

Cullen and Hall note that the Galerkin formulation is conservative and therefore avoids aliasing errors associated with the nonlinear terms. This permits the forecasts to retain accuracy over a longer time period, say the order of 3 days (also see section 5.4.1).

3.6. Closure

In this chapter we have examined the Galerkin finite-element method. As with the traditional Galerkin method, the solution is continuous throughout the domain. As indicated in section 3.4, the accuracy of the method typically relates directly to the closeness of the interpolation inherent in the trial solution.

Simple examples were considered in section 3.2 to clarify the steps required to implement the Galerkin finite-element method (also see appendix 2). These examples also permit a crude appreciation of the accuracy of the method.

Like the finite-difference method, the Galerkin finite-element method generates algebraic equations that are local in character; the corresponding matrix of equations is sparse. In one dimension, with the aid of linear trial functions, algebraic formulae can be generated that have an immediate (and familiar) finite-difference interpretation. However, in more than one dimension, and with higher-order elements, the correspondence is lost. Generally the integral nature of the Galerkin formulation produces smoother and more accurate solutions than the purely local finite-difference formulations.

The Galerkin method is so well established theoretically that it has, in a sense, encouraged the use of the weak formulation in preference to the original (strong) partial-differential-equation formulation. Thus theoretical analysis (e.g. to establish solution uniqueness) of particularly complex equations, such as the Navier–Stokes equations (Temam, 1979b), is often made directly on the weak formulation, and related by extension to the strong form. Of course, the weak formulation is only half of the Galerkin method. The other half is the introduction of an appropriate trial solution.

In section 3.4 the main sources of error in the solution were discussed and, where appropriate, specific error estimates given. Finally, in section 3.5 we have looked at specific "real" problems, of elliptic, parabolic, and hyperbolic character, to see how the method is used in practice and to see what other techniques can be introduced to enhance its range of applicability.

References

Adam, Y. *J. Comp. Phys.* **24**, 10–22 (1977).

Arakawa, A. J. *Comp. Phys.* **1**, 119–143 (1960).

Astley, R. J., and Eversman, W. *J. Sound Vib.* **74**, 103–121 (1981).

Astley, R. J., and Eversman, W. "Acoustic Transmission in Lined Ducts", in *Finite Elements in Fluids* (ed. R. H. Gallagher), Vol. 4, Wiley, London (1982).

Baker, A. J., and Manhardt, P. D. *AIAA J.* **16**, 807–814 (1978).

Baker, A. J., and Soliman, M. O. *J. Comp. Phys.* **32**, 289–324 (1979).

Bathe, K.-J., and Wilson, E. L. *Numerical Methods in Finite Element Analysis*, Prentice-Hall, Englewood Cliffs, NJ (1976).

Birkhoff, G., and de Boor, C. R. In *Approximations of Functions* (ed. H. L. Garabedian), pp. 164–190 Elsevier (1965).

Bramble, J. H., and Schatz, A. H. *Math. Comp.* **31**, 94–111 (1977).

Brebbia, C. A. *The Boundary Element Method for Engineers*, Pentech Press, London (1978).

Ciment, M., Leventhal, S. H., and Weinberg, B. C. *J. Comp. Phys.* **28**, 135–166 (1978).

Culham, W. E., and Varga, R. S. *Soc. Pet. Eng. J.* **11**, 374–388 (1971).

Cullen, M. J. P., and Hall, C. D. *Quart, J. R. Met. Soc.* **105**, 571–592 (1979).

Cullen, M. J. P., and Morton, K. W. *J. Comp. Phys.* **34**, 245–267 (1980).

Cushman, J. H. *I. J. Num. Meth. Eng.* **14**, 1643–1651 (1979).

Cushman, J. H. *I. J. Num. Meth. Eng.* **17**, 975–989 (1981).

Davis, R. T., and Rubin, S. G. *Comp. and Fluids* **8**, 101–132 (1980).

Dennis, S. C. R., and Walker, J. D. A. *J. Fluid Mech.* **48**, 771–789 (1971).

Dougalis, V. A., and Serbin, S. M. *SIAM J. Num. Anal.* **17**, 431–446 (1980).

Douglas, J., and Dupont, T. *Numer. Math.* **22**, 99–109 (1974).

Dupont, T. *SIAM J. Num. Anal.* **10**, 880–889 (1973).

Ergatoudis, J. G., Irons, B. M., and Zienkiewicz, O. C. *Int. J. Solids Structures* **4**, 31–42 (1968).

Fairweather, G. *Finite Element Galerkin Methods for Differential Equations*, Dekker, New York (1978).

Fletcher, C. A. J. The Application of the Finite Element Method to Two-Dimensional Inviscid Flow. WRE-TN-1606, Salisbury, South Australia (1976).

Fletcher, C. A. J. Improved Integration Techniques for Fluid Flow Finite Element Formulations. WRE-TR-1810, Salisbury, South Australia (1977).

Fletcher, C. A. J. "Burgers' Equation: A Model for All Reasons", in *Numerical Solution of Partial Differential Equations* (ed. J. Noye), pp. 139–225, North-Holland, Amsterdam (1982a).

Fletcher, C. A. J. "A Comparison of the Finite Element and Finite Difference Methods for Computational Fluid Dynamics", in *Finite Element Flow Analysis* (ed. T. Kawai), pp. 1003–1010, Univ. of Tokyo Press, (1982b).

Gartling, D. K. *Comp. Meth. App. Mech. Eng.* **12**, 365–382 (1977).

Gartling, D. K. "A Finite Element Analysis of Volumetrically Heated Fluids in an Axisymmetric Enclosure". *3rd Finite Element in Flow Problems Conference*, Banff, Canada, pp. 174–182 (1980).

Gray, W. G., and Pinder, G. F. *I. J. Num. Meth. Eng.* **10**, 893–923 (1976).

Grotkop, G. *Comp. Meth. Appl. Mech. Eng.* **2**, 147–157 (1973).

Heubner, K. H. *The Finite Element Method for Engineers*, Wiley, New York (1975).

Hughes, T. J. R., Taylor, R. L., and Levy, J. F. In *Finite Elements in Fluids* (ed. R. H. Gallagher et al.), Vol. 3, pp. 55–72, Wiley, London (1978).

Hughes, T. J. R., Liu, W. K., and Brooks, A. *J. Comp. Phys.* **30**, 1–60 (1979).

Irons, B. M. *Int. J. Num. Meth. Eng.* **2**, 5–32 (1970).

Irons, B. M., and Razzaque, A. In *The Mathematical Foundations of the Finite Element Method with Applications to Partial Differential Equations* (ed. A. K. Aziz), pp. 557–587, Academic Press, (1972).

Irons, B., and Ahmad, S. *Techniques of Finite Elements*, Wiley, Chichester (1980).

Isaacson, E., and Keller, W. B. *An Analysis of Numerical Methods*, Wiley, New York (1966).

Jennings, A. *Matrix Computation for Engineers and Scientists*, Wiley, London (1977).

Jespersen, D. C. *J. Comp. Phys.* **16**, 383–390 (1974).

Kawahara, M., Takeuchi, N., and Yoshida, T. *Int. J. Num. Meth. Eng.* **12**, 331–351 (1978).

Khosla, P. K., and Rubin, S. G. *J. Eng. Math.* **13**, 127–141 (1979).

Mikhlin, S. G. *Variational Methods in Mathematical Physics*, Pergamon, Oxford (1964).

Mitchell, A. R., and Wait, R. *The Finite Element Method in Partial Differential Equations*, Wiley, London (1977).

Norrie, D., and de Vries, G. *Finite Element Bibliography*, Plenum, New York (1976).

Oden, J. T. *Finite Elements of Nonlinear Continua*, McGraw-Hill, New York (1972).

Oden, J. T., and Reddy, J. N. *An Introduction to the Mathematical Theory of Finite Elements*, Wiley, New York (1976a).

Oden, J. T., and Reddy, J. N. *SIAM J. Num. Anal.* **13**, 393–404 (1976b).

Oden, J. T. "Penalty Methods and Selective Reduced Integration for Stokesian Flows", in *3rd Finite Element in Flow Problems Conference*, Banff, Canada, pp. 140–145 (1980).

Phuoc, H. B., and Tanner, R. I. *J. Fluid Mech.* **98**, 253–271 (1980).

Pinder, G. F., and Gray, W. G. *Finite Element Simulation in Surface and Subsurface Hydrology*, Academic Press, New York (1977).

Richtmyer, R., and Morton, K. W. *Difference Methods for Initial-Value Problems*, Wiley, New York, 2nd Edn (1967).

Rubbert, P. E., and Saaris, G. R. Review and Evaluation of a Three-Dimensional Lifting Potential Flow Analysis Method for Arbitrary Configurations. AIAA Paper 72-188 (1972).

Schreker, G. O., and Maus, J. R. Noise Characteristics of Jet-flap Type Exhaust Flows. NASA CR-2342 (1974).

Schultz, M. *Spline Analysis*, Prentice-Hall, Englewood Cliffs, NJ (1973).

Showalter, R. E. *SIAM J. Num. Anal.* **12**, 456–463 (1975).

Staniforth, A. N. "A Review of the Applications of the Finite Element Method to Meteorological Flows," in *Finite Element Flow Analysis* (ed. T. Kawai) Univ. of Tokyo Press, Tokyo (1982), pp. 835–842.

Staniforth, A. N., and Daley, R. W. *Mon. Weath. Rev.* **107**, 107–121 (1979).

Strang, G., and Fix, G. J. *An Analysis of the Finite Element Method*, Prentice-Hall, Englewood Cliffs, NJ (1973).

Swartz, B., and Wendroff, B. *Math. Comp.* **23**, 37–50 (1969).

Taylor, C., and Davis, J. *Comp. and Fluids* **3**, 125–148 (1975).

Taylor, C. "The Utilisation of the F. E. M. in the Solution of Some Free Surface Problems", in *3rd Finite Element in Flow Problems Conference*, Banff, Canada, pp. 54–81 (1980).

Temam, R., and Thomasset, F. "Numerical Solution of the Navier-Stokes Equations by a Finite Element Method", in *2nd Finite Element in Flow Problems Conference*, St. Margharita Ligure, Italy (1976).

Temam, R. "Some Finite Element Methods in Fluid Flow", in *Lecture Notes in Physics*, No. 90, pp. 34–55 Springer-Verlag, Berlin, (1979a).

Temam, R. *Navier–Stokes Equations*, North-Holland, Amsterdam (1979b).

Thomée, V. *Math. Comp.* **31**, 652–660 (1977).

Tuann, S.-Y., and Olson, M. D. *J. Comp. Phys.* **29**, 1–19 (1978).

Wellford, L. C., and Oden, J. T. *Comp. Meth. App. Mech. Eng.* **5**, 83–96 (1975).

Wheeler, M. F. *SIAM J. Num. Anal.* **10**, 723–759 (1973).

Whiteman, J. R. *A Bibliography for Finite Elements*, Academic Press (1975).

Zienkiewicz, O. C. *The Finite Element Method*, McGraw-Hill, London, 3rd Edn. (1977).

CHAPTER 4

Advanced Galerkin Finite-Element Techniques

In this chapter we examine some of the techniques that have been combined with the Galerkin method to overcome specific deficiencies and to give an improved performance. The important category of generating algorithms for convection-dominated flows will be dealt with in chapter 7.

For parabolic problems the time (or timelike) derivatives are often represented by finite differences and the solution algorithm is most often implicit. To avoid an expensive matrix factorization at every time step it is advantageous to introduce a splitting of the spatial operator. This is considered in section 4.1.

It has been found in practice that using a reduced order of numerical integration can produce a more accurate solution. Reduced integration and the related concept of least-squares residual fitting are demonstrated in section 4.2.

For some problems the local behavior (e.g. close to a corner) has a different character to that in the rest of the domain. The overall solution can be obtained most efficiently by introducing special trial functions locally. This idea is explored in section 4.3.

Most of the problems to which Galerkin methods have been applied are governed by partial differential equations. However, some problems are better described by integral equations. In section 4.4 we consider such problems and examine, in particular, boundary-integral methods.

4.1. Time Splitting

Here we discuss the solution of partial differential equations that contain time-dependent terms, for example

$$\frac{\partial T}{\partial t} = \alpha_x \frac{\partial^2 T}{\partial x^2} + \alpha_y \frac{\partial^2 T}{\partial y^2}. \tag{4.1.1}$$

The problem in question may be genuinely time dependent. Transient heat conduction in a two-dimensional region is governed by the above equation. Alternatively time may be used as an iteration parameter. If only the steady-state solution is of interest, time-dependent terms may be introduced to give

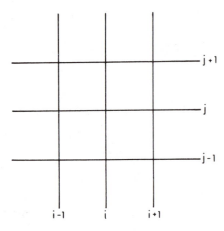

Figure 4.1. Grid notation

a more stable or more efficient path to the steady-state solution. We will refer to this as the *pseudotransient* approach.

Application of the Galerkin finite-element method with linear rectangular elements to eq. (4.1.1) produces the following system of ordinary differential equations:

$$m_y \otimes m_x \frac{\partial T}{\partial t} = \frac{\alpha_x}{\Delta x^2} m_y \otimes \delta_x^2 T + \frac{\alpha_y}{\Delta y^2} m_x \otimes \delta_y^2 T. \qquad (4.1.2)$$

m_x, m_y, δ_x^2, and δ_y^2 are operators that have the same interpretation on a uniform grid (Fig. 4.1) as given in section 3.3, namely

$$\delta_x^2 T_{i,j} = T_{i-1,j} - 2T_{i,j} + T_{i+1,j}, \qquad (4.1.3)$$

$$m_y T_{i,j} = \tfrac{1}{6} T_{i,j-1} + \tfrac{2}{3} T_{i,j} + \tfrac{1}{6} T_{i,j+1}. \qquad (4.1.4)$$

The *tensor product* $\otimes$ can be illustrated by noting that

$$m_y \otimes m_x I = \begin{bmatrix} \tfrac{1}{6} \\ \tfrac{2}{3} \\ \tfrac{1}{6} \end{bmatrix} \begin{bmatrix} \tfrac{1}{6} & \tfrac{2}{3} & \tfrac{1}{6} \end{bmatrix} I = \begin{bmatrix} \tfrac{1}{36} & \tfrac{1}{9} & \tfrac{1}{36} \\ \tfrac{1}{9} & \tfrac{4}{9} & \tfrac{1}{9} \\ \tfrac{1}{36} & \tfrac{1}{9} & \tfrac{1}{36} \end{bmatrix}.$$

If a finite-difference representation for $\partial T/\partial t$ is introduced into eq. (4.1.2), the following is obtained:

$$m_y \otimes m_x(T^{n+1} - T^n) = (r_x m_y \otimes \delta_x^2 + r_y m_x \otimes \delta_y^2)((1 - \theta)T^n + \theta T^{n+1}), \qquad (4.1.5)$$

where

$$r_x = \alpha_x \frac{\Delta t}{\Delta x^2}, \qquad r_y = \alpha_y \frac{\Delta t}{\Delta y^2}.$$

If $\theta = 0$, the right-hand side of eq. (4.1.5) is evaluated at time level n. Then the time step Δt must be restricted to obtain solutions to the global system derived from eq. (4.1.5). If $\theta = 1$, the global system of equations derived from eq. (4.1.5) must be factorized at every time step, but there is no stability restriction on the time step. Matrix factorization is generally computationally expensive.

The purpose of *time splitting* is to replace equations like (4.1.5) with a sequence of equations in which the implicit terms can be efficiently factorized. This can be clarified by manipulating eq. (4.1.5) to give

$$(m_y \otimes m_x - \theta r_x m_y \otimes \delta_x^2 - \theta r_y m_x \otimes \delta_y^2) \Delta T^{n+1} = (r_x m_y \otimes \delta_x^2 + r_y m_x \otimes \delta_y^2) T^n$$

$$= \Delta t R^n,$$

(4.1.6)

where

$$\Delta T^{n+1} = T^{n+1} - T^n,$$

and R^n is the spatial residual. If $\theta = \frac{1}{2}$, the algorithm (4.1.6) has a truncation error of $O(\Delta t^2, \Delta x^2, \Delta y^2)$. If a higher-order term

$$\theta^2 r_x r_y \delta_y^2 \otimes \delta_x^2 \Delta T^{n+1}$$

is added to the left-hand side of eq. (4.1.6), it can be written

$$(m_y - \theta r_y \delta_y^2) \otimes (m_x - \theta r_x \delta_x^2) \Delta T^{n+1} = \Delta t \, R^n.$$

(4.1.7)

Eq. (4.1.7) can be split into two one-dimensional problems (at the implicit level),

$$(m_y - \theta r_y \delta_y^2) \Delta T^* = \Delta t \, R^n$$

(4.1.8)

and

$$(m_x - \theta r_x \delta_x^2) \Delta T^{n+1} = \Delta T^*.$$

(4.1.9)

For the first half time step, eq. (4.1.8) is applied to all nodes on a grid line in the y direction (constant i in Fig. 4.1), generating a tridiagonal system of equations that can be efficiently solved for ΔT^* using the Thomas algorithm (Isaacson and Keller, 1966, p. 55). This process is repeated along all grid lines in the y direction. During the second half time step eq. (4.1.9) is applied to all nodes on each grid line in the x direction (constant j in Fig. 4.1).

The splitting (4.1.8), (4.1.9) implies shape functions that can be split to give the contributions m_x, m_y, and so on. This is possible with Lagrange shape functions but not with serendipity shape functions. Using quadratic Lagrange shape functions would lead to a pentadiagonal system of equation to be solved for on each grid line. However, this can be carried out efficiently.

Gourlay (1977) discusses time splitting in a general context that is equally applicable to finite-difference or finite-element formulations. Fairweather (1978) provides error estimates for schemes similar to eqs. (4.1.8) and (4.1.9). Hayes (1979) has developed a multilevel scheme that avoids iteration at each time step when solving nonlinear time-dependent problems. Deconinck and Hirsch (1981) have applied a scheme similar to eqs. (4.1.8) and (4.1.9) to transonic inviscid flow.

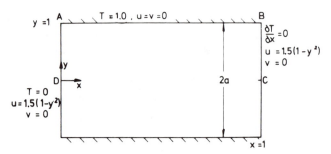

Figure 4.2. Domain and boundary conditions for thermal entry problem

4.1.1. Thermal entry problem

Here we consider a simple time-splitting scheme to obtain the *steady-state* solution. Time is introduced into this problem to generate a pseudotransient path to the steady-state solution. The computational domain and boundary conditions are shown in Fig. 4.2. The governing equation for the temperature distribution in the entrance of a two-dimensional duct can be written

$$\frac{\partial T}{\partial t} + \frac{\partial}{\partial x}(uT) + \frac{\partial}{\partial y}(vT) = \alpha_x \frac{\partial^2 T}{\partial x^2} + \alpha_y \frac{\partial^2 T}{\partial y^2}, \qquad (4.1.10)$$

where T is a nondimensional temperature and u and v are nondimensional velocity components that are known. Thus the problem is linear in T.

Eq. (4.1.10) can be reduced to a system of algebraic equations by applying the Galerkin finite-element method in space with linear rectangular elements and a finite-difference interpretation in time. In applying the Galerkin method the same trial functions have been used to interpolate T and the *groups* (uT) and (vT). Representing groups in this way produces a more economical solution procedure Fletcher (1983). This technique has been used previously with a least-squares finite-element formulation for subsonic compressible flow (Fletcher, 1979b). The group representation is referred to as a "product approximation" by Christie et al. (1981).

The application of the *alternating-direction implicit* (ADI) method to the system of algebraic equations gives, for the first half time step,

$$m_y \otimes m_x(T^* - T^n) + 0.5\lambda_x m_y \otimes \delta_x(uT)^* + 0.5\lambda_y m_x \otimes \delta_y(vT)^n$$
$$= 0.5 r_x m_y \otimes \delta_x^2 T^* + 0.5 r_y m_x \otimes \delta_y^2 T^n, \qquad (4.1.11)$$

and for the second half time step,

$$m_y \otimes m_x(T^{n+1} - T^*) + 0.5\lambda_x m_y \otimes \delta_x(uT)^* + 0.5\lambda_y m_x \otimes \delta_y(vT)^{n+1}$$
$$= 0.5 r_x m_y \otimes \delta_x^2 T^* + 0.5 r_y m_x \otimes \delta_y^2 T^{n+1}. \qquad (4.1.12)$$

In Eqs. (4.1.11) and (4.1.12) δ_x and δ_y are the same as the centered-difference first-derivative operators, such as

$$\delta_x T_{i,j} = \frac{T_{i+1,j} - T_{i-1,j}}{2},$$

and λ_x and λ_y are given by

$$\lambda_x = \frac{\Delta t}{\Delta x} \quad \text{and} \quad \lambda_y = \frac{\Delta t}{\Delta y}.$$

For grid lines in the x direction (constant j in Fig. 4.1), eq. (4.1.11) is used, with terms at the intermediate time level (*) treated implicitly. For grid lines in the y direction (constant i in Fig. 4.1) eq. (4.1.12) is used, with terms at time level $n + 1$ treated implicitly.

The operators m_y in eq. (4.1.11) and m_x in eq. (4.1.12) produce a non-tridiagonal structure in the system of equations. However, a tridiagonal scheme can be obtained if the terms on either side of the jth grid line (for the first half time step) are treated explicitly. This reduces the accuracy of the transient solution, but that is not important with a pseudotransient approach, since it does not affect the steady-state solution.

However, such a scheme is only conditionally stable. The maximum time step is shown in Table 4.1 (Scheme 1). The maximum time step was predicted by a von Neumann stability analysis and confirmed by the numerical results. The numerical results were obtained on a uniform 21×21 grid covering the domain shown in Fig. 4.2.

An examination of the algorithm in Scheme 1 indicates that the main contribution to the lack of stability is the small weight at the i, j node arising from the "mass matrix" $m_y \otimes m_x I$. This weight is $\frac{4}{9}$. Since the mass matrix plays no useful role in the pseudotransient approach, it is convenient to lump all the contributions from $m_y \otimes m_x$ onto the i, j node. This gives a weight of unity. It can be seen from Table 4.1 that a larger time step is possible (Scheme 2), but the scheme is unstable if too large a time step is used.

A further modification is possible to generate an unconditionally stable scheme. In eq. (4.1.11) part of the term $\delta_y^2 T$ lies on the jth grid line. This term and a corresponding contribution from $\delta_x^2 T$ in eq. (4.1.12) have been treated

Table 4.1. Comparison of schemes for thermal entry problem

Scheme	Description	Δt_{max}	Number of time steps	Relative execution time	Rms difference
1	ADIFEM	0.0003	1250	7.73	0.0502
2	ADIFEM, fully lumped	0.002	247	1.49	0.0502
3	ADIFEM, fully lumped, fully implicit	—	40	0.21	0.0502
4	ADI finite difference	—	332	1.00	0.0601

implicitly in Scheme 3. To complete the comparison, an equivalent ADI finite difference is included (Scheme 4). This scheme is also unconditionally stable.

The rms difference shown in Table 4.1 is based on the difference between the steady-state solution and an "exact" solution on the centerline (DC in Fig. 4.2). The results presented in Table 4.1 indicate that the progressive increase in the maximum allowable time step leads to a smaller number of time steps to reach the steady state and a shorter execution time. It was found that too large a time step for the finite-difference method slowed down the rate of convergence. For this problem the finite-element method is both more economical and more accurate than the finite-difference method (Fletcher, 1980b).

4.1.2. Viscous compressible flow

This problem is more complicated than the previous problem because a system of nonlinear equations is required to determine the flow behavior. This system of equations can be written

$$\frac{\partial \mathbf{q}}{\partial t} + \frac{\partial \mathbf{F}}{\partial x} + \frac{\partial \mathbf{G}}{\partial y} = \frac{\partial^2 \mathbf{R}}{\partial x^2} + \frac{\partial^2 \mathbf{S}}{\partial x \partial y} + \frac{\partial^2 \mathbf{T}}{\partial y^2}, \qquad (4.1.13)$$

where

$$\mathbf{q}^t = \{\rho, \rho u, \rho v\},$$

$$\mathbf{F}^t = \{\rho u, \rho u^2 + p, \rho u v\},$$

$$\mathbf{G}^t = \{\rho v, \rho u v, \rho v^2 + p\},$$

$$\mathbf{R}^t = \left\{\varepsilon_\rho, \frac{4u}{3\,\mathrm{Re}}, \frac{v}{\mathrm{Re}}\right\}, \qquad (4.1.14)$$

$$\mathbf{S}^t = \left\{0, \frac{v}{3\,\mathrm{Re}}, \frac{u}{3\,\mathrm{Re}}\right\},$$

$$\mathbf{T}^t = \left\{\varepsilon_\rho, \frac{u}{\mathrm{Re}}, \frac{4v}{3\,\mathrm{Re}}\right\}.$$

In the above equation, u, v are the two velocity components and ρ is the density. The pressure p can be eliminated from eqs. (4.1.14) by using

$$1 + \gamma M_\infty^2 p = \rho\{1 + 0.5(\gamma - 1)M_\infty^2(1 - u^2 - v^2)\}, \qquad (4.1.15)$$

where M_∞ is the freestream Mach number and γ is the specific-heat ratio. In this problem only the steady-state solution is of interest.

Time splitting for this problem can be illustrated by considering the flow past a rectangular obstacle shown in Fig. 4.3. For this problem the following boundary conditions have been imposed:

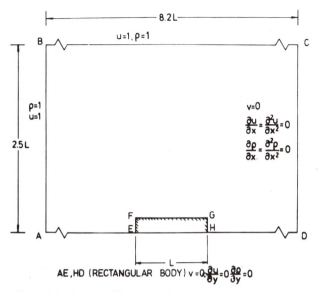

Figure 4.3. Compressible viscous flow past a rectangular obstacle

on AB,	$u = 1$,	$\rho = 1$,	
on BC,	$u = 1$,	$\rho = 1$,	

$$\text{on } CD, \qquad v = 0, \qquad \frac{\partial u}{\partial x} = \frac{\partial^2 u}{\partial x^2} = \frac{\partial \rho}{\partial x} = \frac{\partial^2 \rho}{\partial x^2} = 0, \qquad (4.1.16)$$

on $EFGH$, $\qquad u = v = 0$,

$$\text{on } AE \text{ and } HD \qquad v = 0, \qquad \frac{\partial u}{\partial y} = \frac{\partial \rho}{\partial y} = 0,$$

The Galerkin finite-element method with linear rectangular elements has been applied to the system of equations (4.1.13). In a manner similar to the previous problem, groups of terms, such as $\rho u v$, have been interpolated linearly.

If a finite-difference representation is introduced for $\partial \mathbf{q}/\partial t$ and mass lumping is applied, the following ADI scheme can be generated:

$$\varDelta \mathbf{q}^* + 0.5 \lambda_x m_y \otimes \delta_x \mathbf{F}^* + 0.5 \lambda_y m_x \otimes \delta_y \mathbf{G}^n$$
$$= 0.5 r_x m_y \otimes \delta_x^2 \mathbf{R}^* + 0.5 r_{xy} \delta_{xy} \mathbf{S}^n + 0.5 r_y m_x \otimes \delta_y^2 \mathbf{T}^n \qquad (4.1.17)$$

and

$$\varDelta \mathbf{q}^{n+1} + 0.5 \lambda_x m_y \otimes \delta_x \mathbf{F}^* + 0.5 \lambda_y m_x \otimes \delta_y \mathbf{G}^{n+1}$$
$$= 0.5 r_x m_y \otimes \delta_x^2 \mathbf{R}^* + 0.5 r_{xy} \delta_{xy} \mathbf{S}^* + 0.5 r_y m_x \otimes \delta_y^2 \mathbf{T}^{n+1}, \qquad (4.1.18)$$

where $\Delta\mathbf{q}^* = \mathbf{q}^* - \mathbf{q}^n$ and can be interpreted as a correction to the solution at the nth time level. Other new terms in eqs. (4.1.17) and (4.1.18) are defined by

$$r_{xy} = \frac{\Delta t}{4 \Delta x \Delta y},$$

$$\delta_{xy}\mathbf{S} = (\mathbf{S}_{i+1,j+1} - \mathbf{S}_{i-1,j+1}) - (\mathbf{S}_{i+1,j-1} - \mathbf{S}_{i-1,j-1}).$$

$$(4.1.19)$$

The $\mathbf{F}$ and $\mathbf{G}$ terms in eqs. (4.1.17) and (4.1.18) are nonlinear, and where they appear implicitly, must be linearized to generate a tridiagonal system of equations. This can be done by expanding about the nth time-level. For example,

$$(\rho uv)^{n+1} = (\rho uv)^n + (uv)^n \Delta\rho^{n+1} + (\rho v)^n \Delta u^{n+1} + (\rho u)^n \Delta v^{n+1}. \quad (4.1.20)$$

Upon rearrangement eqs. (4.1.17) and (4.1.18) can be written

$$(\mathbf{H}^n + \mathbf{A}_x^n) \Delta\mathbf{p}^* = -0.5 \Delta t \, \mathbf{R}^n \quad (4.1.21)$$

and

$$(\mathbf{H}^* + \mathbf{A}_y^*) \Delta\mathbf{p}^{n+1} = -0.5 \Delta t \, \mathbf{R}^*, \quad (4.1.22)$$

where $\Delta\mathbf{p}^t = \{\Delta\rho, \Delta u, \Delta v\}$. $\mathbf{H}$ contains the contributions from $\Delta\mathbf{q}$. $\mathbf{A}_x$ and $\mathbf{A}_y$ contain contributions from the spatial terms. $\mathbf{R}$ is the residual of all the spatial terms. The steady-state solution corresponds to $\mathbf{R} = 0$. This also implies that $\Delta\mathbf{p} = 0$ at the steady state. Consequently it is possible to modify the terms multiplying $\Delta\mathbf{p}$ in eqs. (4.1.21) and (4.1.22) without altering the steady-state solution.

A von Neumann stability analysis applied to eqs. (4.1.21) and (4.1.22) indicates that the scheme is only conditionally stable for large values of Re, even if Scheme 3 in section 4.1.1 is used. An improvement can be made if eqs. (4.1.21) and (4.1.22) are replaced by

$$(\mathbf{H}^n + \mathbf{A}_{fd,\,x}^n) \Delta\mathbf{p}^* = -0.5 \Delta t \, \mathbf{R}^n, \quad (4.1.23)$$

$$(\mathbf{H}^* + \mathbf{A}_{fd,\,y}^*) \Delta\mathbf{p}^{n+1} = -0.5 \, \Delta t \, \mathbf{R}^*, \quad (4.1.24)$$

where $\mathbf{A}_{fd}$ represents the terms that would arise if finite-difference expressions were introduced on the left-hand side of eqs. (4.1.21) and (4.1.22). It may be noted that the right-hand side is still evaluated using the full finite-element form. The treatment of the left-hand side is equivalent, with linear elements, to eqs. (4.1.8) and (4.1.9) with mass lumping. The stability restrictions for this scheme are shown in Table 4.2. It is found (Fletcher, 1982) that it is the finite-element treatment of the *convective terms* that causes the lack of unconditional stability.

Numerical solutions using the above scheme have been obtained for the flow past a rectangular obstacle shown in Fig. 4.3. These were obtained with a 101×49 grid which had been stretched to cluster more points close to the body. The smallest step size was $\Delta x = \Delta y = 0.025$, where the length of the body is unity. A typical velocity-vector plot is shown in Fig. 4.4. The length of

Table 4.2. Stability restrictions for implicit scheme (4.1.23), (4.1.24) with $\Delta x = 0.10$, $\Delta y = 0.10$, $M_\infty = 0.90$

Δt \ Re	5	20	100	1000	10,000
0.004	Stable	Stable	Stable	Stable	Stable
0.02	Stable	Stable	Stable	Stable	Stable
0.10	Stable	Stable	Stable	Stable	Stable
0.40	Stable	Stable	Stable	Unstable	Unstable
0.80	Stable	Stable	Stable	Unstable	Unstable

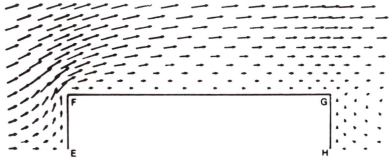

Figure 4.4. Velocity-vector plot for compressible flow past a rectangular obstacle

the shaft of the arrow indicates the magnitude of the velocity vector. Velocity vectors in Fig. 4.4 correspond to alternate grid points.

The pressure distribution on the same body is shown in Fig. 4.5. There is some evidence that the pressure solution is oscillatory adjacent to the upstream salient F. Ideally a very refined grid is required in this area, or an analytic solution could be introduced locally (see section 4.3.1).

4.2. Least-squares Residual Fitting

When numerical integration is required to evaluate the inner products associated with applying the finite-element method (e.g. eq. (3.2.44)), it is often found (Zienkiewicz and Hinton, 1976) that using a lower-order Gauss quadrature formula *(reduced integration)* produces superior results to carrying out the integration exactly.

In a sense the use of reduced integration relaxes the constraints on the finite-element solution imposed by the form of the trial functions. This can be illustrated as follows. A trial solution for an unknown, q, is introduced by

$$q(x,y) = \sum_j \phi_j(x,y)\bar{q}_j. \tag{4.2.1}$$

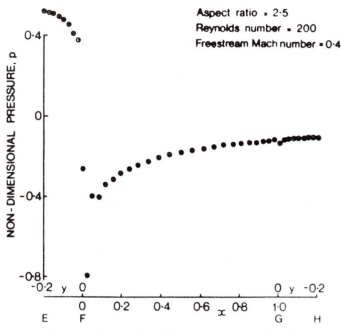

Figure 4.5. Pressure distribution on a rectangular obstacle

Substitution into the governing equation produces a residual $R(\bar{q}_j, x, y)$. Application of the Galerkin method requires

$$(\phi_k, R(\bar{q}_j, x, y)) = 0. \qquad (4.2.2)$$

If the inner product is evaluated numerically, eq. (4.2.2) is replaced with

$$\sum_{i=1}^{M} W_i \phi_k(x_i, y_i) R_i(\bar{q}_j, x_i, y_i) = 0, \qquad k = 1, \ldots, N, \qquad (4.2.3)$$

where W_i is the weight (determined by the particular quadrature formula) associated with the ith function-evaluation point. For linear problems

$$R_i = \sum_{j=1}^{T} L(\phi_j(x_i, y_i))\bar{q}_j, \qquad (4.2.4)$$

where T is the total number of nodal parameters $\bar{q}_j$ (some of these will be known). Eq. (4.2.4) gives one linear relationship among the N unknowns $\bar{q}_j$. Each additional evaluation point (x_i, y_i) introduces another linear relationship among the unknowns $\bar{q}_j$. Clearly N evaluation points would provide sufficient independent relationships to solve for the $\bar{q}_j$'s.

Substitution of eq. (4.2.4) into eq. (4.2.3) gives

$$\sum_{i=1}^{M} W_i N_k(x_i, y_i) \sum_{j=1}^{T} L(\phi_j(x_i, y_i))\bar{q}_j = 0, \qquad k = 1, \ldots, N. \qquad (4.2.5)$$

This permits the M function-evaluation points to determine the solution. If the integration is carried out exactly, M will be two or three times N. M is typically the product of the number of contributing elements and the number of function-evaluation points per element. The large number of function-evaluation points follows from the constraints placed on the finite-element solution by the form of eq. (4.2.1). The use of reduced integration relaxes some of these constraints, but introduces additional errors by not performing the integration exactly.

Gauss quadrature formulae have the interesting property of sampling at points where an integrand of products of order p coincides with a least-squares fit of the products of order $p + 1$. Thus using reduced Gauss integration to evaluate eq. (4.2.2) is equivalent to the exact numerical integration of

$$((\phi_k)_{ls}, (R)_{ls}) = 0, \tag{4.2.6}$$

where $(R)_{ls}$ is a least-squares fit of the residual over the element.

Clearly $(R)_{ls}$ resembles R over each element. The above interpretation suggests formally replacing eq. (4.2.2) with

$$(\phi_k, (R)_{ls}) = 0. \tag{4.2.7}$$

We will illustrate this idea with two of the problems that have been utilized previously. First Duncan's problem (previously analysed in section 1.2.1),

$$\frac{dy}{dx} - y = 0, \tag{4.2.8}$$

will be considered in the region $0 \le x \le 1$, with the boundary condition $y = 1$ at $x = 0$. If the region is split into two one-dimensional quadratic elements as shown in Fig. 4.6, the problem has five nodal parameters, $\bar{y}_i$ of which four are unknown. Substitution of a quadratic trial solution into eq. (4.2.8) produces a quadratic residual in each element. Evaluation of $(R)_{ls}$ for each element and evaluation of eq. (4.2.7) produces four algebraic equations. The solution of

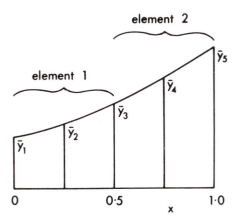

Figure 4.6. Two-element representation of $dy/dx - y = 0$

Table 4.3. Two-element solution of $dy/dx - y = 0$

x	Exact numerical integration $\bar{y}_{ei}$	Element least-squares residual fit $\bar{y}_{ls}$	Reduced integration $\bar{y}_{ri}$	Exact solution $y = e^x$
0	1.0000	1.0000	1.0000	1.0000
0.25	1.2707	1.2838	1.2838	1.2840
0.50	1.6403	1.6486	1.6486	1.6487
0.75	2.0990	2.1165	2.1165	2.1170
1.00	2.6938	2.7180	2.7180	2.7183
$\|\bar{y} - y_{ex}\|_{2,d}$	0.0342	0.0006	0.0006	

these equations is shown as $\bar{y}_{ls}$ in Table 4.3. Also shown in Table 4.3 are the solutions obtained via exact integration, $\bar{y}_{ei}$, and reduced integration, $\bar{y}_{ri}$. It is apparent from the results shown in Table 4.3 that the solution obtained from a least-squares fit of the residual over each element coincides with the solution obtained by reduced integration, and both solutions are considerably more accurate than the use of exact integration.

A second problem will be used to investigate the least-squares residual fit formulation (4.2.7). The problem to be considered is that of incompressible, inviscid flow about a two-dimensional cylinder. This problem was considered in section 3.2.3. The solution is sought in a quarter plane (Fig. 3.16). After application of the Galerkin finite-element method the following algebraic equations are obtained:

$$\sum_j a_{kj}\bar{u}_j + \sum_j b_{kj}\bar{v}_j = 0, \qquad k = 1, \ldots, N, \qquad (4.2.9)$$

and

$$\sum_j b_{kj}\bar{u}_j - \sum_j a_{kj}\bar{v}_j = 0, \qquad k = 1, \ldots, N. \qquad (4.2.10)$$

If Green's theorem is not applied (see section 3.2.3), a_{kj}, b_{kj} have the following form:

$$a_{kj} = \int\int N_k \frac{\partial N_j}{\partial x} dx\, dy, \qquad (4.2.11)$$

and

$$b_{kj} = \int\int N_k \frac{\partial N_j}{\partial y} dx\, dy. \qquad (4.2.12)$$

After application of the isoparametric mapping, eqs. (4.2.11) and (4.2.12) become, at an element level,

$$(a_{kj})_e = \sum_{l=1}^{Le} c_{jkl}\bar{y}_l \qquad (4.2.13)$$

and

$$(b_{kj})_e = -\sum_{l=1}^{Le} c_{jkl}\bar{x}_l, \qquad (4.2.14)$$

where $(\bar{x}_l, \bar{y}_l)$ are the coordinates of the nodes of the element, and the coefficients c_{jkl} are given by

$$c_{jkl} = \int\int N_k \left\{ \frac{\partial N_j}{\partial \eta} \frac{\partial N_l}{\partial \xi} - \frac{\partial N_j}{\partial \xi} \frac{\partial N_l}{\partial \eta} \right\} d\xi \, d\eta. \qquad (4.2.15)$$

Logically the least-squares residual fit (4.2.7) should be made in the physical plane. That is, if N_k and N_j in eqs. (4.2.11) and (4.2.12) are quadratic serendipity shape functions, it would be feasible to replace eqs. (4.2.11) and (4.2.12) with

$$(a_{kj})_{1s} = \int\int N_k F \, dx \, dy,$$
$$(b_{kj})_{1s} = \int\int N_k G \, dx \, dy, \qquad (4.2.16)$$

where

$$F = a_0 + a_1 x + a_2 y + a_3 xy \qquad (4.2.17)$$

and a_0 to a_3 are chosen so that

$$\int\int_e \left[\frac{\partial N_j}{\partial x} - F(x,y) \right]^2 dx \, dy \text{ is a minimum,}$$

and similarly for G. This requires a separate integration over every element to evaluate eqs. (4.2.16). This is less efficient than evaluating eqs. (4.2.13) and (4.2.14) for each element after evaluating eq. (4.2.15) just once.

Alternatively F can be expressed as

$$F = a_0 + a_1 \xi + a_2 \eta + a_3 \xi\eta \qquad (4.2.18)$$

and the integrations carried out in the (ξ, η) plane. The solution obtained by this means is more accurate than the use of exact integration, but is very sensitive to element size (Fletcher, 1979a).

Examination of eqs. (4.2.11) to (4.2.15) indicates that the $\partial N_j/\partial \eta$ and $\partial N_j/\partial \xi$ terms in eq. (4.2.15) come from the equation residual. Consequently the least-squares residual fit is applied in the (ξ, η) plane by replacing eq. (4.2.15) with

$$c_{jkl} = \int\int N_k \left\{ \left(\frac{\partial N_j}{\partial \eta} \right)_{1s} \frac{\partial N_l}{\partial \xi} - \left(\frac{\partial N_j}{\partial \xi} \right)_{1s} \frac{\partial N_l}{\partial \eta} \right\} d\xi \, d\eta, \qquad (4.2.19)$$

where $(\partial N_j/\partial \eta)_{1s}$ etc. are represented by

$$\left(\frac{\partial N_j}{\partial \eta} \right)_{1s} = a_0 + a_1 \xi + a_2 \eta + a_3 \xi\eta. \qquad (4.2.20)$$

Solutions obtained with this scheme are shown as case 1 in Table 4.4.

In Table 4.4 σ is the rms difference between the finite-element solution for the tangential body-surface velocity, q_{Ta}, and the exact solution for the tangential body-surface velocity, q_{Te}. That is,

Table 4.4. Comparison of different Galerkin finite-element formulations

Case	Comments	σ
1	Exact numerical integration, eq. (4.2.19)	0.017
2	Exact numerical integration, eq. (4.2.15)	0.034
3	Reduced numerical integration, eq. (4.2.15)	0.005

$$\sigma = \left\{ \frac{1}{N_B} \sum_{l=1}^{N_B} (q_{Ta} - q_{Te})_l^2 \right\}^{1/2}, \qquad (4.2.21)$$

where N_B is the number of nodal points on the body. Also shown in Table 4.4 are the solutions corresponding to the exact and reduced integration of eq. (4.2.15).

The results shown in Table 4.4 were obtained with 25 quadratic rectangular serendipity elements and 149 nodal unknowns. The present results used eq. (4.2.10) on the line DC (Fig. 3.16). The results given in section 3.2.3 were based on the use of eq. (4.2.9) on the line DC. An examination of the results shown in Table 4.4 indicates that the least-squares residual fit procedure produces more accurate result than the use of exact numerical integration. However, the use of reduced integration is more accurate, for this problem, than either of the other procedures.

It has been found (Fletcher, 1980a) that least-squares residual fitting and reduced integration give no significant improvement in accuracy for triangular elements applied to this problem.

4.3. Special Trial Functions

The (Galerkin) finite-element method is effective partly because it makes use of piecewise polynomial approximation (Strang and Fix, 1973, Preface). However, we need to qualify this statement by adding the phrase "assuming we know nothing about the solution".

In contrast, an important feature of the traditional Galerkin method is to choose the nature of the trial functions so that an accurate solution can be obtained with as few trial functions as possible. Any knowledge we have about the solution should be exploited.

If the same strategy is followed with the Galerkin finite-element method, we generally find that the introduction of special functions is very disruptive of the rest of the solution procedure. Therefore the benefits of supplementing the polynomial trial functions with special functions must be significant. Here we examine three situations where the benefits justify the extra effort.

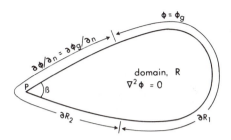

Figure 4.7. Singularities in elliptic
boundary-value problems

4.3.1. Singularities

Elliptic boundary-value problems are characteristically smooth in the interior.
Singularities (discontinuities) can only occur at the boundary. Two types of
singularities can be identified:

(i) geometric singularities,
(ii) discontinuities in the boundary conditions.

The different types of singularity are illustrated in Fig. 4.7 in relation to the
Laplace equation

$$\nabla^2 \phi = 0 \quad \text{in } R \qquad (4.3.1)$$

with

$$\phi = \phi_g \quad \text{on } \partial R_1,$$

$$\frac{\partial \phi}{\partial n} = \frac{\partial \phi_g}{\partial n} \quad \text{on } \partial R_2. \qquad (4.3.2)$$

In Fig. 4.7 a corner (a geometric singularity) occurs at P. At the junction between ∂R_1 and ∂R_2 the boundary conditions change discontinuously.

A particularly severe singularity is associated with the classical fracture-
mechanics problem, where a crack runs into the domain of interest. The cross
section of a cracked beam in torsion is shown in Fig. 4.8. The crack is represented by the line $P_1 P$ terminating at P. This problem is governed by the
equation

$$\nabla^2 \phi + f = 0 \quad \text{in } R \qquad (4.3.3)$$

with boundary conditions

$$\phi = 0 \quad \text{on } PP_1, P_2 P_3, \text{ and } P_4 P_5 \qquad (4.3.4)$$

and

$$\frac{\partial \phi}{\partial n} = 0 \quad \text{on } P_1 P_2, P_3 P_4, \text{ and } P_5 P_1. \qquad (4.3.5)$$

Fix et al. (1973) have sought finite-element solutions to this problem by
considering

$$\phi_a - \sum_j c_j \psi_j, \qquad (4.3.6)$$

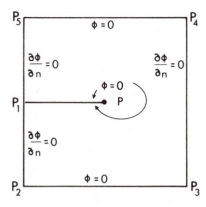

Figure 4.8. Idealization of a cracked beam in torsion

where ϕ_a is the global finite-element solution. ψ_j are analytic functions which are solutions of the governing equation *in the immediate neighborhood of the singularity*. The functions ψ_j are defined in an (r, θ) coordinate system centered at P, as

$$\psi_j(r, \theta) = \begin{cases} r^{v_j} \sin v_j\theta & \text{if } 0 \leq r \leq r_0, \\ P_j(r) \sin v_j\theta & \text{if } r_0 \leq r \leq r_1, \\ 0 & \text{if } r_1 \leq r, \end{cases} \qquad (4.3.7)$$

where $v_j = j + \frac{1}{2}$ and $P_j(r)$ is a blending function to allow ψ_j to go to zero smoothly when $r \geq r_1$. The form of the representation for the ψ_j in eq. (4.3.7) coincides with the leading terms of an analytic expression developed by Lehman (1959).

For the global finite-element solution Fix et al. considered both linear and cubic elements. With linear elements one singular term was included (eq. (4.3.6)). With cubic elements three terms were included. The number of terms is such as to permit the composite finite-element solution (4.3.6) to have the same convergence properties as the finite-element solution ϕ_a would have in the absence of any singularities.

Numerical results for this problem indicate that without special analytic functions, bicubic Hermite shape functions give less accurate solutions than bilinear shape functions (Strang and Fix, 1973, p. 271). The introduction of the analytic functions produces a very large increase in accuracy, with the cubic combination significantly more accurate than the linear combination (Fix et al., 1973).

The introduction of special analytic functions does add to the complexity of formulating the problem from a finite-element perspective. In particular, the matrix of algebraic equations becomes close to linearly dependent if high-order terms (large j in eq. (4.3.7)) are included. Thus the ill-conditioned stiffness matrix may well require a higher-precision computation to retain sufficient accuracy in the solution.

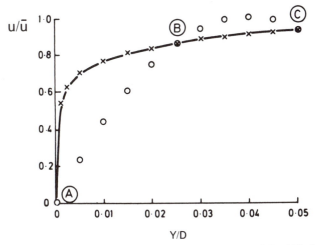

Figure 4.9. Turbulent velocity profile near wall of pipe:——Re = 4.5 × 10⁵, fine mesh; ○ parabolic interpolation through ABC; × logarithmic interpolation through ABC (after Taylor et al., 1977; reprinted with permission of Pergamon Press)

Second, the use of the special functions disturbs the banded structure of the stiffness matrix. Fix et al. (1973) present a procedure in which equations formed near the singularity are considered last. By an appropriate matrix factorization it is possible to solve the global system of equations with only a small operational overhead to account for the special functions.

4.3.2. Near-wall turbulent flows

For turbulent flows in pipes the longitudinal velocity increases very rapidly, over a small radial distance, from zero at the wall to an almost constant value across the radius. A large number of conventional elements would be required, close to the wall, to represent the velocity profile accurately.

Taylor et al. (1977) avoid this problem by introducing a logarithmic element, in place of a quadratic serendipity element, adjacent to the wall. The relative merits of logarithmic interpolation and polynomial interpolation for this problem can be appreciated from Fig. 4.9.

The velocity profile close to the wall is represented closely by the function

$$S_k(\eta) = \log_{M_k}[M_k - 0.5(M_k - 1)(1 + \eta)], \qquad (4.3.8)$$

which satisfies the conditions (Fig. 4.10)

$$S_k = \begin{cases} 0 & \text{at } \eta = +1 \\ 1 & \text{at } \eta = -1. \end{cases}$$

The parameter M_k is solution dependent and can be obtained as follows. First, $S_k(O)$ is related to the nodal velocities (Fig. 4.10) by

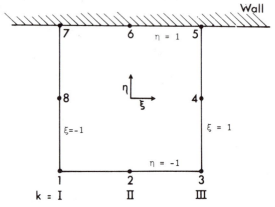

Figure 4.10. Near-wall element

$$S_I(O) = L_I = u_8/u_1,$$
$$S_{III}(O) = L_{III} = u_4/u_3,$$
$$\qquad\qquad (4.3.9)$$
$$S_{II}(O) = L_{II} = 0.5(L_I + L_{III}).$$

Then M_k is chosen to satisfy

$$2M_k^{L_k} - M_k - 1 = 0, \qquad k = \text{I, II, III.} \qquad (4.3.10)$$

The above functions have been used to generate the logarithmic interpolation shown in Fig. 4.9. Clearly, on a coarse grid, logarithmic interpolation is considerably more accurate than quadratic polynomial interpolation.

The logarithmic shape functions are expressed in the product form

$$J^k(\eta) N_i(\xi, \eta),$$

where $N_i(\xi, \eta)$ is a conventional quadratic serendipity shape function (section 3.1.2) and $J^k(\eta)$ is given by

$$J^k(\eta) = \frac{\log_{10}[M_k - 0.5(M_k - 1)(1 + \eta)]}{[(1 - \eta^2)L_k - 0.5(\eta - \eta^2)]\log_{10} M_k} \qquad (4.3.11)$$

and

$$k = \begin{cases} \text{I} & \text{if } i = 1, 8, \\ \text{II} & \text{if } i = 2, \\ \text{III} & \text{if } i = 3, 4. \end{cases}$$

Due to the complex nature of the shape functions (4.3.11), Gauss quadrature is required to evaluate the usual Galerkin inner products. Due to the boundary conditions on u, no shape functions associated with nodes 5, 6, and 7 in Fig. 4.10 are required.

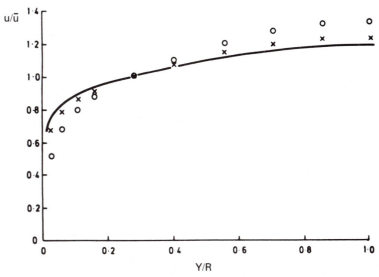

Figure 4.11. Longitudinal velocity profile for developing pipe flow, $x/D = 60$, $Re = 1 \times 10^5$: ——fine mesh; o quadratic coarse grid; × logarithmic coarse grid (after Taylor et al., 1977; published with permission of Pergamon Press)

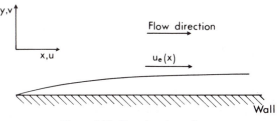

Figure 4.12. Boundary-layer flow

Typical results, using a coarse mesh, are shown in Fig. 4.11. The results with one logarithmic element adjacent to the wall show excellent agreement with the fine-mesh results.

4.3.3. Dorodnitsyn boundary-layer formulation

For the solution of attached laminar and turublent boundary-layer flows, Dorodnitsyn (1960) introduced a general formulation that converts the problem of seeking solutions for $u(x, y)$, $v(x, y)$ to the problem of seeking solutions for $\Theta(\xi, u)$. Here u and v are the longitudinal and transverse velocity components (Fig. 4.12), and Θ and ξ are defined by

$$\Theta = \frac{1}{\partial u/\partial \eta}, \qquad \eta = \mathrm{Re}^{1/2} u_e y, \qquad u' = \frac{u}{u_e}$$

and

$$\xi = \int_0^x u_e(x') \, dx'.$$

$u_e(x)$ is the given velocity distribution at the outer edge of the boundary layer. The governing equation for Θ is (with the prime dropped from u')

$$\frac{\partial}{\partial \xi} \int_0^1 f_k(u) u \Theta \, du = \frac{\dot{u}_e}{u_e} \int_0^1 f_k'(u)(1 - u^2)\Theta \, du$$

$$+ \int_0^1 f_k'(u) \frac{\partial}{\partial u} \left\{ \left(1 + \frac{v_T}{v}\right) \mathcal{T} \right\} du. \tag{4.3.12}$$

The derivation of this equation is given in section 5.5. Here $f_k(u)$ is a general weighting function that is to be chosen. Also

$$f_k'(u) \equiv \frac{df_k(u)}{du}, \qquad \dot{u}_e \equiv \frac{du_e}{d\xi}, \quad \text{and} \quad \mathcal{T} \equiv \frac{1}{\Theta}. \tag{4.3.13}$$

v is the kinematic viscosity, and v_T is an effective viscosity introduced to represent the Reynolds stresses in turbulent boundary layers.

Eq. (4.3.12) can be given a finite-element interpretation by introducing the following trial solution:

$$\Theta(\xi) = \frac{\theta(\xi)}{1 - u} = \sum_{j=1}^{N_T} \frac{N_j(u)}{1 - u} \theta_j(\xi). \tag{4.3.14}$$

It is known that $\partial u/\partial \eta \to 0$ at the outer edge of the boundary layer, where $u = 1$. The factor $1/(1 - u)$ is introduced to account for this behavior. $N_j(u)$ are conventional one-dimensional linear or quadratic shape functions (section 3.1.1). θ is an auxiliary function that is interpolated from its nodal values θ_j. However, Θ is obtained in terms of the trial functions $N_j(u)/(1 - u)$. A comparable trial solution is introduced for $\mathcal{T}$ by

$$\mathcal{T} = (1 - u)\tau(\xi) = \sum_{j=1}^{N_T} (1 - u)N_j(u)\tau_j(\xi). \tag{4.3.15}$$

As in eq. (4.3.14), the factor $(1 - u)$ is introduced to ensure the correct behavior at the outer edge of the boundary layer. The condition $\mathcal{T} = 1/\Theta$ (eq. (4.3.13)) is preserved for nodal values of $\mathcal{T}(\tau)$ and $\Theta(\theta)$:

$$\tau_j = 1/\theta_j. \tag{4.3.16}$$

But it is not preserved for the interpolated solution. However, in the limit of $N_T \to \infty$ the relationship $\mathcal{T} = 1/\Theta$ is recovered everywhere.

The general weighting function $f_k(u)$ is chosen to be

$$f_k(u) = (1 - u)N_k(u). \tag{4.3.17}$$

To obtain eq. (4.3.12) it is necessary that $f_k(1) = 0$. Clearly eq. (4.3.17) satisfies this condition.

Substitution of eqs. (4.3.14), (4.3.15), and (4.3.17) into eq. (4.3.12) indicates that we have a *modified Galerkin* finite-element formulation. Evaluation of the various terms gives the following system of ordinary differential equations:

$$\mathbf{CC\dot{\theta}} = \frac{\dot{u}_e}{u_e}\mathbf{EF\theta} + \mathbf{AA\tau'}, \tag{4.3.18}$$

where an element of $\dot{\theta}$ is $d\theta_j/d\xi$ and an element of τ' is $(1 + v_T/v)_j\tau_j$. Elements of $\mathbf{CC}$, $\mathbf{EF}$, and $\mathbf{AA}$ are given by

$$CC_{kj} = \int_0^1 N_k N_j u \, du,$$

$$EF_{kj} = \int_0^1 N_j \left(\frac{dN_k}{du}(1 - u) - N_k\right)(1 + u) \, du, \tag{4.3.19}$$

$$AA_{kj} = \int_0^1 \left(\frac{dN_j}{du}(1 - u) - N_j\right)\left(\frac{dN_k}{du}(1 - u) - N_k\right) du.$$

A very efficient algorithm for solving eqs. (4.3.18) can be obtained by introducing a Crank–Nicolson integration scheme in the ξ direction, that is, the x direction. Then eq. (4.3.18) becomes

$$\sum_j \left[CC_{kj} - 0.5 \, \Delta x \left\{\frac{\dot{u}_e}{u_e}EF_{kj} - AA_{kj}\left(1 + \frac{v_T}{v}\right)_j \tau_j^2\right\}\right]^n \Delta\theta_j^{n+1}$$

$$= \Delta x \left[\frac{\dot{u}_e}{u_e}\sum_j EF_{kj}\theta_j + \sum_j AA_{kj}\left(1 + \frac{v_T}{v} + 0.5\frac{\Delta v_T}{v}\right)_j \tau_j\right]^n. \tag{4.3.20}$$

Here $\Delta\theta_j^{n+1}$ is the correction to θ_j at the nth time level,

$$\Delta\theta_j^{n+1} = \theta_j^{n+1} - \theta_j^n. \tag{4.3.21}$$

If linear shape functions are used in eqs. (4.3.19), then eq. (4.3.20) is a tridiagonal system of equations that can be solved for $\Delta\theta_j^{n+1}$. Such a scheme has a truncation error of $O(\Delta\xi, \Delta u^2)$. The above scheme has been utilized to solve both laminar and turbulent boundary-layer flows with linear and quadratic shape functions (Fletcher and Fleet, 1982). A typical result, showing the surface-stress variation at the surface of a circular cylinder, is given in Fig. 4.13. The efficiency of the present formulation comes, in part, from the special trial functions used in eqs. (4.3.14) and (4.3.15).

The Dorodnitsyn boundary-layer formulation can also be given a spectral interpretation. This is described in section 5.5.

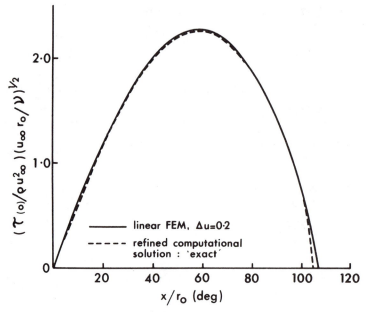

Figure 4.13. Shear-stress variation on the surface of a circular cylinder

4.4. Integral Equations

The traditional Galerkin method has been applied to integral equations (e.g. section 1.6.4), but the use of Galerkin finite-element methods with integral equations has not been substantial. However, the finite-element method in a collocation form is being actively developed. Typically this is referred to as the *boundary-element* (Brebbia, 1978) or *boundary-integral-equation* (Zienkiewicz, 1977) method.

Integral equations arise, most commonly, from boundary-value problems of *potential theory*. Potential theory is applicable if a particular problem can be described by

$$\nabla^2 \phi(\mathbf{x}_p, \mathbf{x}_q) = -4\pi\delta(\mathbf{x}_p - \mathbf{x}_q), \qquad (4.4.1)$$

where δ is the Dirac delta function located at $\mathbf{x}_q$. $\mathbf{x}_p$ is a general field point, so that $\phi(\mathbf{x}_p, \mathbf{x}_q)$ is the value of the potential at $\mathbf{x}_p$ due a unit "source" at $\mathbf{x}_q$. Eq. (4.4.1) reduces to the Laplace equation if $\mathbf{x}_q$ lies outside the domain of interest.

Potential theory is useful in describing inviscid, incompressible flow, seepage problems in porous media, steady heat-conduction problems, and electrical-conductor problems, and so on (Sneddon, 1966).

Integral-equation formulations of potential theory rest on the ready availability of simple solutions of eq. (4.4.1). For instance, if we consider the inviscid, incompressible flow around a two-dimensional isolated airfoil (section 1.6.4), the velocity potential ϕ is given at any point away from the airfoil by

$$\phi(\mathbf{x}_p) = \int_s \log|\mathbf{x}_p - s|\sigma(s)\,ds, \tag{4.4.2}$$

where a distribution of *sources* of density σ is placed on the surface of the airfoil. The factor $\log|\mathbf{x}_p - s|$ satisfies eq. (4.4.1) and describes the influence of the source distribution on the rest of the flowfield.

The source density σ is determined to satisfy the boundary condition of no flow through the body surface. The normal velocity is defined by

$$v_n = \frac{\partial\phi}{\partial n}(\mathbf{x}_p) = \int_s \frac{\partial}{\partial n}\log|\mathbf{x}_p - s|\sigma(s)\,ds + \pi\sigma(\mathbf{x}_p), \tag{4.4.3}$$

where $\mathbf{x}_p$ is now any point on the surface and v_n is known in terms of the local surface slope. Eq. (4.4.3) is a *Fredholm integral equation of the second kind*, for which computational solution techniques, including the Galerkin method, are well established (Atkinson, 1976). We return to the computational aspects below. Once σ has been obtained from solving eq. (4.4.3), the solution ϕ follows from eq. (4.4.2).

An alternative integral-equation formulation for potential theory is possible (Jaswon and Symm, 1977) via Green's theorem. This is obtained by starting with two potentials, ϕ and ψ, which satisfy

$$\nabla^2\phi = 0 \quad \text{and} \quad \nabla^2\psi(\mathbf{x}_q) = -4\pi\delta(\mathbf{x}_q - \mathbf{x}_p). \tag{4.4.4}$$

Green's theorem then gives

$$\int_R (\phi\nabla^2\psi - \psi\nabla^2\phi)\,dv = \int_s \left(\psi\frac{\partial\phi}{\partial n} - \phi\frac{\partial\psi}{\partial n}\right)ds = -4\pi\phi(\mathbf{x}_p) \tag{4.4.5}$$

in a general domain. If the potential problem is considered in a two-dimensional domain, then

$$\psi = \log|\mathbf{x}_p - s|, \tag{4.4.6}$$

and eq. (4.4.5) is replaced by

$$\int_s \log|\mathbf{x}_p - s|\frac{\partial\phi(s)}{\partial n}\,ds - \int_s \frac{\partial}{\partial n}\log|\mathbf{x}_p - s|\phi(s)\,ds = 2\pi\phi(\mathbf{x}_p). \tag{4.4.7}$$

If the field point $\mathbf{x}_p$ is restricted to the surface s, eq. (4.4.7) becomes

$$\int_s \log|\mathbf{x}_p - s|\frac{\partial\phi(s)}{\partial n}\,ds - \int_s \frac{\partial}{\partial n}\log|\mathbf{x}_p - s|\phi(s)\,ds = \beta\phi(\mathbf{x}_p). \tag{4.4.8}$$

If the boundary has a corner at $\mathbf{x}_p$, then β is the angle of the corner. Where the boundary is smooth, $\beta = \pi$ (Fig. 4.7). Eq. (4.4.8) represents a general com-

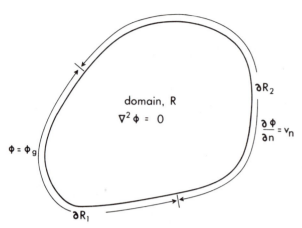

Figure 4.14. Domain for internal potential problems

patibility condition between $\phi(s)$ and $\partial\phi(s)/\partial n$. The way in which eq. (4.4.8) is used to obtain the solution is indicated in section 4.4.1.

Eq. (4.4.8) is well suited to the solution of internal problems depicted in Fig. 4.14, which are governed by

$$\nabla^2 \phi = 0 \quad \text{in } R,$$

$$\phi = \phi_g \quad \text{on } \partial R_1, \tag{4.4.9}$$

$$\partial\phi/\partial n = v_n \quad \text{on } \partial R_2.$$

If the generalized form of eq. (4.4.3) is applied to internal problems, an additional condition,

$$\int_s \frac{\partial\phi(s)}{\partial n}\, ds = 0, \tag{4.4.10}$$

must be imposed. For this and other reasons Jaswon and Symm (1977) consider the formulation (4.4.8) to be superior to the use of the generalized form (4.4.3) for internal potential problems. For external problems, neither formulation has a specific advantage. However, eq. (4.4.8) is the preferred starting point for boundary-element (boundary-integral-equation) methods.

Jaswon and Symm (1977) show that the Green's-theorem formulation can be recast as a generalized "source" method, eq. (4.4.3). It can be seen directly that if $\partial\phi(s)/\partial n$ is given everywhere on the boundary, eq. (4.4.8) also becomes a Fredholm integral equation of the second kind.

Therefore we briefly consider the solution of Fredholm integral equations of the second kind via the Galerkin (finite-element) method. Following Atkinson (1976), we will start with the following standard form:

$$\lambda h(x) - \int_a^b K(x,s)h(s)\, ds = y(x), \quad a \le x \le b, \tag{4.4.11}$$

where $y(x)$ and λ are given and $h(x)$ is to be solved for. The similarity to eq. (4.4.3) is clear.

The Galerkin method proceeds in the conventional manner by introducing a trial solution

$$h(x) = \sum_{j=1}^{N_T} \alpha_j N_j(x) \qquad (4.4.12)$$

and by defining a residual

$$R = y(x) - \lambda \sum_{j=1}^{N} \alpha_j \left\{ N_j(x) + \int_a^b K(x,s) N_j(s)\, ds \right\}. \qquad (4.4.13)$$

The Galerkin method forms algebraic equations to evaluate the coefficients α_j by the repeated evaluation of

$$(R, N_k) = 0. \qquad (4.4.14)$$

This leads to the linear system of equations

$$\mathbf{K}\boldsymbol{\alpha} = \mathbf{Y}, \qquad (4.4.15)$$

where an element of $\mathbf{Y}$ is

$$Y_k = \int_a^b y(x) N_k(x)\, dx \qquad (4.4.16)$$

and an element of $\mathbf{K}$ is

$$K_{kj} = \lambda \int_a^b N_j(x) N_k(x)\, dx - \int_a^b \int_a^b K(x,s) N_j(s) N_k(x)\, ds\, dx. \qquad (4.4.17)$$

The double integral appearing in eq. (4.4.17) makes the Galerkin formulation computationally expensive. However, Wendland (1979) and Wendland et al. (1979) have introduced a modified Galerkin finite-element method which appears to be very efficient. Error estimates comparable to those given in section 3.4.3 are also given by Wendland et al.

4.4.1. Boundary-element method

We now turn our attention to the boundary-element method. This method could be applied by starting from eq. (4.4.8). However, a more "natural" (strictly familiar) approach is to start with the method of weighted residuals. Previously we have considered interior methods of weighted residuals (section 1.3) in which the boundary residual, R_b in eq. (1.3.2), is satisfied exactly and the governing equation is satisfied approximately.

Now we consider a boundary method of weighted residuals in which the governing equation is satisfied exactly but the boundary conditions are satisfied approximately. It is easier to describe the method by considering a specific

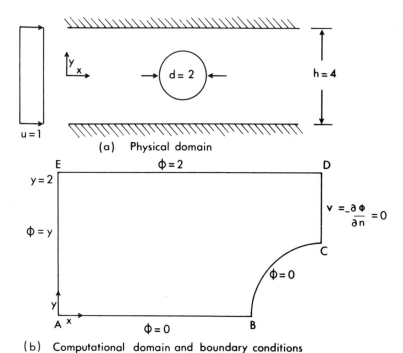

(a)   Physical domain

(b)   Computational domain and boundary conditions

Figure 4.15. Inviscid flow past a circular cylinder in a channel

example: inviscid, incompressible flow past a circular cylinder in a channel
(Fig. 4.15). This problem is of historical interest, since it was the first fluid-flow
problem solved by the conventional finite-element method that appeared in
the literature (Martin, 1969). We will solve the problem with a mixed stream-
function (ϕ) and velocity (u, v) formulation. The velocity components are
connected to the stream function by

$$u = \frac{\partial \phi}{\partial y}, \qquad v = -\frac{\partial \phi}{\partial x}. \qquad (4.4.18)$$

To maintain a consistent notation we have departed from the use of the symbol
ψ for the steam function. The governing equation for this problem is

$$\nabla^2 \phi = 0. \qquad (4.4.19)$$

The boundary conditions are shown in Fig. 4.15. The definition of the velocity
components, eq. (4.4.18), indicates that the tangential component of the
velocity at the boundary, v_T, is equal to the normal derivative of the stream
function.

The problem will be generalized by specifying the boundary conditions as

$$\phi = \phi_g \quad \text{on } DAC \qquad (4.4.20)$$

and

$$v_T = v_{Tg} \quad \text{on } CD. \tag{4.4.21}$$

The following weighted residuals can be constructed:

$$\iint \psi \nabla^2 \phi \, dx \, dy = R_E,$$

$$\int_{DAC} (\phi - \phi_g) \frac{\partial \psi}{\partial n} ds = R_{DAC}, \tag{4.4.22}$$

$$\int_{CD} (v_T - v_{Tg}) \psi \, ds = R_{CD},$$

where ψ is a weight function that satisfies

$$\nabla^2 \psi = 0 \tag{4.4.23}$$

except at a singular point, $\mathbf{x}_q$. In two dimensions a solution of eq. (4.4.23), with $\mathbf{x}_q$ restricted to a surface s, is

$$\psi = \log|\mathbf{x}_p - s|, \tag{4.4.24}$$

that is, eq. (4.4.6). The residuals, eqs. (4.4.22), will be combined in the following way:

$$R_E = R_{CD} - R_{DAC}. \tag{4.4.25}$$

Substituting the expressions (4.4.22) and differentiating the left-hand side by parts twice gives

$$\iint \phi \nabla^2 \psi \, dx \, dy = - \int_{CD} v_{Tg} \psi \, ds - \int_{DAC} v_T \psi \, ds$$
$$+ \int_{CD} \phi \frac{\partial \psi}{\partial n} ds + \int_{DAC} \phi_g \frac{\partial \psi}{\partial n} ds. \tag{4.4.26}$$

Substitution of eq. (4.4.23) and evaluation of the left-hand side gives

$$2\pi\phi(\mathbf{x}_p) + \int_{CD} \phi(s) \frac{\partial \psi}{\partial n} ds + \int_{DAC} \phi_g(s) \frac{\partial \psi}{\partial n} ds$$

$$= \int_{CD} v_{Tg}(s) \psi \, ds + \int_{DAC} v_T(s) \psi \, ds. \tag{4.4.27}$$

If eq. (4.4.24) were substituted into eq. (4.4.27), the result would be eq. (4.4.7). However, here we have split up the boundary integral to allow the boundary conditions to appear explicitly. This is convenient in constructing a computational algorithm. Eq. (4.4.27) is applicable to any point $\mathbf{x}_p$ in the computational domain (Fig. 4.15). On the boundary s, this equation becomes

$$\beta\phi(\mathbf{x}_p) + \int_{CD} \phi(s)\frac{\partial\psi}{\partial n} ds + \int_{DAC} \phi_g(s)\frac{\partial\psi}{\partial n} ds$$

$$= \int_{CD} v_{Tg}(s)\psi\, ds + \int_{DAC} v_T(s)\psi\, ds. \quad (4.4.28)$$

As with the conventional Galerkin method, a trial solution is introduced for ϕ and v_T:

$$\phi = \sum_j N_j(\xi)\bar{\phi}_j \quad (4.4.29)$$

and

$$v_T = \sum_j N_j(\xi)\bar{v}_{Tj}. \quad (4.4.30)$$

Here $N_j(\xi)$ are linear shape functions defined over one-dimensional elements (section 3.1.1), in element coordinates ξ on the boundary s. Since ψ and $\partial\psi/\partial n$ are known, eq. (4.4.28) can be evaluated with $\mathbf{x}_p$ corresponding to each node in turn, after substitution of eqs. (4.4.29) and (4.4.30). The result is a linear system of algebraic equations

$$\mathbf{K}\boldsymbol{\phi} = \mathbf{Y}, \quad (4.4.31)$$

where $\boldsymbol{\phi}$ is a vector representing the nodal unknowns, $\bar{\phi}_j$ in eq. (4.4.29) and $\bar{v}_{Tj}$ in eq. (4.4.30). An element of $\mathbf{Y}$ is given by

$$Y_p = \sum_{j=1}^{L}\left(\int_{CD} \psi(\bar{\mathbf{x}}_p, s)N_j(\xi)\, ds\right)\bar{v}_{Tgj}$$

$$- \sum_{j=L+1}^{N_T}\left(\int_{DAC} \frac{\partial\psi}{\partial n}(\bar{\mathbf{x}}_p, s)N_j(\xi)\, ds + \delta_{pj}\beta\right)\bar{\phi}_{gj}, \quad (4.4.32)$$

and an element of $\mathbf{K}$ is given by

$$K_{pj} = \begin{cases} \delta_{pj}\beta + \int_{CD} \dfrac{\partial\psi}{\partial n}(\mathbf{x}_p, s)N_j(\xi)\, ds & \text{if } \mathbf{x}_j \text{ is on } CD, \\[2mm] - \int_{DAC} \psi(\mathbf{x}_p, s)N_j(\xi)\, ds & \text{if } \mathbf{x}_j \text{ is on } DAC. \end{cases} \quad (4.4.33)$$

In eqs. (4.4.32) and (4.4.33), $\delta_{pj} = 1$ if $\mathbf{x}_j = \mathbf{x}_p$ and $\delta_{pj} = 0$ otherwise.

The above formulation is a *collocation* method. In contrast to the conventional Galerkin method (4.4.17), the evaluation of eq. (4.4.33) involves only a single integral. Even with piecewise polynomials as trial functions, the matrix $\mathbf{K}$ in eq. (4.4.31) is typically full. Consequently the solution of eq. (4.4.31) is an $O(N^3)$ process. Clearly this puts an effective upper limit on the number of equations that can be solved economically. However, since ψ is a solution of the governing equations, very accurate solutions are typically obtained with a very small value of N_T.

This is the main strength of the method. A second strong point is that no grid need be constructed in the computational domain, as in the conventional

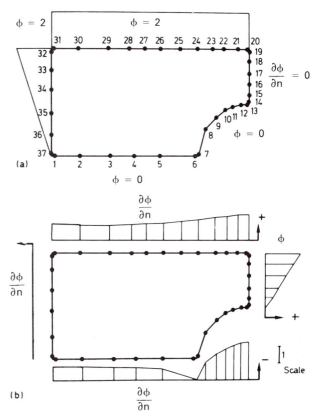

Figure 4.16. Velocity distributions on the flowfield boundary (after Brebbia, 1978; reprinted with permission of Pentech Press)

finite-element method. This follows from the fact that all the integrations in eq. (4.4.11) are carried out over the boundary only. A relative weakness of the boundary-integral method is that it is only applicable where general solutions, like eq. (4.4.6), exist. In practice this restricts the method to linear partial differential equations like the Laplace equation.

A typical result using linear elements is shown in Fig. 4.16 for the problem depicted in Fig. 4.15. Brebbia (1978) reports that this solution is superior to the conventional finite-element solution obtained by Martin (1969) for the same problem. The tangential velocity on ABC (Fig. 4.16) shows the correct qualitative behavior of dropping to zero at B and rising to a maximum at C.

A problem arises with the boundary-element method when the boundary slope changes discontinuously. Thus at points A, B, C, D, and E in Fig. 4.15, $\partial\phi/\partial n$ is multivalued. Brebbia (1978) tackles this problem by defining two nodal values of $\partial\phi/\partial n$ (or ϕ) which correspond to approaching the relevant corner from the two boundary directions.

In general a discontinuity can arise in the Green's theorem version of the integral equation formulation at the junction between the specification of ϕ and $\partial\phi/\partial n$ (Fig. 4.14). A singularity occurs in either formulation if the boundary over which the integral is defined has a geometric discontinuity. Since the solution in the domain for the integral-equation formulation depends on the solution obtained on the boundary, singularities degrade the accuracy of integral equation formulations to a greater extent than for differential equations solved over the domain.

A more satisfactory approach to the problem of singularities occuring on the boundary is to introduce local analytic functions to account for the singularities (as in section 4.3). This has been done, in the boundary-element context, by Wendland and Stephan (1980) and Wendland (1981). The techniques and error estimates given in these papers are essentially extensions of the modified Galerkin finite-element method described in Wendland et al. (1979).

4.5. Closure

In this chapter a number of techniques have been introduced to extend the Galerkin finite-element method.

The idea of obtaining solutions to elliptic-boundary value problems by constructing an equivalent parabolic problem and adopting a pseudotransient approach is very powerful. This technique has been used extensively with finite-difference methods. Time splitting plays a central role in the pseudotransient approach. We expect to see such techniques applied in a finite-element context to an increasing extent in the future.

Reduced integration and least-squares residual fitting are often very effective techniques in practice. However, at the present time, there is no a priori guarantee that solutions of superior accuracy will be produced in the general situation.

It seems likely that for many problems, more efficient solutions could be obtained by the introduction of special trial functions. However, if the manual effort of modifying an existing computer program to incorporate special functions is included, then it may turn out that a very refined grid adjacent to the singularity is to be preferred. But if the code containing the special trial functions is to be used extensively, then the benefits of introducing local analytic functions are substantial.

Integral equations do not appear to be well suited to the conventional Galerkin finite-element method. This is because the integral nature of the formulation prevents the use of test functions of small support, and consequently prevents the generation of a sparse stiffness matrix. However, the modified Galerkin method of Wendland et al. (1979) looks promising.

References

Atkinson, K. E. *A Survey of Numerical Methods for the Solution of Fredholm Integral Equations of the Second Kind*, SIAM, Philadelphia (1976).

Brebbia, C. A. *The Boundary Element Method for Engineers*, Pentech Press, London (1978).

Christie, I., Griffiths, D. F., Mitchell, A. R., and Sanz-Serna, J. M. "Product Approximation for Nonlinear Problems in the Finite Element Method", *Inst. Math. Appl. J. Num. Anal.* **1**, 253–266 (1981).

Deconinck, H., and Hirsch, C. *Lecture Notes in Physics*, Vol. 141, pp. 138–143, Springer-Verlag, Berlin, (1981).

Dorodnitsyn, A. A. In *Advances in Aeronautical Sciences*, Vol. 3, Pergamon, New York (1960).

Fairweather, G. *Finite Element Galerkin Methods for Differential Equations*, Dekker, New York (1978).

Fix, G. J., Gulati, S., and Wakoff, G. I. *J. Comp. Phys.* **13**, 209–228 (1973).

Fletcher, C. A. J. "On the Application of an Improved Finite Element Formulation with Isoparametric Elements", in *3rd International Conference on Finite Elements in Engineering in Australia*, Sydney, pp. 671–681 (1979a).

Fletcher, C. A. J. *J. Comp. Phys.* **33**, 301–312 (1979b).

Fletcher, C. A. J. *Comp. Meth. App. Mech. Eng.* **24**, 251–267 (1980a).

Fletcher, C. A. J. "On the Application of Alternating Direction Implicit Finite Element Methods to Flow Problems", in *3rd Finite Element for Flow Problems Conference*, Banff, Canada (1980b).

Fletcher, C. A. J. *Comp. Meth. App. Mech. Eng.* **30**, 307–322 (1982).

Fletcher, C. A. J. *Comp. Meth. App. Mech. Eng.* **37**, 225–243 (1983).

Fletcher, C. A. J., and Fleet, R. W. "A Dorodnitsyn Finite Element Boundary Layer Formulation", in *8th International Conference on Numerical Methods in Fluid Dynamics*, Aachen (1982).

Gourlay, A. R. In *The State of the Art in Numerical Analysis*, (ed. D. Jacob), pp. 757–796, Academic Press (1977).

Hayes, L. J. "Finite Element Patch Approximations and Alternating-Direction Methods", in *Advances in Computing Methods for Partial Differential Equations III* (eds. R. Vichnevetsky and R. S. Stepleman), p. 162–166, IMACS (1979).

Isaacson, E., and Keller, H. B. *Analysis of Numerical Methods*, Wiley, New York (1966).

Jaswon, M. A., and Symm, G. T. *Integral Equation Methods in Potential Theory and Elastostatics*, Academic Press, London (1977).

Lehman, S. *J. Math. Mech.* **8**, 727–760 (1959).

Martin, H. C. "Finite Element Analysis of Fluid Flows", in *Proceedings 2nd Conference on Matrix Methods in Structural Mechanics*, publ. as AFFDL-TR-68-150 (1969).

Sneddon, I. N. *Mixed Boundary Value Problems in Potential Theory*, North Holland, Amsterdam (1966).

Strang, G., and Fix, G. J. *An Analysis of the Finite Element Method*, Prentice-Hall, Englewood Cliffs, NJ (1973).

Taylor, C., Hughes, T. G., and Morgan, K. *Comp. and Fluids* **5**, 191–204 (1977).

Wendland, W. L. *Elliptic Systems in the Plane*, Pitman, London (1979).

Wendland, W. L., Stephan, E., and Hsiao, G. C. *Math. Meth. Appl. Sci.* **1**, 265–321 (1979).

Wendland, W. L., and Stephan, E. "Boundary Integral Methods for Mixed Boundary Value Problems", in *Innovative Numerical Analysis for the Engineering Sciences* (eds. R. Shaw et al.), pp. 543–554, Univ. Press of Virginia, Charlottesville (1980).

Wendland, W. L. "On the Asymptotic Convergence of Some Boundary Element Methods", in *Conference on Mathematics of Finite Elements and Applications*, MAFELAP 1981, Brunel Univ. (1981).

Zienkiewicz, O. C., and Hinton, E. "Reduced Integration, Function Smoothing and Non-conformity in Finite Element Analysis", in *2nd International Conference on Finite Elements in Engineering in Australia*, Adelaide (1976).

Zienkiewicz, O. C. *The Finite Element Method*, McGraw-Hill, London, 3rd Edn., (1977).

CHAPTER 5

Spectral Methods

In chapter 2 it was indicated that modern computational Galerkin methods have developed in two directions. The first direction leads to finite-element methods, which were discussed in chapters 3 and 4. Finite-element methods are characterized by the use of local, low-order polynomials as test and trial functions in subdomains called finite elements.

In this chapter we will pursue the second direction mentioned in chapter 2. This will lead to spectral methods, which make use of global, orthogonal test and trial functions. As with the traditional Galerkin method, the accuracy, and hence efficiency, of spectral methods is very dependent on making the correct choice for the test and trial functions. This aspect of spectral methods will be considered in section 5.1.

In section 5.2 we provide a couple of simple examples to introduce the Galerkin spectral method and to draw attention to some of its features. Specific techniques for improving the efficiency of spectral methods are discussed in section 5.3. The collocation (or pseudospectral) and tau methods are closely related to the Galerkin spectral method, and these are discussed in section 5.4.

In section 5.5 we present an example where a classical solution technique can be made more efficient by recasting it as a spectral method. Finally, in section 5.6 applications of spectral methods are considered, particularly in the areas of direct simulation of turbulence and global weather prediction.

5.1. Choice of Trial Functions

As with traditional Galerkin methods, the nature of the trial (and test) functions has a significant impact on the accuracy of spectral methods. However, we begin by considering a typical general error estimate in a form similar to those given in section 3.4.2. Pasciak (1980) has analysed the inviscid convection equation

$$\frac{\partial u}{\partial t} + \sum_{i=1}^{n} \left(V_i \frac{\partial u}{\partial x_i} + \frac{\partial u}{\partial x_i}(V_i u) \right) = 0,$$

where $V_i(x)$ is known and u is periodic on ∂R. Pasciak used the same techniques that are appropriate to finite-element theory. For an N-term spectral approximation u_a, the following error estimate is obtained:

$$\|u(t) - u_a(t)\|_{H^0, R} \le C_s N^{-k} \|u_0\|_{H^{k+1}, R}, \tag{5.1.1}$$

where u_0 is the initial data and C_s is a positive constant independent of u_0 and N. Clearly the spectral method is capable of giving very rapid convergence (and by implication, high accuracy) if the initial data (and boundary conditions) are sufficiently smooth.

Orszag (1980) indicates that a comparable convergence rate is possible for a spectral solution, based on a Fourier sine series, of the heat-conduction equation

$$\frac{\partial u}{\partial t} = \alpha \frac{\partial^2 u}{\partial x^2} \tag{5.1.2}$$

with boundary conditions

$$u(0, t) = u(\pi, t) = 0$$

and initial conditions

$$u(x, 0) = u_0(x), \qquad 0 \le x \le \pi,$$

if $\|u\|_{H^{k+1}, R}$ is bounded and if

$$\frac{\partial^{2r} u}{\partial x^{2r}} = 0 \quad \text{at } x = 0 \text{ and } \pi \quad \text{for } r \le k. \tag{5.1.3}$$

Based on the previous application of spectral methods, we can construct a hierarchy of the more common choices for the trial functions. These are shown in Table 5.1.

If eigenfunctions of a closely related problem are available, accurate solutions with relatively few terms in the trial solution are possible. Surface spherical harmonics are used as the trial functions in global weather prediction (Bourke et al., 1977). Surface spherical harmonics are eigenfunctions of the Laplace equation written in spherical coordinates. The Laplacian forms a central part of the equations obtained from a stream-function–vorticity formulation. However, the effectiveness of the eigenfunction expansion (in achieving a very high convergence rate) typically requires that the exact solution must be infinitely differentiable and that precise, typically homogeneous,

Table 5.1. Hierarchy of trial functions

Trial function	Comments
Eigenfunctions	Suggested by a related problem
Fourier series	Periodic boundary conditions; infinitely differentiable
Legendre polynomials	Good wavelength resolution; nonperiodic
Chebyshev polynomials	Very robust; nonperiodic; minimax

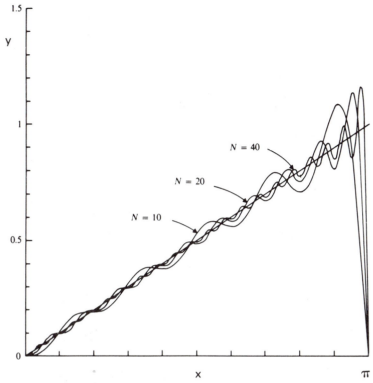

Figure 5.1. N-term Fourier sine series for linear function $y = x/\pi$ (after Gottlieb and Orszag, 1977; reprinted with permission of the Society for Industrial and Applied Mathematics)

boundary conditions must hold. If the various conditions do not hold, the convergence rate with N, the number of terms in the trial solution, will be much less.

As noted by Orszag (1980), Fourier series are capable of giving a convergence rate that is greater than any power of $1/N$ as $N \to \infty$ if sufficient homogeneous boundary conditions, like eq. (5.1.3), can be imposed and if the exact solution is infinitely differentiable.

It is noted in Table 5.1 that Fourier series are suitable if the solution, and hence the boundary conditions, are periodic. If the underlying problem is not periodic, the use of a Fourier series will convert the problem into a periodic problem with a discontinuity at one boundary. A general Fourier series will produce a constant overshoot in the neighborhood of the discontinuity as $N \to \infty$. This is called the Gibbs phenomenon.

The Gibbs phenomenon can be illustrated by fitting the linear function $y = x/\pi$ with a sine series. The result is shown in Fig. 5.1. The Gibbs phenomenon occurs at $x = \pi$. The effect of the discontinuity at $x = \pi$ reduces the convergence rate to $O(N^{-1})$ throughout the domain $0 \leq x \leq \pi$. If a cosine series is used, there is no Gibbs phenomenon, but the rate of convergence is still reduced. The rate of convergence is typically $O(N^{-2})$ except close to $x = 0$ and π, where it is $O(N^{-1})$.

Discontinuities in the interior are equally disruptive of the convergence rates suggested by eq. (5.1.1). Majda et al. (1978) apply a Fourier-series trial solution to the model hyperbolic equation

$$\frac{\partial u}{\partial t} - \frac{\partial u}{\partial x} = 0 \tag{5.1.4}$$

in the interval $-\pi \leq x \leq \pi$. The problem is periodic, and the initial conditions are smooth except for a jump discontinuity at $x = 0$. Majda et al. found that the global convergence rate (i.e. even away from the discontinuity) was only $O(N^{-2})$.

Gottlieb and Orszag (1977) consider the model hyperbolic problem

$$\frac{\partial u}{\partial t} + \frac{\partial u}{\partial x} = x + t \tag{5.1.5}$$

with initial and boundary conditions

$$u(x, 0) = 0 \quad \text{for } 0 \leq x \leq \pi$$

and

$$u(0, t) = 0 \quad \text{for } t \geq 0.$$

This problem has the exact solution $u = xt$. As might be expected, if a Galerkin procedure is applied with a Fourier sine series, the discontinuity at $x = \pi$ causes a very poor rate of convergence. However, it is also found that the Galerkin solution does not converge to the exact solution. If an even number of terms are included in the Fourier sine series, the following asymptotic solution is generated:

$$u_a = \begin{cases} xt & \text{if } x \geq t, \\ \pi(x - t) + xt & \text{if } x < t. \end{cases} \tag{5.1.6}$$

The Galerkin solution $u_a(x, 1)$ with $N = 100$ is shown in Fig. 5.2. If an odd number of terms are included, the following asymptotic solution is obtained:

$$u_a = \begin{cases} xt & \text{if } x \geq t, \\ \pi(t - x) + xt & \text{if } x < t. \end{cases} \tag{5.1.7}$$

The source of the problem is that the assumed trial solution, the sine series, is incomplete after the differentiation operation (5.1.5) has been applied. This effect does not produce the spurious solutions (5.1.6) and (5.1.7) if the problem is periodic.

The problem of nonperiodic boundary conditions can be overcome by supplementing the Fourier-series trial solution by a few terms of a polynomial. However, the convergence rate is then determined by the number of terms

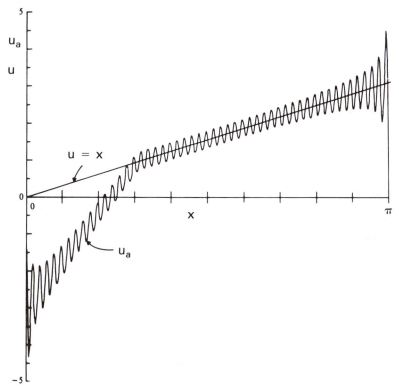

Figure 5.2. 100-term Fourier-sine-series solution of eq. (5.1.5) at $t = 1$ (after Gottlieb and Orszag, 1977; reprinted with permission of the Society for Industrial and Applied Mathematics)

included in the polynomial expansion (Gottlieb and Orszag, 1977). Consequently, for nonperiodic problems it is better to introduce orthogonal, nonperiodic polynomials, such as Legendre or Chebyshev polynomials, in the first instance.

Legendre polynomials are orthogonal over the range $-1 \le x \le 1$. If the problem being solved has an exact solution which is infinitely differentiable, the use of Legendre polynomials will give the same order of convergence as a Fourier series as $N \to \infty$. Gottlieb and Orszag indicate that a very rapid rate of convergence is achieved if at least π terms are included in the trial solution for every complete wave expected in the exact solution.

If an internal discontinuity occurs, the solution close to the discontinuity behaves like a Fourier series representation. The global convergence rate is then $O(N^{-1})$ except near the boundaries, where it drops to $O(N^{-1/2})$. This is a disadvantage of Legendre polynomials.

Chebyshev polynomials are included in Table 5.1 because they are often used in spectral methods. If they are used in a conventional Galerkin formulation, Chebyshev polynomials produce off-diagonal contributions to the mass matrix, since they are not strictly orthogonal (e.g. see section 1.3.5). However, if the Galerkin method is generalized, as in section 1.3.6, the weighted orthogonality of Chebyshev polynomials can be taken advantage of. In this case the trial solution is constructed from Chebyshev polynomials $T_j(x)$, but the test functions are chosen to be $T_k(x)/(1 - x^2)^{1/2}$.

If the underlying exact solution is sufficiently smooth, then a trial solution of Chebyshev polynomials will generate convergence rates that are better than algebraic—indeed, comparable to those of Fourier series trial solutions— as $N \to \infty$. If an internal discontinuity occurs, the convergence rate drops to $O(N^{-1})$ as with Legendre polynomials. However due to the weight factor $(1 - x^2)^{-1/2}$, Chebyshev polynomials maintain the $O(N^{-1})$ convergence at the boundaries as well as in the interior. The solutions obtained with Chebyshev polynomials generally have close to the minimum possible maximum error (minimax principle; Hamming, 1973).

With reference to Table 5.1, the higher an item is on the list, the more restrictions are required (on differentiability and boundary conditions) to achieve a high rate of convergence. However, if the restrictions are met, the convergence rate will be higher for the same number of terms in the trial solution. Conversely, the lower an item on the list, the more robust will be the accuracy and convergence of the solution in relation to initial- and boundary-condition specification.

Due to the global nature of the trial functions, the accuracy and convergence rate of any of the trial functions indicated in Table 5.1 will be higher, if the exact solution is sufficiently smooth, than for the purely local trial functions used with the finite-element method.

However, if discontinuities occur, either at the boundary or in the interior, the convergence rate is typically $O(N^{-1})$ or $O(N^{-2})$, about the same as for a low-order finite-difference method.

5.2. Examples

Here we will examine two examples to illustrate the mechanics and some of the characteristics of spectral methods. The first example is the linear heat-conduction equation which was considered previously in sections 1.2.4 and 3.2.4. In the second the spectral method will be applied to Burgers' equation, which is nonlinear. This is the same example that was considered previously in sections 1.2.5 and 3.2.5. For both examples the spectral method will be introduced to represent the spatial behavior only.

5.2.1. Unsteady heat conduction

As in section 1.2.4, we consider the transient temperature distribution in a rod (Fig. 1.7). With an appropriate nondimensionalization (section 1.2.4) the governing equation is

$$\frac{\partial \theta}{\partial t} - \frac{\partial^2 \theta}{\partial x^2} = 0. \tag{5.2.1}$$

The initial condition is

$$\theta(x, 0) = \sin \pi x + x, \tag{5.2.2}$$

and the boundary conditions are

$$\theta(0, t) = 0 \quad \text{and} \quad \theta(1, t) = 1. \tag{5.2.3}$$

A trial solution is introduced as

$$\theta_a(x, t) = \theta_0(x) + \sum_{j=1}^{N} a_j(t) \sin j\pi x \tag{5.2.4}$$

where

$$\theta_0(x) = \sin \pi x + x.$$

Since $\theta_0(x)$ satisfies the initial and boundary conditions, the trial functions $\sin j\pi x$ satisfy homogeneous boundary conditions.

Introduction of the trial solution (5.2.4) into the governing equation (5.2.1) produces a residual R. A system of ordinary differential equations for the coefficients a_j in eq. (5.2.4) is created by the repeated evaluation of the inner product

$$(R, \sin k\pi x) = 0. \tag{5.2.5}$$

The ensuing ordinary differential equations can be written

$$\frac{da_k}{dt} + (k\pi)^2 a_k + r_k = 0, \qquad k = 1, 2, \ldots, N, \tag{5.2.6}$$

where

$$r_k = \begin{cases} \pi^2 & \text{if } k = 1, \\ 0 & \text{if } k \neq 1. \end{cases} \tag{5.2.7}$$

Because the trial functions in eq. (5.2.4) are orthogonal over the domain $0 \leq x \leq 1$, *explicit* ordinary differential equations (5.2.6) are obtained. This may be contrasted with the situation for the traditional Galerkin formulation. In that case a *system* of coupled ordinary differential equations was generated: eq. (1.2.49).

Examination of eqs. (5.2.6) and (5.2.7) indicates that

$$a_1 = \exp(-\pi^2 t) - 1,$$
$$a_k = 0, \qquad k > 1. \tag{5.2.8}$$

Substitution into eq. (5.2.4) gives the following for the trial solution:

$$\theta_a(x, t) = \sin \pi x \exp(-\pi^2 t) + x. \tag{5.2.9}$$

Table 5.2. Solution of eq. (5.2.1) by spectral method

t value $(x = 0.50)$	Approximate solution θ_a			Exact solution θ	Approximate solution θ_b
	$N = 1$	$N = 3$	$N = 5$		
0	1.5000	1.5000	1.5000	1.5000	1.5000
0.02	1.3143	1.3466	1.3384	1.3408	1.3404
0.04	1.1620	1.1993	1.1911	1.1943	1.1931
0.06	1.0371	1.0752	1.0670	1.0707	1.0690
0.08	0.9347	0.9729	0.9647	0.9686	0.9667
0.10	0.8507	0.8889	0.8807	0.8847	0.8828
0.12	0.7819	0.8201	0.8118	0.8158	0.8139
0.14	0.7254	0.7636	0.7553	0.7592	0.7574
0.16	0.6791	0.7173	0.7090	0.7128	0.7111
0.18	0.6411	0.6793	0.6710	0.6747	0.6731
0.20	0.6099	0.6481	0.6399	0.6434	0.6420

But this is the exact solution. This has occurred because the trial solution "contains" the exact solution. A more interesting situation develops if the following initial conditions are considered:

$$\theta(x,0) = 5x - 4x^2. \tag{5.2.10}$$

The ordinary differential equations are developed as before, except that r_k, given by eq. (5.2.7), is replaced by

$$r_k = \begin{cases} 32/k\pi & \text{if } k = 1, 3, 5, \ldots, \\ 0 & \text{if } k = 2, 4, 6, \ldots. \end{cases} \tag{5.2.11}$$

Solutions of eqs. (5.2.6) and (5.2.11) are shown in Table 5.2 for various values of N. These solutions have been obtained by numerically integrating eq. (5.2.6) as in eq. (1.2.55) with $\Delta t = 0.001$. Also shown in Table 5.2 are the exact solution θ and an approximate solution θ_b, which contains errors solely due to the numerical integration scheme. Consequently, to assess the accuracy of the spectral representation, the approximate solution θ_a should be compared with θ_b. For further comments see section 1.2.4.

The results shown in Table 5.2 indicate that the accuracy increases rapidly with N. For the same value of N the accuracy of the spectral method is comparable to that of the traditional Galerkin method (Table 1.7). However, for the spectral method the expensive matrix factorization (1.2.54) is avoided.

5.2.2. Burgers' equation

To ascertain how the spectral method copes with a nonlinear problem, we will apply it to Burgers' equation, which governs the propagating-"shock" problem. This problem was considered in section 1.2.5. The governing equation is

$$\frac{\partial u}{\partial t} + u\frac{\partial u}{\partial x} - \frac{1}{Re}\frac{\partial^2 u}{\partial x^2} = 0 \qquad (5.2.12)$$

with boundary conditions

$$u(-1,t) = 1, \qquad u(1,t) = 0 \qquad (5.2.13)$$

and initial conditions

$$u(x,0) = \begin{cases} 1, & \text{if } x \leq 0, \\ 0, & \text{if } x > 0. \end{cases} \qquad (5.2.14)$$

Compared with eq. (5.2.1), Burgers' equation includes the additional complication of the nonlinear convective term $u\,\partial u/\partial x$. However, a spectral formulation will be introduced for eq. (5.2.12) without making any special provision for the nonlinear term.

The following trial solution is used:

$$u(x,t) = \sum_{j=0}^{N} a_j(t)P_j(x), \qquad (5.2.15)$$

where $P_j(x)$ are Legendre polynomials (Dahlquist et al., 1974). Legendre polynomials are orthogonal over the interval $-1 \leq x \leq 1$. Following the conventional Galerkin procedure (e.g. as in section 1.2.5), we obtain the following system of ordinary differential equations:

$$\mathbf{M\dot{A}} + (\mathbf{B} + \mathbf{C})\mathbf{A} = 0, \qquad (5.2.16)$$

where an element of $\mathbf{M}$ is

$$m_{kk} = \frac{2}{2k+1},$$
$$m_{jk} = 0, \qquad j \neq k. \qquad (5.2.17)$$

Thus $\mathbf{M}$ is purely diagonal, as we would expect from the use of orthogonal functions. An element of $\mathbf{B}$ is given by

$$b_k = \sum_{i=1}^{N} a_i \left(P_j \frac{dP_i}{dx}, P_k \right). \qquad (5.2.18)$$

The nonlinearity of the convective term manifests itself by the dependence of b_{kj} on the solution a_i.

An element of $\mathbf{C}$ is given by

$$c_{kj} = \frac{1}{Re}\left(\frac{\partial P_j}{\partial x}, \frac{\partial P_k}{\partial x} \right). \qquad (5.2.19)$$

As in section (1.2.5), the initial values of a_j are obtained by applying the Galerkin method to the initial conditions. The result is the following system of algebraic equations:

$$\mathbf{MA} = \mathbf{D}, \qquad (5.2.20)$$

where an element of $\mathbf{D}$ is given by

$$d_k = \int_{-1}^{0} P_k(x)\, dx. \tag{5.2.21}$$

The two coefficients a_N and a_{N-1} in eq. (5.2.15) have been chosen so that the boundary conditions (5.2.13) are satisfied. Strictly, this makes the present method a *tau* method (see section 5.4.2).

The system of equations (5.2.16) has been integrated using a variable-time-step, fourth-order Runge–Kutta scheme. The time steps have been kept sufficiently small so that any error in the solution is due to the spatial representation (5.2.15). After a specific time, typically $t = 0.80$, the spectral solution is compared with the exact solution (1.2.61). Solutions at $\mathrm{Re} = 10$ and various values of N are shown in Fig. 5.3 and Table 5.3. It can be seen that the accuracy of the solution increases rapidly with N. A comparison of Tables 1.8 and 5.3 indicates that the accuracy of the traditional Galerkin method and spectral method is comparable. However, the spectral method requires less computational work to obtain the solution.

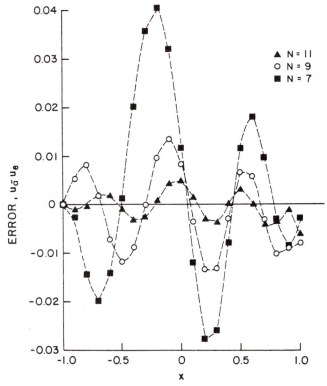

Figure 5.3. Error distribution for the solution of Burgers' equation, $\mathrm{Re} = 10$

Table 5.3. Solution of Burgers' equation by spectral method: $t = 0.80$, Re $= 10$

x	Approximate solution			Exact solution
	$N = 7$	$N = 9$	$N = 11$	
-1.0	1.0000	1.0000	1.0000	1.0000
-0.9	0.9974	1.0054	0.9986	1.0000
-0.8	0.9856	1.0082	0.9997	0.9999
-0.7	0.9799	1.0020	1.0017	0.9998
-0.6	0.9855	0.9926	1.0012	0.9996
-0.5	1.0000	0.9870	0.9978	0.9988
-0.4	1.0176	0.9882	0.9937	0.9972
-0.3	1.0290	0.9934	0.9904	0.9934
-0.2	1.0258	0.9947	0.9857	0.9852
-0.1	1.0007	0.9816	0.9730	0.9687
0	0.9491	0.9442	0.9422	0.9372
0.1	0.8696	0.8758	0.8829	0.8816
0.2	0.7643	0.7748	0.7887	0.7920
0.3	0.6390	0.6463	0.6611	0.6648
0.4	0.5025	0.5011	0.5109	0.5104
0.5	0.3657	0.3546	0.3573	0.3542
0.6	0.2405	0.2233	0.2224	0.2225
0.7	0.1377	0.1213	0.1236	0.1280
0.8	0.0653	0.0561	0.0646	0.0684
0.9	0.0259	0.0241	0.0330	0.0344
1.0	0.0136	0.0078	0.0102	0.0164
$\|u - u_e\|_{rms}$	0.0200	0.0086	0.0031	

For a value $N = 9$ solutions at different Reynolds numbers Re are shown in Fig. 5.4 and Table 5.4. As the Reynolds number is increased, the "shock" profile becomes steeper. For a finite number of terms, N, in the trial solution, the accuracy deteriorates with increasing Re. This is seen most easily by examining the value of the error in the rms norm, $\|u_a - u_e\|_{rms}$, given in Table 5.4. At Re $= 100$ a sufficiently large value of N would suppress the unphysical "wiggles". The trend with increasing values of Re is the same as for the traditional Galerkin method (section 1.2.5).

A computer program, BURG1, to obtain spectral solutions to the propagating-shock problem is described and listed in appendix 1.

Gottlieb and Orszag (1977) have considered a linearised Burgers' equation—namely, eq. (5.2.12) with $u_0 \, \partial u/\partial x$ replacing $u \, \partial u/\partial x$—with boundary conditions

$$u(-1, 0) = 0 \quad \text{and} \quad u(1, 0) = 0 \qquad (5.2.22)$$

and initial condition

$$u(x, 0) = g(x). \qquad (5.2.23)$$

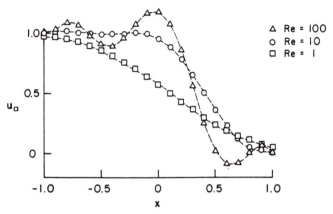

Figure 5.4. 9-term Legendre polynomial solution of Burgers' equation, various Reynolds numbers

Table 5.4. Solution of Burgers' equation by spectral method for various Reynolds numbers: $N = 9$

x	Re = 1.0, $t = 0.19$		Re = 10, $t = 0.80$		Re = 100, $t = 0.66$	
	Approx.	Exact	Approx.	Exact	Approx.	Exact
−1.0	0.9718	0.9706	1.0000	1.0000	1.0000	1.0000
−0.9	0.9593	0.9574	1.0054	1.0000	1.0052	1.0000
−0.8	0.9426	0.9395	1.0082	0.9999	1.0830	1.0000
−0.7	0.9198	0.9162	1.0020	0.9998	1.0646	1.0000
−0.6	0.8898	0.8865	0.9926	0.9996	0.9760	1.0000
−0.5	0.8521	0.8499	0.9870	0.9988	0.8968	1.0000
−0.4	0.8068	0.8058	0.9882	0.9971	0.8853	1.0000
−0.3	0.7544	0.7545	0.9934	0.9933	0.9515	1.0000
−0.2	0.6957	0.6965	0.9947	0.9849	1.0603	1.0000
−0.1	0.6321	0.6328	0.9816	0.9681	1.1509	1.0000
0	0.5648	0.5651	0.9442	0.9360	1.1634	1.0000
0.1	0.4956	0.4954	0.8758	0.8793	1.0609	1.0000
0.2	0.4262	0.4258	0.7748	0.7882	0.8436	0.9986
0.3	0.3585	0.3585	0.6463	0.6514	0.5508	0.8218
0.4	0.2943	0.2955	0.5011	0.5040	0.2503	0.0297
0.5	0.2355	0.2383	0.3546	0.3479	0.0165	0.0002
0.6	0.1834	0.1879	0.2233	0.2174	−0.0967	0.0000
0.7	0.1393	0.1449	0.1213	0.1244	−0.0813	0.0000
0.8	0.1037	0.1092	0.0561	0.0662	0.0095	0.0000
0.9	0.0762	0.0804	0.0241	0.0331	0.0718	0.0000
1.0	0.0555	0.0578	0.0078	0.0157	0.0000	0.0000
$\|u - u_e\|_{rms}$	0.0029		0.0086		0.1175	

A Legendre spectral formulation was stable and convergent. However, a Chebyshev spectral formulation would not converge for $N < N_{crit}$, where N_{crit} depends on the value of $\operatorname{Re} u_0$. The solution develops a narrow boundary layer adjacent to the boundary $x = 1$ for large values of $\operatorname{Re} u_0$. Attempting to satisfy the outflow boundary condition with small values of N causes the lack of convergence.

The examples of the spectral method given in this section have been restricted to small values of N to facilitate comparison with the traditional Galerkin examples of section 1.2. The use of spectral methods for "real" problems (see section 5.6) implies a large number of terms, N, in the trial solution. As foreshadowed in chapter 2, the method becomes very uneconomical, particularly for nonlinear problems, unless special techniques are introduced. Some of the special techniques are described in the next section.

5.3. Techniques for Improved Efficiency

The spectral method, as described in section 5.2, is more efficient than the traditional Galerkin method through use of orthogonal test and trial functions. However, for parabolic problems

$$\frac{\partial u}{\partial t} = L(u),$$

after application of the Galerkin method (e.g. eq. (5.2.16)), the execution time will be dominated by the evaluation of the spatial terms at each time step. For spectral methods each linear term requires a summation over N terms. In contrast, for finite-difference or finite-element methods, a summation over only three or five terms is typically required. For a problem involving a quadratic nonlinearity, like Burgers' equation, a double summation over N terms must be evaluated with the spectral method. These summations must be repeated for each unknown coefficient in the trial solution.

Specific techniques have been developed to make the evaluation of the spatial terms more efficient. Some of these are discussed in sections 5.3.1 and 5.3.2. The nonsparse nature of the matrix equations in a spectral formulation also affects the time-integration schemes. This is discussed in section 5.3.3.

5.3.1. Recurrence relations

The evaluation of the derivative $\partial u/\partial x$, after substitution of a general trial solution

$$u(x, t) = \sum_j a_j(t)\phi_j(x), \qquad (5.3.1)$$

results in

$$\frac{\partial u}{\partial x} = \sum_j a_j(t) \frac{\partial \phi_j}{\partial x}. \tag{5.3.2}$$

However, the evaluation of the series is often more efficient if $\partial u/\partial x$ is defined in terms of the same trial functions $\phi_j(x)$ as appear in eq. (5.3.1). Thus let

$$\frac{\partial u}{\partial x} = \sum_j b_j(t) \phi_j(x). \tag{5.3.3}$$

If eqs. (5.3.1) and (5.3.3) are Fourier series, then relations between a_j and b_j are available directly. If eqs. (5.3.1) and (5.3.3) are based on Chebyshev or Legendre polynomials, Gottlieb and Orszag (1977) give relationships between b_j and a_j for many different operators $L(u)$. In particular, for $L(u) = \partial u/\partial x$ and if the trial functions ϕ_j are Chebyshev polynomials,

$$b_j = 2 \sum_{\substack{P=j+1 \\ P+j\,\text{odd}}}^{N} P a_P, \qquad j = 1, \ldots, N-1,$$

$$b_0 = \sum_{\substack{P=1 \\ P\,\text{odd}}}^{N} P a_P. \tag{5.3.4}$$

For $L(u) = \partial^2 u/\partial x^2$,

$$b_j = \sum_{\substack{P=j+2 \\ P+j\,\text{even}}}^{N} P(P^2 - j^2) a_P, \qquad j = 1, \ldots, N-2,$$

$$b_0 = 0.5 \sum_{\substack{P=2 \\ P\,\text{even}}}^{N} P^3 a_P. \tag{5.3.5}$$

If the trial functions ϕ_j are Legendre polynomials and $L(u) = \partial u/\partial x$, then

$$b_j = (2j + 1) \sum_{\substack{P=j+1 \\ P+j\,\text{odd}}}^{N} a_P, \qquad j = 0, \ldots, N-1, \tag{5.3.6}$$

and if $L(u) = \partial^2 u/\partial x^2$, then

$$b_j = (j + \tfrac{1}{2}) \sum_{\substack{P=j+2 \\ P+j\,\text{even}}}^{N} [P(P+1) - j(j+1)] a_P, \qquad j = 0, \ldots, N-2. \tag{5.3.7}$$

An advantage of the form (5.3.3) over (5.3.2) is that, in applying the Galerkin method with orthogonal functions, a contribution is obtained from b_k only. A further advantage comes from combining equations like eqs. (5.3.4) to (5.3.7) with appropriate recurrence relationships. Thus for Chebyshev polynomials,

$$2T_j = \frac{1}{j+1}\frac{dT_{j+1}}{dx} - \frac{1}{j-1}\frac{dT_{j-1}}{dx}. \qquad (5.3.8)$$

This equation can be used to obtain the following recurrence relationship for b_j in eq. (5.3.4):

$$b_j = b_{j+2} + 2(j+1)a_{j+1}, \qquad 1 \le j \le N-1,$$
$$b_0 = 0.5b_2 + a_1, \qquad\qquad\qquad (5.3.9)$$

and $b_N = b_{N+1} = 0$. To evaluate all the b_k's that arise in applying the Galerkin method to $\partial u/\partial x$ requires $O(N^2)$ operations to evaluate eqs. (5.3.4) or (5.3.2) but only $O(N)$ to evaluate eq. (5.3.9).

Comparable recurrence relationships can be obtained for other operators $L(u)$ and trial functions.

5.3.2. Nonlinear terms

The evaluation of quadratically nonlinear terms in the direct manner of eq. (5.2.18) is an $O(N^3)$ process. This essentially limited the use of the spectral method to small values of N until Orszag (1969) introduced the *transform* technique. More recently, Orszag (1980) has expressed the spirit of transform techniques in the following way: "Transform freely between the physical and spectral representations, evaluating each term in whatever representation that term is most accurately, and simply, evaluated."

Implicit in the above remarks is the requirement that a typical transformation such as

$$u(x_l) = \sum_{j=1}^{N} a_j \phi_j(x_l), \qquad l = 1, \ldots, N, \qquad (5.3.10)$$

and inverse transformation (assuming ϕ_j are orthonormal functions)

$$a_k = \int_R u\phi_k(x)\,dx, \qquad k = 1, \ldots, N, \qquad (5.3.11)$$

can be evaluated very economically. A direct evaluation of eq. (5.3.10), or (5.3.11) using numerical quadrature, would require an $O(N^2)$ process. However, if the trial and test functions ϕ_j are members of a Fourier series, the fast Fourier transform will require only $O(N\log_2 N)$ operations.

The fast Fourier transform was given prominence by Cooley and Tukey (1965) and is discussed in detail by Brigham (1974). An essential feature of the fast Fourier transform is to repeatedly factorize the summation in eq. (5.3.10) into the product of two summations. If N is a multiple of 2 (i.e. $N = 2^p$), then the N^2 process is replaced by a $N(2 + 2 + 2 + \cdots)$, or $O(N\log_2 N)$, process.

The use of the fast Fourier transform with the Galerkin method is described by Orszag (1971c). However, Orszag (1980) notes that comparable fast transforms are possible for other orthogonal trial functions. We will use the expression "fast transform" to imply a technique that will evaluate systems of equations like (5.3.10) or (5.3.11) in $O(N \log N)$ operations rather than $O(N^2)$ operations.

Orszag (1980) points out that the main advantage of transform methods comes from their ability to split up multidimensional transforms into a sequence of one-dimensional transforms. An example is given of solving the Navier–Stokes equations for three-dimensional incompressible flow with periodic boundary conditions. A trial solution with 128 unknown coefficients in each direction is used. The evaluation of all the nonlinear terms in spectral space requires about 5×10^5 sec per time step on a CRAY-1 computer. Using a fast transform to physical space permits an evaluation in 20 sec per time step. However, the fast transform provides a speedup by a factor of 2, and the conversion to a sequence of one-dimensional transformations provides the rest of the speedup, a factor $\sim 10^4$.

We will now describe, conceptually, the integration of one time step of the spectral formulation of Burgers' equation using some of the above ideas. We assume that a_j^n are known at time level n. The following sequence is required:

(1) evaluate

$$u_a^n(x_l) = \sum_j a_j^n \phi_j(x_l), \quad l = 1, \ldots, 2N, \qquad \text{F.T.:} \ O(2N \log 2N);$$

(2) evaluate

$$b_j^{(1)n} \text{ from } a_j^n \ \text{ by recurrence:} \qquad O(2N);$$

(3) evaluate

$$\frac{\partial u_a^n(x_l)}{\partial x} = \sum_j b_j^{(1)n} \phi_j(x_l), \quad l = 1, \ldots, 2N, \qquad \text{F.T.:} \ O(2N \log 2N);$$

(4) evaluate

$$w^n(x_l) = u_a^n(x_l) \frac{\partial u_a^n(x_l)}{\partial x}, \quad l = 1, \ldots, 2N: \qquad O(2N);$$

(5) evaluate

$$d_k^n = \int_R w^n \phi_k \, dx, \qquad \text{F.T.:} \ O(2N \log 2N);$$

(6) evaluate

$$s_k^n = \sum_j \left(\phi_k, \frac{\partial^2 \phi_j}{\partial x^2} \right) a_j^n \ \text{ by recurrence:} \qquad O(N);$$

(7) evaluate

$$\frac{da_k^{n+1}}{dt} = d_k^n - \frac{s_k^n}{\text{Re}}, \qquad\qquad O(N);$$

(8) evaluate

$$a_k^{n+1} = a_k^n + f\left(\frac{da_k^{n+1}}{dt}\right): \qquad\qquad O(N).$$

In the above, d_k^n would be evaluated by a quadrature scheme using the values of $w(x_l)$ evaluated at step (4). It is necessary to evaluate $w(x_l)$ at $2N$ points to avoid *aliasing errors*. This will be pursued in section 5.4.1.

For large values of N it is apparent that the above sequence is substantially more economical than the elementary procedure described in section 5.2.2.

5.3.3. Time differencing

Spectral methods have been applied, most often, to parabolic problems. Generally a low-order finite-difference scheme is used for time differencing and integration. After application of a Galerkin spectral method the resulting equations can be written in the following time-differenced form:

$$a_k^{n+1} - a_k^n = \Delta t(1 - \theta)L_k(a_k^n) + \Delta t\,\theta L_k(a_k^{n+1}), \qquad (5.3.12)$$

where θ is a parameter chosen to control the degree of implicitness and hence the stability of the scheme. Gottlieb and Orszag (1977) show that eq. (5.3.12) is stable if $\theta \geq 0.5$.

For $\theta = 0$, eq. (5.3.12) is completely explicit. For problems with periodic boundary conditions and a Fourier-series trial solution the restriction on the time step, when $\theta = 0$, is comparable to that found with (spatial) finite-difference formulations. For the model hyperbolic problem

$$\frac{\partial u}{\partial t} + \frac{\partial u}{\partial x} = 0, \qquad (5.3.13)$$

Δt_{max} is $O(N^{-1})$, which is equivalent to $O(\Delta x)$ for a (spatial) finite-difference formulation. For the model parabolic problem,

$$\frac{\partial u}{\partial t} = \frac{\partial^2 u}{\partial x^2}, \qquad (5.3.14)$$

Δt_{max} is $O(N^{-2})$ when $\theta = 0$.

If Chebyshev polynomials are used for the trial functions, with $\theta = 0$, the restriction on the time step is more severe. For the model hyperbolic problem (5.3.13), Δt_{max} is $O(N^{-2})$. The cause of this restriction is the property of the Chebyshev collocation points, $\cos(j\pi/N)$, being spaced at a distance of $O(N^{-2})$ near the boundaries. For the model parabolic problem (5.3.14), Δt_{max} is $O(N^{-4})$.

Gottlieb and Orszag (1977) discuss a number of semiimplicit schemes that for the hyperbolic problem

$$\frac{\partial u}{\partial t} + b(x)\frac{\partial u}{\partial x} = 0, \qquad -1 \le x \le 1, \qquad (5.3.15)$$

restore the time-step restriction for Chebyshev trial functions; that is, Δt_{max} is $O(N^{-1})$. To streamline the notation, we recall that the trial solution u_a is given by

$$u_a = \sum_j a_j \phi_j(x),$$

and we define L_a^{max} as being the Chebyshev spectral approximation to $0.5 b_{max} \partial/\partial x$, where $b_{max} = \max(b(x))$ and $b(x) > 0$. A semiimplicit scheme can be written

$$(I - \Delta t\, L_a^{max})u_a^{n+1} = u_a^n + \Delta t\,(L_a - L_a^{max})u_a^n. \qquad (5.3.16)$$

Gottlieb and Orszag note that these equations are tridiagonal and so can be solved efficiently.

More recently, Gottlieb and Turkel (1980) have developed unconditionally stable explicit schemes. For eq. (5.3.15) with periodic boundary conditions and a complex Fourier series as a trial solution, that is,

$$u_a(x_l) = \sum_{j=-N}^{N} a_j \exp\left(\frac{ijl\pi}{N}\right), \qquad l = 0, \ldots, 2N - 1, \qquad (5.3.17)$$

the Galerkin procedure gives

$$\frac{d\mathbf{A}}{dt} + i\mathbf{B}\mathbf{A} = 0, \qquad (5.3.18)$$

where an element of $\mathbf{A}$ is a_j and an element of $\mathbf{B}$ is given by

$$b_{mn} = \begin{cases} (n - N - 1)\hat{b}_{m-n} & \text{if } |m - n| \le N \\ 0 & \text{if } |m - n| > N \end{cases}, \qquad m, n = 1, \ldots, 2N + 1. \qquad (5.3.19)$$

Here $\hat{b}$ are the coefficients of an expansion for $b(x)$ in eq. (5.3.15):

$$b(x_l) = \sum_{j=-N}^{N} \hat{b}_j \exp\left(\frac{ijl\pi}{N}\right), \qquad l = 0, \ldots, 2N - 1. \qquad (5.3.20)$$

Gottlieb and Turkel solve eq. (5.3.18) with a modified leapfrog scheme

$$\mathbf{A}^{n+1} = \mathbf{A}^{n-1} - i\mathbf{C}\mathbf{A}^n, \qquad (5.3.21)$$

where an element of $\mathbf{C}$ is given by

$$c_{mn} = \begin{cases} \sin[\sigma(n - N - 1)\Delta t]\hat{b}_{m-n} & \text{if } |m - n| \le N, \\ 0 & \text{if } |m - n| > N, \end{cases} \qquad (5.3.22)$$

where $\sigma = \alpha\rho(B)$; $\rho(B)$ is the spectral radius of B. α is chosen so that $\sigma \geq \rho(B)$ $\geq \max |b(x)|$.

As indicated above, implicit schemes for eq. (5.3.12) are stable. However, if $\theta \geq 0.5$, the structure of $L_k(a_k^{n+1})$ for the general case implies that a full matrix of equations must be solved at each time step. Orszag (1980) discusses the possibility of doing this economically using iterative schemes based on approximately representing $L_k(a_k^{n+1})$ by a low-order finite-difference scheme.

In more than one dimension, time-splitting techniques (comparable to those described in section 4.1) are possible. These have been used in a spectral context by Bourke et al. (1977) and Orszag and Kells (1980).

5.4. Alternative Spectral Methods

So far we have focused our discussion of spectral methods on Galerkin methods. However, if we define a spectral method as a method that incorporates a global trial solution of orthogonal trial functions, then other spectral methods are possible. In principle we could use any of the methods of weighted residuals (section 1.3). In fact, the collocation or *pseudospectral* method has achieved considerable popularity, since it is well suited to transform techniques (section 5.3.2). We will discuss the pseudospectral method in section 5.4.1.

Another spectral method, which could have been included in section 1.3, has been used occasionally. This is the *tau* method. This method is similar to the Galerkin method and will be described in section 5.4.2.

5.4.1. Pseudospectral method

This method was introduced by Orszag (1971c) as a viable alternative to the Galerkin spectral method when nonlinearities force fast-transform techniques to be used.

We illustrate the pseudospectral method for a nonlinear diffusion equation given by Orszag (1980):

$$\frac{\partial u}{\partial t} = e^u \frac{\partial^2 u}{\partial x^2}.$$ (5.4.1)

A trial solution is introduced by

$$u_a(x, t) = \sum_{j=1}^{N} a_j(t)\phi_j(x).$$ (5.4.2)

A sequence of steps to integrate the pseudospectral solution of eq. (5.4.1) over one time step can be set down as follows (we assume that $u_a^n(x_l)$, $l = 1, \ldots, N$, are known at time-level n):

(1) evaluate

$$a_j^n(t) \text{ to satisfy eq. (5.4.2) at } x_l, \quad \text{F.T.:} \quad O(N \log N);$$

(2) evaluate

$$\frac{\partial^2 u_a^n}{\partial x^2} = \sum \frac{\partial^2 \phi_j}{\partial x^2} a_j^n \quad \text{by recurrence:} \quad O(N);$$

(3) evaluate

$$\frac{\partial u_a^{n+1}(x_j, t)}{\partial t} = \exp\{u_a(x_j, t)\} \frac{\partial^2 u_a(x_j, t)}{\partial x^2}, \quad l = 1, \ldots, N, \quad O(N);$$

(4) evaluate

$$u_a^{n+1}(x_l) = u_a^n(x_l) + f\left(\frac{\partial u_a^{n+1}}{\partial t}\right), \quad l = 1, \ldots, N: \quad O(N).$$

As with the collocation method in section 1.3.2, step (3) above forces the governing equation to be satisfied at the N collocation points x_l. To uniquely define the N values of a_j, only N values of $u(x_l)$ are required. In the Galerkin spectral method $2N$ points were used to avoid aliasing errors.

Aliasing errors are a disadvantage of the pseudospectral method. *Aliasing* is the phenomenon where high frequencies of the solution on a discrete grid appear as low frequencies. Hamming (1973) cites the example of two trigonometric functions

$$\cos(\pi(n + \varepsilon)t + \phi) \quad \text{and} \quad \cos(\pi(n - \varepsilon)t + \phi),$$

which are plotted in Fig. 5.5. If the values of the following frequency and so on are to be determined from the equally spaced sample points shown in Fig. 5.5, it is clear that it is not possible to differentiate between a frequency of $\pi(n - \varepsilon)$ and $\pi(n + \varepsilon)$.

In a typical nonlinear problem with a general Fourier-series trial solution the following expression might need to be evaluated in spectral space:

$$w(k) = \sum_{\substack{p+q=k \\ \|p\|, \|q\| < N/2}} u(p)v(q), \tag{5.4.3}$$

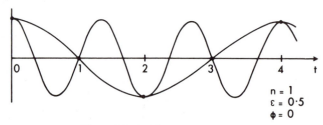

Figure 5.5. An illustration of aliasing

where k is a wave vector. If a transformation is made to physical space to evaluate the corresponding local products $\hat{w}_j = \hat{u}_j \hat{v}_j$ at N points, and an inverse transform $\hat{w}(k)$ is computed, it is found that

$$\hat{w}(k) = w(k) + w(k + N) + w(k - N). \tag{5.4.4}$$

$w(k)$ is the true inverse transform, and the extra terms $w(k + N)$ and $w(k - N)$ are the aliases due to sampling at N discrete points x_j. Thus it is not possible to distinguish between wave vectors k, $k \pm N$, $k \pm 2N$, ... Orszag (1971c) demonstrates that if at least $2N$ sample points are used in physical space, aliasing will be eliminated.

Historically aliasing was considered a major cause of nonlinear instability for the long-time integration of the equations governing global atmospheric circulation. For many problems (e.g. turbulence modeling) there is sufficient dissipation that any errors introduced by aliasing have a negligible effect on the stability of the time integration (Orszag, 1972). Therefore, for situations where aliasing is not critical the pseudospectral method is to be preferred, since it is more economical than the Galerkin spectral method.

Fornberg and Whitham (1978) apply a pseudospectral method to the Korteweg–de Vries equation with a modified leapfrog time-differencing scheme similar to eqs. (5.3.21) and (5.3.22).

Morchoisne (1979) applies a pseudospectral method to Burgers' equation and the incompressible Navier–Stokes equations. The approach of Morchoisne is interesting because an equation splitting is introduced and because the pseudospectral representation is applied in time as well as space. The Morchoisne formulation is discussed by Fletcher (1982).

5.4.2. The tau method

The tau method is described at length by Lanczos (1956), who originated the method in 1938. The idea behind the method is to introduce a small perturbation to the given problem so that an exact solution can be obtained to the perturbed problem. If the perturbation is small, then the exact solution of the perturbed problem will be an approximate solution of the given problem.

As a modification to the Galerkin method, the tau method can be introduced as follows. Suppose we wish to solve

$$\frac{du}{dt} = L(u) + f, \tag{5.4.5}$$

with boundary conditions

$$B(u) = 0. \tag{5.4.6}$$

A trial solution based on orthonormal functions $\phi_j(x)$, chosen from a complete set, is introduced as

$$u_a(x, t) = \sum_{j=1}^{N+l} a_j(t)\phi_j(x). \tag{5.4.7}$$

Orthonormal functions are orthogonal functions normalized so that the orthogonality condition (5.5.9) equals unity.

N equations for the a_j's are obtained from

$$\frac{da_k}{dt} = (\phi_k, L(u_a)) + (\phi_k, f), \qquad k = 1, \ldots, N, \tag{5.4.8}$$

and l equations are obtained from imposing the boundary conditions

$$\sum_{j=1}^{N+l} a_j(t) B(\phi_j) = 0. \tag{5.4.9}$$

If $N + l = \infty$ in eq. (5.4.7), then the trial solution is exact, from the definition of a complete set of functions. Therefore the solution by the tau method, u_a, is the exact solution of the perturbed problem

$$\frac{du_a}{dt} = L(u_a) + f + \sum_{p=1}^{\infty} \tau_p(t)\phi_{N+p}(x). \tag{5.4.10}$$

The coefficients τ_1 to τ_l come from eq. (5.4.9). The other τ coefficients are given by

$$\tau_p = -(\phi_{N+p}, L(u_a) f), \qquad p = l + 1, \ldots. \tag{5.4.11}$$

In practice, of course, eqs. (5.4.10) and (5.4.11) are not used.

The essential difference from the Galerkin method is that in the tau method, the trial functions are not required to satisfy the boundary conditions individually. With complicated boundary conditions this aspect can make the Galerkin method cumbersome and computationally less efficient, since the inner products are more expensive to evaluate. In the tau method enough of the coefficients in the trial solution are chosen to ensure exact satisfaction of the boundary conditions.

Strictly the solutions of Burgers' equation described in sections 1.2.5 and 5.2.2 were obtained by the tau method. For the propagating-shock problem the last two coefficients in the trial solution were chosen to satisfy the boundary conditions at $x = \pm 1$.

Gottlieb and Orszag (1977) give numerous examples of the tau spectral formulation. The general character of the tau method is similar to that of the Galerkin method, as might be expected. For the model hyperbolic problem (5.3.13) and a Chebyshev trial solution, Gottlieb and Orszag find that the Galerkin method is more accurate than the tau method for relatively inaccurate approximations (small N). For accurate approximations (large N) the reverse is true. Used with a Legendre trial solution, the tau method produces errors near the boundaries that are much smaller than the errors with a Chebyshev trial solution.

5.5. Orthonormal Method of Integral Relations

In this section we will start with the method of integral relations, which is a member of the class of methods of weighted residuals, and show that it can be upgraded into a Galerkin spectral method, that is, *the orthonormal method of integral relations*. Like the traditional Galerkin method, the method of integral relations endeavors to use special test and trial functions so that an accurate solution can be obtained with relatively few unknown coefficients. The essential feature of the orthonormal method of integral relations is to take the special test functions and construct orthonormal functions from them with respect to a weight function suggested by the problem at hand.

The method will be described in relation to two-dimensional laminar boundary-layer flow, which is shown schematically in Fig. 4.12. The governing equations can be written, in nondimensional form, as

$$\frac{\partial u}{\partial x} + \frac{\partial v}{\partial y} = 0 \qquad (5.5.1)$$

and

$$u\frac{\partial u}{\partial x} + v\frac{\partial u}{\partial y} = u_e\frac{du_e}{dx} + \frac{1}{\text{Re}}\frac{\partial^2 u}{\partial y^2}, \qquad (5.5.2)$$

where u and v are the velocity components shown in Fig. 4.12, $u_e(x)$ is a given velocity distribution at the outer edge of the boundary layer, and Re is the Reynolds number.

The following variables are defined:

$$\xi = \int_0^x u_e(x')\,dx', \qquad \eta = \text{Re}^{1/2}u_e y, \quad \text{and} \quad u' = \frac{u}{u_e}, \qquad (5.5.3)$$

and they are introduced into eqs. (5.5.1) and (5.5.2) in place of x and y. We will call the resulting equations (5.5.1)$_{\xi,\eta}$ and (5.5.2)$_{\xi,\eta}$. The method of integral relations was applied to them by Dorodnitsyn (1960) in the following way. The composite equation

$$f_k(u') \times \left[\text{eq. (5.5.1)}_{\xi,\eta}\right] + \frac{df_k}{du'}(u') \times \left[\text{eq. (5.5.2)}_{\xi,\eta}\right] = 0$$

is formed and integrated across the boundary layer with respect to η. If $f_k(u') = 0$ at $\eta = \infty$ ($u' = 1$), it is found that the normal velocity v does not appear in the composite equation.

The integration variable is changed to u', and the final form of the equation to be solved is as follows (dropping the prime from u'):

$$\frac{d}{d\xi}\int_0^1 \Theta u f_k(u)\,du = \frac{1}{u_e}\frac{du_e}{d\xi}\int_0^1 \Theta(1 - u^2)\frac{df_k}{du}(u)\,du$$
$$- \left[\frac{df_k}{du}(u)\mathscr{T}\right]_{u=0} - \int_0^1 \frac{d^2 f_k}{du^2}\mathscr{T}\,du. \qquad (5.5.4)$$

This is a partial differential equation for $\Theta = \Theta(\xi, u)$, where $\Theta = 1/\tau = \partial u/\partial \eta$. A trial solution for Θ is introduced as follows:

$$\Theta = \frac{1}{1 - u}\left(a_0 + \sum_{j=1}^{N-1} a_j u^j\right). \tag{5.5.5}$$

The factor $1 - u$ is introduced to ensure the correct behavior of Θ as $u \to 1$ (the outer edge of the boundary layer). Dorodnitsyn (1960) used the following form for the weight function f_k:

$$f_k(u) = (1 - u)^k. \tag{5.5.6}$$

Substitution of eqs. (5.5.5) and (5.5.6) into eq. (5.5.4) and evaluation of the various integrals generates a system of ordinary differential equations for the a_j's that can be written

$$\mathbf{M\dot{A}} = \mathbf{B}. \tag{5.5.7}$$

The method encounters the same problem as the traditional Galerkin method if solutions are sought for large N. In particular $\mathbf{M}$ becomes progressively more ill conditioned with increasing N. For small values of N the method of integral relations is very effective and has been applied to a wide range of problems (Holt, 1977).

The orthonormal method of integral relations is based on introducing orthonormal functions $g_j(u)$ instead of the test functions $f_k(u)$ and instead of the trial functions u^j. The ultimate result is a diagonal form for $\mathbf{M}$ in eq. (5.5.7) and a formulation that can be extended to large N. Thus the high accuracy of a spectral method is combined with the inherent economy of the method of integral relations.

The orthonormal functions $g_j(u)$ are generated from the Dorodnitsyn test functions $f_k(u)$ by

$$g_j = \sum_{k=1}^{j} e_{kj} f_k(u), \tag{5.5.8}$$

where e_{kj} are evaluated via the Gram–Schmidt orthonormalization process (Isaacson and Keller, 1966) so that the functions g_j satisfy the conditions

$$(g_i, g_j) = \begin{cases} 1 & \text{if } i = j, \\ 0 & \text{if } i \neq j, \end{cases} \tag{5.5.9}$$

where (g_i, g_j) is a general inner product,

$$(g_i, g_j) = \int_0^1 g_i(u) g_j(u) w(u)\, du. \tag{5.5.10}$$

The function $w(u)$ depends on the problem being solved. The appropriate form for $w(u)$ for the present example will be indicated below.

Once the orthonormal functions have been generated, the following trial solution is introduced:

$$\Theta = \frac{1}{1-u}\left(b_0 + \sum_{j=1}^{N-1} b_j g_j(u)\right).$$ (5.5.11)

The leading coefficient b_0 is retained to ensure that Θ behaves properly at $u = 1$. If eq. (5.5.11) is substituted into eq. (5.5.4) with g_k replacing f_k, the result is

$$\frac{d}{d\xi}\int_0^1 \left(b_0 + \sum_{j=1}^{N-1} b_j g_j\right) g_k \frac{u}{1-u}\, du = c_k', \qquad k = 1, \ldots, N,$$ (5.5.12)

where

$$c_k' = \frac{u_{e\xi}}{u_e}\int_0^1 \Theta \frac{dg_k}{du}(1 - u^2)\, du - \left[\frac{dg_k}{du}\mathscr{T}\right]_{u=0} - \int_0^1 \frac{d^2 g_k}{du^2}\mathscr{T}\, du.$$

Comparing eqs. (5.5.10) and (5.5.12) indicates that eq. (5.5.9) will be satisfied if

$$w(u) = \frac{u}{1-u}.$$ (5.5.13)

Making use of eq. (5.5.9) permits eq. (5.5.12) to be written

$$\frac{db_0}{d\xi}v_k + \frac{db_k}{d\xi} = c_k', \qquad k = 1, \ldots, N - 1,$$ (5.5.14)

where

$$v_k = \int_0^1 g_k(u) w(u)\, du.$$ (5.5.15)

When $k = N$,

$$\frac{db_0}{d\xi} = \frac{c_N'}{v_N}.$$ (5.5.16)

Consequently the equations (5.5.14) are replaced by

$$\frac{db_k}{d\xi} = c_k' - c_N' \frac{v_k}{v_N}.$$ (5.5.17)

v_k, given by eq. (5.5.15), is evaluated once and for all. Eqs. (5.5.16) and (5.5.17) can be integrated conveniently using a fourth-order Runge–Kutta scheme.

Fletcher and Holt (1975) considered three *Falkner–Skan* solutions corresponding to a favorable, a neutral, and an unfavorable pressure gradient. For each case solutions at various N for the orthonormal and conventional methods of integral relations were compared with the exact solution. The results for the favorable-pressure-gradient case ($\beta = 0.5$) are shown in Fig. 5.6. The parameter β is defined by

$$\beta = (2\xi)^{1/2} u_\xi / u.$$ (5.5.18)

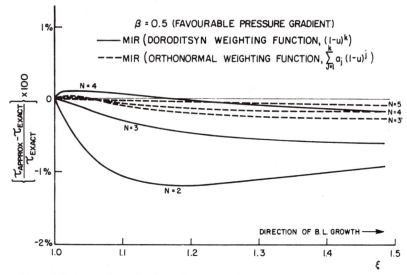

Figure 5.6. Comparison of MIR surface stresses for Falkner–Skan solution, $\beta = 0.5$

The exact solution provides the starting data at $\xi = 1$. The general trend is for the approximate solutions to diverge initially from the exact solution and subsequently to reapproach it. It can be seen that increasing the order reduces the maximum error and that for the same order the orthonormal method of integral relations gives a more accurate answer than the conventional method of integral relations.

The orthonormal method of integral relations has been applied to boundary-layer growth on an inclined cone in a supersonic flow by Fletcher and Holt (1976). This problem is governed by the continuity equation, the longitudinal and circumferential momentum equations, and the enthalpy equation. After application of the Howarth, Mangler, and Crocco transformations the problem reduces to a function of two variables (ζ, u). Here u is the longitudinal velocity component and ζ is a similarity coordinate analogous to ξ in eq. (5.5.4).

The governing equations equivalent to (5.5.12) can be written

$$\frac{\partial}{\partial \zeta} \int_0^1 g_k \frac{w}{\tau} du = c_k^{(1)},$$

$$\frac{\partial}{\partial \zeta} \int_0^1 g_k \frac{w^2}{\tau} du = c_k^{(2)}, \qquad (5.5.19)$$

$$\frac{\partial}{\partial \zeta} \int_0^1 g_k \frac{sw}{\tau} du = c_k^{(3)},$$

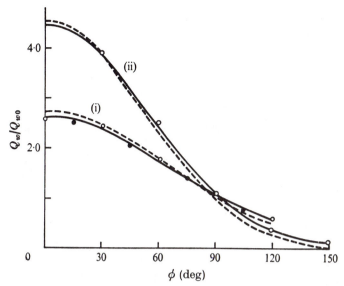

Figure 5.7. Surface heat-transfer variation for inclined cone: ——— including displacement thickness; – – – – excluding displacement thickness; ● experimental data (Tracy, 1963); ○ computational solution (Lubard and Helliwell, 1974)

where w is a nondimensional circumferential velocity and s is a nondimensional enthalpy. The following trial solutions are introduced:

$$y_1 = \frac{w}{\tau} = \frac{1}{1-u}\left\{b_{01} + \sum_{j=1}^{N-1} b_{j1} g_j(u)\right\}, \tag{5.5.20}$$

$$y_2 = \frac{w^2}{\tau} = \frac{1}{1-u}\left\{b_{02} + \sum_{j=1}^{N-1} b_{j2} g_j(u)\right\}, \tag{5.5.21}$$

$$y_3 = \frac{sw}{\tau} = \frac{1}{1-u}\left\{b_{03} + \sum_{j=1}^{N-1} b_{j3} g_j(u)\right\}. \tag{5.5.22}$$

Here $g_j(u)$ are orthonormal functions with respect to $w(u) = 1/(1-u)$ in eq. (5.5.10). Because of the form of the left-hand side of eq. (5.5.19), it is convenient to introduce the trial solutions for groups of terms.

After substitution of eqs. (5.5.20) to (5.5.22) into eqs. (5.5.19), explicit ordinary differential equations for $b_{01}, b_{11}, \ldots$ are obtained. After numerical integration of these equations, substitution of the values of $b_{01}, b_{11}, \ldots$ into eqs. (5.5.20) to (5.5.22) gives $y_1(\zeta, u)$, y_2 and y_3. Algebraic manipulation then provides the more physically meaningful variables such as velocity, temperature, and skin friction. A typical circumferential variation of surface heat transfer is shown in Fig. 5.7. The results were obtained at $M_\infty = 7.95$, $\theta_b = 10°$, (i) $\alpha = 12°$, (ii) $\alpha = 24°$, $\mathrm{Re}_x = 4.2 \times 10^5$, and $T_w/T_0 = 0.41$. θ_b is the cone half angle, and α is the cone incidence. T_w and T_0 are the wall and stagnation temperatures respectively; Re_x is the Reynolds number, and M_∞ is the freestream Mach number.

The results are compared with experimental data (filled circles) due to Tracy (1963) and a finite-difference solution of the parabolized Navier–Stokes equations (open circles) due to Lubard and Helliwell (1974). The dashed line corresponds to an inviscid solution obtained about the cone surface (Fletcher, 1975), whereas the solid line corresponds to the inviscid solution about the cone surface plus displacement thickness.

The orthonormal method of integral relations has been applied to turbulent boundary-layer flows by Yeung and Yang (1981) and Fleet and Fletcher (1982).

5.6. Applications

Spectral methods have been utilized in many areas, with the major emphasis occurring in two: global atmospheric modeling and fundamental turbulence studies. Most applications have been to time-dependent mixed initial-boundary-value problems with finite-difference schemes to provide time differencing and integration. Here we will examine global atmospheric modeling in section 5.6.1, direct turbulence simulation in section 5.6.2, and other applications in section 5.6.3.

5.6.1. Global atmospheric modeling

The first application of a spectral method to a meteorological flow was due to Silberman (1954), who considered the vorticity equation in a spherical coordinate system. Lorenz (1960) established that a truncated spectral representation of nondivergent barotropic flow conserved the mean squared kinetic energy and mean squared vorticity. Platzman (1960) showed that this property would prevent nonlinear instability from developing.

In the early spectral methods all nonlinear terms were evaluated in spectral space via interaction coefficients. Necessarily the order N of the trial solution was limited. Orszag (1970) demonstrated the transform technique on the vorticity equation and showed that the $O(N^5)$ operations per time step associated with the interaction coefficient formulation could be reduced to $O(N^3)$ operations per time step if a transform method was used.

Subsequently Bourke (1972) applied a transform Galerkin spectral method to the divergent barotropic vorticity equations in two dimensions (latitude and longitude). However, in extending the method to three dimensions Bourke (1974) used a finite-difference formulation in the vertical direction, whereas Machenhauer and Daley (1972) used a spectral method in all three directions.

Bourke et al. (1977) illustrate the application of a Galerkin spectral method to the nondivergent vorticity equation in the following form:

$$\frac{\partial}{\partial t} \nabla^2 \psi = \frac{1}{a \cos^2 \phi} \left[\frac{\partial}{\partial \lambda} (u \nabla^2 \psi) + \cos \phi \frac{\partial}{\partial \phi} (v \nabla^2 \psi) \right] - 2\Omega \frac{v}{r}, \quad (5.6.1)$$

where ϕ is the latitude, u and v are latitude-scaled horizontal velocity components, λ is the longitude, r is the earth's radius, and Ω the earth's rotation rate. ψ is the stream function, which is related to the vorticity ζ by

$$\zeta = \nabla^2 \psi. \quad (5.6.2)$$

Bourke et al. consider eq. (5.6.1) to be in a form well suited to the application of the transform technique to products like $u \nabla^2 \psi$. A trial solution for ψ is introduced as

$$\psi(\lambda, \phi, t) = r^2 \sum_{m=-N}^{N} \sum_{l=|m|}^{|m|+N} \psi_l^m(t) Y_l^m(\lambda, \phi), \quad (5.6.3)$$

where $\psi_l^m(t)$ are the unknown coefficients. The trial functions are given by

$$Y_l^m(\lambda, \phi) = P_l^m(\sin \phi) e^{im\lambda}, \quad (5.6.4)$$

where $P_l^m(\sin \phi)$ is a Legendre polynomial of the first kind, normalized to unity:

$$\int_{-\pi/2}^{\pi/2} P_l^m(\sin \phi) P_l^m(\sin \phi) \cos \phi \, d\phi = 1. \quad (5.6.5)$$

Equivalent trial solutions are introduced for u and v.

Bourke (1972) compared the interaction coefficient technique and the transform technique for the divergent barotropic spectral model, on the basis of computational efficiency. The variation of the time per time step with wavenumber truncation (N in eq. (5.6.3)) is shown in Fig. 5.8. For the interaction-coefficient technique the effect of increasing values of N is potentially ruinous.

Figure 5.8. Variation of computation time with spectral resolution (after Bourke et al., 1977; reprinted with permission of Academic Press)

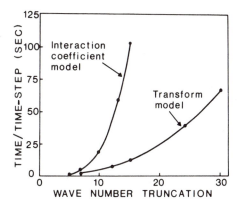

Bourke et al. (1977) subsequently describe an operational model based on a wave-number truncation of $N = 15$. Clearly, from Fig. 5.8, the economy associated with the transform technique is substantial. However to suit a cutoff of $N = 15$, Bourke et al. find it necessary to smooth the topography of the earth's surface.

They use a semiimplicit time-integration scheme of a leapfrog type. For a typical equation, the vorticity equation, this scheme is written

$$\zeta^{n+1} = \zeta^{n-1} + 2\,\Delta t\,a + 2\,\Delta t\,V\zeta^{n+1}. \tag{5.6.6}$$

a contains the nonlinear terms, which are evaluated at time level n (the convection terms) and time level $n - 1$ (the horizontal diffusion terms). V represents the linearized internal vertical diffusion. The semiimplicit scheme requires the solution of full matrices by Gaussian elimination at each time step, but permits a maximum time-step increase from 10 min to 1 hr with only a 3% computational overhead at $N = 15$. Bourke et al. also discuss the possibility of time splitting.

Bourket et al. describe an operational model designed to give 48-hr predictions in 1-hr computing time with 40 K of main memory. To accommodate these restrictions they use a cutoff of $N = 15$ and seven vertical levels. Com-

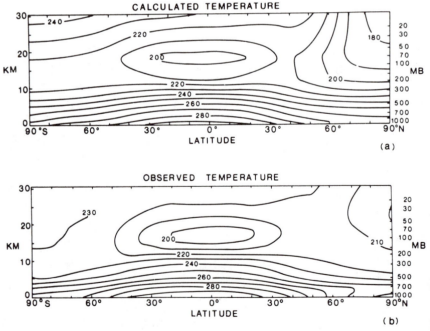

Figure 5.9. Comparison of latitude–height temperature distribution (after Bourke et al., 1977; reprinted with permission of Academic Press)

parisons are made for the Southern Hemisphere, and it is shown that the spectral model, with 256 $((N + 1)^2)$ degrees of freedom at each level, produces better predictions for a 24-hr forecast than does a finite-difference model with approximately 3000 degrees of freedom.

Bourke et al. (1977) also present results obtained for general circulation with an $N = 15$, nine-vertical-level spectral model. Time integration has been continued to 86 days using a temporal stepsize $\Delta t = 30$ min. A comparison of the average temperature distribution (over days 52 to 86) is shown in Fig. 5.9. The predictions are considered reasonable except for the rather

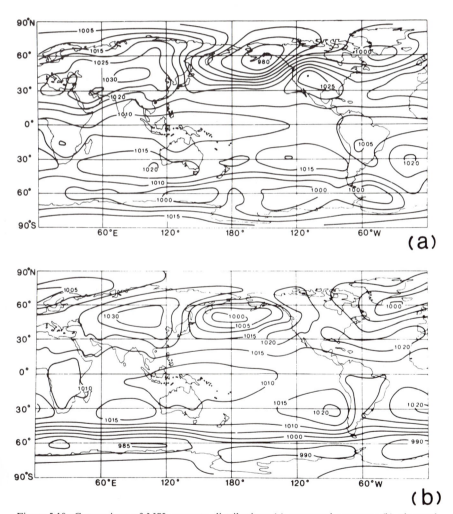

Figure 5.10. Comparison of MSL pressure distribution: (a) computed pressure; (b) observed pressure (after Bourke et al., 1977; reprinted with permission of Academic Press)

low temperatures at 30-km altitude at 90°N. The observations are due to Newell et al. (1972). A comparison of the mean sea-level pressure (averaged over days 52 to 86) is made with the observed data of Schutz and Gates (1971) in Fig. 5.10. The agreement is considered acceptable.

Current operational models (Puri, 1981) use nine vertical levels and wave-number cutoffs in the range $N = 21$ to 31. Projected operational models (Puri, 1982) are expected to approximately double the wave-number cutoff. The main limitation on the useful prediction period appears to be the quality and resolution of the initial data. Currently, reliable four-day and two-day predictions are possible for the Northern and Southern Hemispheres respectively.

5.6.2. Direct turbulence simulation

In contrast to the empirical modeling of the effects of turbulence, it is possible to solve the incompressible unsteady Navier–Stokes equations if the details of the flow can be resolved sufficiently. The governing equations for unsteady incompressible flow are

$$\frac{\partial v_l}{\partial t} + v_j \frac{\partial v_l}{\partial x_j} = \frac{-1}{\rho} \frac{\partial p}{\partial x_l} + v \left\{ \frac{\partial^2 v_l}{\partial x_j \partial x_j} \right\} \qquad (5.6.7)$$

and

$$\frac{\partial v_l}{\partial x_l} = 0, \qquad (5.6.8)$$

where the indices $l, j = 1, 2, 3$ in three dimensions. $v_l(x_j, t)$ are the velocity components, and $p(x_j, t)$ is the pressure. The flow becomes turbulent if the kinematic viscosity is sufficiently small.

However, there is a problem in obtaining sufficient data to compute accurate statistical averages. For homogeneous isotropic turbulence, three-dimensional spatial averages over small bands of wave number have been used (Orszag, 1977). For two-dimensional shear flows, averages are taken over the spanwise coordinate, that is, the coordinate direction over which no change in average properties is expected. The current capability to directly simulate turbulence at high Reynolds numbers is limited by the computing power available. Consequently the theoretically attractive idea of running N independent computational experiments and forming an ensemble average is not popular.

A spectral formulation for the direct simulation of turbulence was given by Orszag and Kruskal (1968). However, it was the introduction of the transform technique to handle nonlinear terms (section 5.3.2) by Orszag (1969) that permitted computational results to be obtained at a reasonable cost.

Early applications of the method included the simulation of three-dimensional homogeneous isotropic turbulence at moderate Reynolds number

(Orszag and Patterson, 1972) and the use of two-dimensional turbulence simulations to provide test data to assess analytic theories of turbulence (Herring et al., 1974).

Homogeneous turbulence can be simulated by solving eqs. (5.6.7) and (5.6.8) with periodic spatial boundary conditions applied to the velocity components. As indicated in section 5.1, a Fourier series is an appropriate trial solution if periodic boundary conditions apply. Thus the velocity components v_l are represented as follows:

$$v_l(\mathbf{x}, t) = \sum_{|\mathbf{k}| < N} u_l(\mathbf{k}, t) e^{i\mathbf{k} \cdot \mathbf{x}}. \tag{5.6.9}$$

A similar trial solution can be introduced for the pressure p. However, p is subsequently eliminated (see Orszag and Kruskal, 1968, for details). Application of the spectral method to eqs. (5.6.7) and (5.6.8) gives

$$\left(\frac{\partial}{\partial t} + vk^2\right) u_l(\mathbf{k}, t) = -ik_m \left(\delta_{lj} - \frac{k_l k_j}{k^2}\right) \sum_{\mathbf{p} + \mathbf{q} = \mathbf{k}} u_j(\mathbf{p}, t) u_m(\mathbf{q}, t). \tag{5.6.10}$$

Here $k = |\mathbf{k}|$ and δ_{lj} is the Kronecker delta.

For two-dimensional homogeneous turbulence simulation a spectral stream-function–vorticity (ψ, ζ) approach has been used by Herring et al. (1974). They report that the pseudospectral formulation is roughly twice as efficient as the Galerkin spectral formulation. Results were presented for values of kinematic viscosity down to $v = 0.001$ and with wave-number cutoffs up to 128.

For the range of Reynolds numbers considered (i.e. up to 1000), it appeared that the energy spectrum function $E(k)$ varied like k^{-4} rather than the theoretical high-Reynolds-number dependence of k^{-3} in the inertial range. However, with an extension of the cutoff to $N = 1024$, solutions for Reynolds numbers up to 2500 have indicated a k^{-3} dependence for $k \leq 50$ (Orszag, 1976).

A simulation of a turbulent free shear layer has been employed to assess a traditional *mixing-length hypothesis*, which relates the Reynolds stress $\overline{uw}$ to the mean velocity $\overline{U}(z)$ in the following way:

$$\overline{uw} = -L^2 \frac{\partial \overline{U}}{\partial z} \left| \frac{\partial \overline{U}}{\partial z} \right|. \tag{5.6.10}$$

A comparison of the direct calculation of $\overline{uw}$ with $-L^2(\partial \overline{U}/\partial z)|\partial \overline{U}/\partial z|$ is shown in Fig. 5.11 for $L = 0.22$ at $t = 2$. The shear layer is directed along the x axis with the z axis normal to the shear layer. This comparison was obtained from a two-dimensional direct simulation.

The testing of simpler, semiempirical turbulence models is an important function of direct turbulence simulation by spectral techniques (Orszag, 1977).

Direct simulation of two different turbulent free shear flows—axisymmetric wake flow and mixing layers—has been reported by Metcalfe and Riley

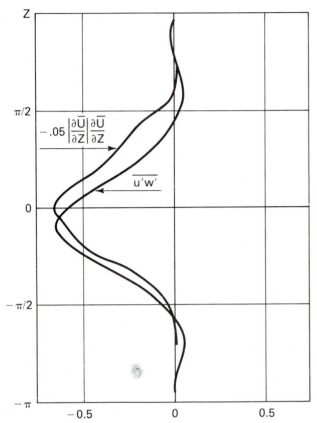

Figure 5.11. Reynolds stress distribution in a turbulent free shear layer (after Orszag, 1977; published with permission of Plenum Press)

(1981). The codes used were based on a three-dimensional 32-mode and 64-mode pseudospectral method. For the wake behind a towed body, good agreement was obtained with experimental data. Metcalfe and Riley noted that the only empiricism in their simulation was in the choice of the initial energy spectrum. Simulations of a turbulent mixing layer demonstrated the same qualitative behavior for mixing-layer thickness and mean-velocity profile development as did the experimental results.

Spectral methods have also been used to predict the *transition* from laminar to turbulent flow. The traditional manner of predicting the transition is to postulate a linear disturbance field on a known solution of the Navier–Stokes equations, which then reduce to the Orr–Sommerfeld equation (1.6.19). The Orr–Sommerfeld equation is then solved to identify any unstable eigenmodes λ, which correspond to exponential disturbance growths with time.

This has been done for two-dimensional channel flow by Orszag (1971b),

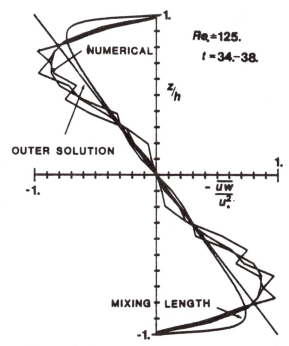

Figure 5.12. Reynolds stress distribution for forced transition in a two-dimensional channel (after Patera and Orszag, 1981; reprinted with permission of Springer-Verlag)

who introduced a Chebyshev trial solution for the disturbance field. The critical Reynolds number (lowest Reynolds number at which disturbances may grow) was found to be 5772.22.

However, it is well known that the transition is very susceptible to three-dimensional disturbances and that it has been observed with a Reynolds number as low as 1000. Orszag and Kells (1980) applied a composite pseudo-spectral, spectral-tau, fractional-step method to plane Couette and Poiseuille flow. The method used, as a trial solution, Fourier series in the plane of the flow and Chebyshev polynomials normal to the flow. It was confirmed computationally that the transition can take place at Reynolds numbers of order 1000 if three-dimensional disturbances are introduced.

Patera and Orszag (1981) have undertaken a direct simulation of a transitional flow in a two-dimensional channel using a 32-mode, three-dimensional pseudospectral method, with both natural transition and forced transition. To hasten development of the turbulent flow, the forced-transition results were obtained by introducing an initial turbulent disturbance. A typical result for the Reynolds stress at a value Re = 125 (based on a friction velocity u^*) is shown in Fig. 5.12. Good agreement is indicated with an asymptotic outer solution and with a mixing-length turbulence model.

5.6.3. Other spectral applications

Taylor and Murdock (1981) have applied a novel pseudospectral method to two-dimensional laminar flow over a flat plate to test the stability of the flow to disturbances in the inflow boundary conditions. An unsteady primitive variable formulation (u, v, p) employs a parabolized x-momentum equation to obtain the longitudinal velocity component u, the continuity equation to obtain the normal velocity component v, and a Poisson equation to obtain the pressure p. Each variable (u, v, p) is represented by a series of Chebyshev polynomials in the longitudinal (x) and normal (y) directions.

As noted previously, Morchoisne (1979) has applied a pseudospectral method in time and space to a developing shock flow governed by Burgers' equation and to two-dimensional flow in a square cavity governed by the Navier–Stokes equations in a stream-function form.

Moin and Kim (1980) discuss the difficulties of providing boundary conditions for a spectral formulation of channel flow when a Poisson equation is to be solved for the pressure. They avoid this problem by making explicit use of the continuity equation. A pseudospectral method is used with trial solutions for velocity components and pressure in terms of a Fourier series in the horizontal direction (with periodic boundary conditions), and Chebyshev polynomials in the normal direction.

Orszag (1971a) has applied a Galerkin spectral method to the inviscid convection of a "cone" and used the results to compare Galerkin methods with finite-difference methods. We will return to this problem in section 6.5.3.

McCrory and Orszag (1980) have investigated the feasibility of applying spectral methods to very distorted regions for diffusion-dominated problems. In this case a pseudospectral method, based on a mixed Fourier-cosine, Chebyshev-polynomial trial solution, was applied to the heat-conduction equation.

Haidvogel et al. (1980) have applied a pseudospectral method to the inviscid vorticity equation that models certain classes of ocean flows. They have used this as a test problem to compare spectral, finite-element, and finite-difference methods. We will return to this problem in section 6.5.4.

5.7. Closure

As with traditional Galerkin methods, the accuracy that can be achieved by spectral methods is typically sensitive to the specific choice for the trial functions. The proper choice depends on the problem and on the nature of the boundary conditions. When an appropriate choice is made, spectral methods provide a very high rate of convergence once sufficient trial functions are included to adequately represent the underlying problem. If an inappropriate choice is made, the result can be poor accuracy or even convergence to a spurious solution.

Elementary examples have demonstrated an accuracy comparable to that of traditional Galerkin methods. The requirement of solving large nonlinear problems has caused the widespread use of fast transforms and a substantial displacement of Galerkin spectral methods by pseudospectral methods. Pseudospectral methods are more efficient, but introduce aliasing errors.

For the solution of time-dependent problems spectral methods have been combined with finite-difference time-integration techniques. Due to the global nature of the trial solution, the generation of efficient implicit algorithms for spectral methods is more difficult than for spatial finite-difference or finite-element formulations.

It has been shown that, through the Gram–Schmidt orthonormalization process, spectral methods can be constructed from test and trial functions appropriate to a specific problem. Most spectral applications to date have been characterized by simple geometries and complex physics—a situation for which spectral methods achieve high accuracy. However, the work of McCrory and Orszag (1980) suggests that complex geometries can also be handled by spectral methods.

References

Bourke, W. *Mon. Weather Rev.* **100**, 683–689 (1972).

Bourke, W. *Mon. Weather Rev.* **102**, 688–701 (1974).

Bourke, W. McAvaney, B., Puri, K., and Thurling, R., *Meth. Comp. Phys.* **17**, 267–325 (1977).

Brigham, E. O. *The Fast Fourier Transform*, Prentice-Hall, Englewood Cliffs, NJ (1974).

Cooley, J. W., and Tukey, J. W. *Math. Comp.* **19**, 297–301 (1965).

Dahlquist, G., Bjorck, A., and Anderson, N. *Numerical Methods*, Prentice-Hall, Englewood Cliffs, NJ (1974).

Dorodnitsyn, A. A. *Adv. in Aero. Sci.*, Vol. 3, Pergamon Press, New York (1960).

Fleet, R. W., and Fletcher, C. A. J. "A Comparison of the Finite Element and Spectral Methods for the Dorodnitsyn Boundary Formulation", in *Fourth International Conference in Australia on Finite Element Methods*, Univ. of Melbourne, August 1982, pp. 59–63 (1982).

Fletcher, C. A. J. *AIAA J.* **13**, 1073–1078 (1975).

Fletcher, C. A. J. "Burgers' Equation: A Model for All Reasons", in *Numerical Solution of Partial Differential Equations* (ed. J. Noye), pp. 139–225, North-Holland (1982).

Fletcher, C. A. J., and Holt, M. *J. Comp. Phys.* **18**, 154–164 (1975).

Fletcher, C. A. J., and Holt, M. *J. Fluid Mech.* **74**, 561–591 (1976).

Fornberg, B., and Whitham, G. B. *Proc. Roy. Soc. A* **289**, 373–404 (1978).

Gottlieb, D., and Orszag, S. A. *Numerical Analysis of Spectral Methods: Theory and Applications*, SIAM, Philadelphia (1977).

Gottlieb, D., and Turkel, E. *Stud. Appl. Math.* **63**, 67–86 (1980).

Haidvogel, D. B., Robinson, A. R., and Schulman, E. E. *J. Comp. Phys.* **34**, 1–53 (1980).

Hamming, R. W. *Numerical Methods for Scientists and Engineers*, McGraw-Hill, New York, 2nd Edn. (1973).

Herring, J. R., Orszag, S. A., Kraichnan, R. H., and Fox, D. G. *J. Fluid Mech.* **66**, 417–444 (1974).

Holt, M. *Numerical Methods in Fluid Dynamics*, Springer-Verlag, Berlin (1977).

Isaacson, E., and Keller, H. B. *Analysis of Numerical Methods*, Wiley, New York (1966).

Lanczos, C. *Applied Analysis*, Prentice-Hall, Englewood Cliffs, NJ (1956).

Lorenz, E. N. *Tellus* **12**, 243–254 (1960).

Lubard, S., and Helliwell, W. S. *AIAA J.* **12**, 965–974 (1974).

Machenhauer, B., and Daley, R. A Baroclinic Primitive Equation Model with Spectral Representation in Three Dimensions, Rep. 4, Inst. Theor. Met., Köbenhavns Universitet, Denmark (1972).

Majda, A., McDonough, J., and Osher, S. *Math. Comp.* **32**, 1041–1081 (1978).

Majda, A., and Osher, S. *Comm. Pure Appl. Math.* **30**, 671–705 (1977).

McCrory, R. L., and Orszag, S. A. *J. Comp. Phys.* **37**, 93–112 (1980).

Metcalfe, R. W., and Riley, J. J. *Lecture Notes in Physics*, Vol. 141, pp. 279–284, Springer-Verlag, Berlin (1981).

Moin, P., and Kim, J. *J. Comp. Phys.* **35**, 381–392 (1980).

Morchoisne, Y. *La Recherche Aerospatiale* **1979-5**, 11–31 (1979).

Newell, R. W., Kidson, J. W., Vincent, D. G., and Boer, G. J. *The General Circulation of the Tropical Atmosphere*, Vol. 1, pp. 17–130, MIT Press, Cambridge, MA (1972).

Orszag, S. A. *Physics of Fluids* **12**, Supplement II, 250–257 (1969).

Orszag, S. A. *J. Atmos. Sci.* **27**, 890–895 (1970).

Orszag, S. A. *J. Fluid Mech.* **49**, 75–112 (1971a).

Orszag, S. A. *J. Fluid Mech.* **50**, 689–703 (1971b).

Orszag, S. A. *Stud. Appl. Math.* **50**, 293–327 (1971c).

Orszag, S. A. *Stud. Appl. Math.* **51**, 253–259 (1972).

Orszag, S. A. *Lecture Notes in Physics*, Vol. 59, pp. 32–51, Springer-Verlag, Berlin (1976).

Orszag, S. A. "Numerical Simulation of Turbulent Flows", in *Handbook of Turbulence* (eds. W. Frost and T. H. Moulden), pp. 281–313, Plenum Press, New York, (1977).

Orszag, S. A. *J. Comp. Phys.* **37**, 70–92 (1980).

Orszag, S. A., and Kells, L. C. *J. Fluid Mech.* **96**, 159–205 (1980).

Orszag, S. A., and Kruskal, M. D. *Physics of Fluids* **11**, 43–60 (1968).

Orszag, S. A., and Patterson, G. S. *Phys. Rev. Lett.* **28**, 76–79 (1972).

Pasciak, J. E. *Math. Comp.* **35**, 1081–1092 (1980).

Patera, A. T., and Orszag, S. A. *Lecture Notes in Physics*, Vol. 141, pp. 329–335, Springer-Verlag, Berlin (1981).

Platzman, G. W., *J. Meteorol.* **17**, 635–644 (1960).

Puri, K., *Mon. Weath. Rev.* **109**, 286–305 (1981).

Puri, K. Private communication (1982).

Schutz, C., and Gates, W. L. Global Climatic Data for Surface, 800 mb: January, Rep. R-915-ARPA, Rand Corp., Santa Monica, CA (1971).

Silberman, I. S. *J. Meteorol.* **11**, 27–34 (1954).

Taylor, T. D., and Murdock, J. W. *Comp. and Fluids* **9**, 255–263 (1981).

Tracy, R. R. Calif. Inst. Techn., Memo No. 69 (1963).

Yeung, W.-S., and Yang, R.-J. "Application of the Method of Integral Relations to the Calculation of Two-dimensional Incompressible Turbulent Boundary Layers", *J. App. Mech.* **48**, 701–706 (1981).

CHAPTER 6

Comparison of Finite-Difference, Finite-Element, and Spectral Methods

In the preceding chapters we have examined the structure and properties of the traditional Galerkin method and observed that its modern developments have gone in two radically different directions. First, finite-element methods use local, low-order polynomial trial functions to generate sparse algebraic equations in terms of meaningful nodal unknowns. Secondly the use of global, orthogonal trial functions permits spectral methods to achieve a high accuracy per degree of freedom.

Whilst the development of Galerkin methods has been continuing, considerably more effort (measured in man–years and computer–years) has been applied to the further advancement of finite-difference techniques.

In a sense, a finite-difference formulation offers a more direct approach to the numerical solution of partial differential equations than does a Galerkin formulation. One simply replaces the derivatives with finite-difference expressions and demands that the resulting algebraic equations be satisfied *exactly* at the grid points. Thus finite-difference formulations can be interpreted as collocation methods without a trial solution. However, difficulties arise in imposing boundary conditions, and low-order finite-difference formulations are often inaccurate, particularly on coarse grids.

In this chapter we propose to compare finite-element, finite-difference, and spectral methods. This will provide an opportunity to reemphasize the strengths and weaknesses of the various methods in relation to solving physically realistic problems—that is, problems that probably will be non-linear, may include severe gradients or discontinuities, and may occur in irregular domains.

We will start (section 6.1) by examining the types of problems and partial differential equations for which the various methods are best suited. In section 6.2 we consider how the various methods handle boundary conditions and complex boundary geometry. The contributions to the computational efficiency of the various methods will be discussed in section 6.3. In section 6.4 the important practical problems of ease of coding and flexibility will be examined. Some of the test cases that have been used to compare the various methods will be considered in section 6.5. In section 6.6 we summarize the main points.

6.1. Problems and Partial Differential Equations

By considering the types of problems for which finite-element and spectral methods are most effective it is possible to conclude that Galerkin methods are compatible with certain types of partial differential equation.

The majority of applications of spectral and finite-element methods have been in the spatial domain. An examination of the spectral method indicates that a small change in an unknown coefficient in the trial solution immediately affects the whole domain. In this sense the spectral method is an elliptic method and well suited to solving elliptic boundary-value problems. The finite-element method is also elliptic. However, since it is a local method, the ability of disturbances in one part of the domain to influence the whole domain is disguised. The influence comes from the coupling between the different equations.

Hyperbolic and parabolic problems have the feature that disturbances can only influence part of the domain. Thus for time-dependent, parabolic problems information can only be carried forward in time. For hyperbolic problems one can define *domains of influence and dependence*. In Fig. 6.1 a disturbance occurring at (x_0, t_0) can only influence the part of the domain bounded by OA and OB.

The parabolic property of only allowing forward-in-time influence can be achieved computationally by using one-sided finite differences. The hyperbolic domains of influence and dependence are readily simulated, for finite-

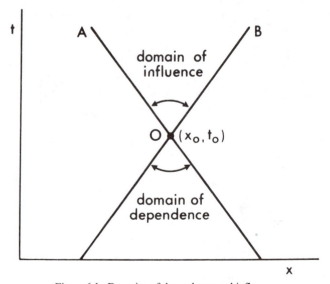

Figure 6.1. Domains of dependence and influence

difference methods, by invoking the *Courant–Friedrichs–Lewy* condition. This requires that the time step Δt be chosen so that the computational domain of dependence contains the physical domain of dependence.

To simulate one-sided differences using the finite-element method requires a special-shaped element (e.g. Fig. 3.22) or a generalized Galerkin approach (chapter 7).

The introduction of finite elements in the time domain is discussed by Zienkiewicz (1977) and by Pinder and Gray (1977). However, although it is possible to reproduce some of the simpler finite-difference schemes (section 3.3), there does not appear to be any special advantage in using finite elements in the time domain. Similarly, no particular advantage was demonstrated by Morchoisne (1979) in applying the spectral method in the time domain.

Parabolic problems do not require the time dimension to be present. Boundary-layer flow is governed by a parabolic partial differential equation with the spatial coordinate in the direction of the boundary-layer development having a timelike role. Consequently application of spectral (Fletcher and Holt, 1975) and finite-element (Fletcher and Fleet, 1982; Soliman and Baker, 1981a, b) methods have combined the Galerkin formulation across the boundary layer with finite differences or ordinary-differential-equation integrators in the timelike direction.

Hyperbolic problems can also occur without time appearing explicitly. Thus supersonic inviscid flow is governed by a hyperbolic partial differential equation. A computationally more difficult situation arises with transonic inviscid flow. A typical example is shown in Fig. 6.2. In the region bounded by the sonic line and the shock, the governing equation is hyperbolic. Outside this region the governing equation is elliptic. Finite-difference methods have been applied to this problem, very effectively, by switching the differencing formulae from centered differences in the elliptic region to one-sided differ-

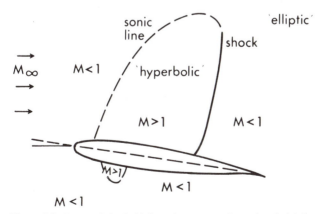

Figure 6.2. Transonic inviscid flow about a two-dimensional airfoil

ences in the hyperbolic region (Jameson, 1978). Finite-element methods have been applied to this problem (Deconinck and Hirsch, 1981), but have not, as yet, reached the level of sophistication of finite-difference formulations (Wigton and Holt, 1981).

We conclude that finite-element and spectral methods are best suited to elliptic problems or to those parts of parabolic and hyperbolic problems that permit "elliptic" behavior. Finite-difference methods are appropriate to all classes of partial differential equations.

6.2. Boundary Conditions and Complex Boundary Geometry

The finite-element method handles boundary conditions either at the problem-formulation level or at the postelement level. Neumann boundary conditions (derivative of the unknown specified) are usually dealt with at the problem-formulation level. Typically the Neumann boundary condition is incorporated into a boundary residual which cancels or permits the evaluation of a surface integral arising from an integration by parts of part of the weighted residual. Dirichlet boundary conditions (nodal unknown specified) are usually substituted after the system of algebraic equations or ordinary differential equations has been formed. No equation is required at the node at which the Dirichlet boundary condition is applied.

Through the availability of the isoparametric formulation (section 2.4), irregular boundary shapes can be handled readily. However, singularities or geometric discontinuities require special treatment. The solution in the neighborhood of the singularity can often be obtained sufficiently accurately by grid refinement. If a more accurate solution is required, a local analytic solution is recommended (section 4.3.1). However, singularities are more disruptive for boundary-element methods.

Complex boundary conditions have a restrictive effect on the Galerkin spectral method, since orthodox application of the Galerkin formulation requires that the trial functions satisfy the boundary conditions individually. However, use of the tau spectral method relaxes this requirement. Even so, the ability of the spectral method to generate solutions of high accuracy for a finite (and relatively small) number of coefficients in the trial solution is very sensitive to the precise form of the boundary conditions (Gottlieb and Orszag, 1977).

If *discontinuities* occur in the exact solution, the very high accuracy of the spectral method is typically reduced to first or second order, globally. This is the case whether the discontinuity occurs at the boundary (Gottlieb and Orszag, 1977) or in the interior (Majda et al., 1978).

In principle this problem can be overcome by a local analytic solution. Thus if the internal discontinuity is due to a shock wave for which the jump conditions are known exactly (*Rankine–Hugoniot* relations), it would be possible to solve the problem by the spectral method in two adjacent regions separated by the shock.

Most applications of spectral methods have been to problems where the boundaries coincide with coordinate lines. Thus for global atmospheric modeling, boundary conditions are applied on spherical surfaces, with the possibility of local perturbations at the earth's surface. In modeling isotropic homogeneous turbulence the boundary conditions are applied at the surface of a "box", which coincides with Cartesian coordinates.

Recently McCrory and Orszag (1980) have shown that complex boundaries can be handled by mapping the distorted physical domain into a regular computational domain. If necessary, the distorted physical domain can be split into a small number of subdomains that can be more easily transformed into regular computational domains. The compatibility conditions between the subdomains do not present any special restrictions (Orszag, 1980). However, it is not clear whether the high accuracy of the spectral method can be maintained when operating in a very distorted physical domain.

Finite-difference methods have some difficulties with boundary conditions in that dummy nodes may need to be introduced outside the computational domain, in order to evaluate derivatives at the boundary. For hyperbolic equations it is possible (Kreiss, 1973) to satisfy the boundary conditions with schemes one order lower than used in the interior without degrading the global accuracy. Often finite-difference formulations requires additional boundary conditions, beyond that required by the partial differential equation. It is mainly the difficulty of imposing boundary conditions satisfactorily that has restricted the wider use of higher-order finite-difference methods.

Traditionally it was considered necessary to apply finite-difference methods on a uniform grid. This created a considerable problem when the boundary did not coincide with a grid line. Since about 1974 (Thompson et al., 1974) there has been considerable development of techniques to permit an irregular physical domain to be mapped into a uniform computational domain. These techniques have often solved the potential equation in the computational domain to obtain the location of an orthogonal mesh in the physical domain. However, the difference equations are obtained and solved in the uniform computational domain. The effect of the distorted grid in the physical domain on the accuracy of the solution does not seem to have been established.

6.3. Computational Efficiency

A number of different ways of defining computational efficiency are discussed by Swartz (1974). We consider Swartz's preferred definition in section 6.5.2. For the moment we are only concerned with the influence of accuracy and economy on computational efficiency (CE). Therefore we will adopt the following simple definition:

$$CE = \frac{k}{\varepsilon \tau},\tag{6.3.1}$$

where ε is the error in the computed solution in some appropriate norm, and τ is the execution (CPU) time or operation count. Thus a solution on a coarse grid that has a large error but a small execution time is as computationally efficient as a solution on a refined grid that produces a small error but requires a large execution time. Since computational efficiency is a property of the method, the relative weight (powers) of ε and τ should be chosen so that the computational efficiency is independent of grid refinement.

The three categories of computational method—finite-difference, finite-element, and spectral methods—can be ranked in that order as far as *economy per degree of freedom* is concerned. This would be expected from the relative sparseness of the algebraic or ordinary differential equations formed by the methods if the same total number of degrees of freedom (unknowns) are involved.

From a consideration of the theoretical error estimates it would be expected that the *accuracy* achieved by the various methods with the same number of degrees of freedom could be ranked as follows: spectral, finite-element, and finite-difference methods.

However the relative computational efficiency of the various methods is not obvious from fundamental considerations. The efficiency will depend on the particular problem being considered, the order of the trial solution, the number of dimensions, and the influence of other aspects of the computational algorithm—for example, whether efficient time-splitting algorithms are available.

The analysis of section 3.1.2 showed that for higher-order, higher-dimensional local methods the algebraic complexity (number of contributing nodes) is much greater for finite-element than for finite-difference formulations. Consequently we would expect the economy to be significantly worse and require the accuracy to be significantly higher if the computational efficiency is to be greater.

A number of test cases will be considered in section 6.5 to refine the comparison, particularly for computational efficiency, of the various methods.

6.4. Ease of Coding and Flexibility

The development of software packages can be approximately quantified. Patterson (1978) quotes the following relationships derived from about sixty IBM software projects:

$$E = 5.2 \, L^{0.91},$$

$$D = 49 \, L^{1.01},\tag{6.4.1}$$

where

L is the number of source-code lines (in 1000s),
E is the total effort in programmer–months, and
D is the number of pages of documentation.

It is apparent that E and D are rather large numbers.

We will interpret *ease of coding* as both the programming effort (i.e. E) and the algebraic manipulation effort required to make the problem ready for programming. Once a computer program has been written to solve a specific problem, a number of changes will be necessary to solve a new problem. We will understand *flexibility* to be an inverse measure of the number of changes. The assessment of the three categories of method in relation to ease of coding and flexibility will necessarily be qualitative.

From the nature of finite-difference methods one would expect them to require relatively little algebraic manipulation and relatively straightforward programming, except that special procedures might be required for particular boundary conditions. Strictly, according to the above criterion, finite-difference methods are inflexible, since we would expect very little of the existing program to be relevant to the next problem. However, the relative ease of coding applies to the second problem just as much as to the first.

Finite-element methods require some preliminary algebraic manipulation and more programming effort than finite-difference methods. However, the modularity of the finite-element method lends itself to efficient programming, and a number of general-purpose (particularly structural) finite-element computer packages exist (NASTRAN etc.). Flexibility is an important feature of the finite-element method. In solving a new problem relatively few changes need be made in an existing computer package.

Spectral methods require substantial preliminary algebraic manipulation and programming if an efficient code (i.e. incorporating the techniques discussed in section 5.3) is to be generated. Also, the solution of a new problem typically requires a new trial solution, new boundary-condition specification, and so on—in short, a complete new program. Consequently spectral methods are difficult to code and inflexible.

However, the importance of these two criteria is a function of subsequent usage. Thus a spectral code developed to provide operational weather predictions on an ongoing basis could afford a substantially higher "ease-of-coding" cost than a code written to solve one particular problem on a limited budget in a limited time.

6.5. Test Cases

In the past, few comparisons have been made of the finite-difference, finite-element, and spectral methods. Inevitably, the more precise the comparison, the simpler the model problem. The extent to which an advantage demon-

Table 6.1. Accuracy comparison for Burgers' equation, $N = 11$, $Re = 10$

Method	Rms error, $\{\|u - u_{ex}\|_{2,d}\}/N^{1/2}$
Spectral, $t = 0.80$	0.0031
Linear finite-element, $t = 0.47$	0.0050
3-pt. finite-difference, $t = 0.81$	0.0215

strated for a linear model problem will persist for a realistic nonlinear problem with complex boundaries is difficult to predict.

Here we will review some of the more interesting test cases that are available. We will be concerned mainly with *accuracy, economy,* and *computational efficiency.* The problems considered in sections 6.5.1 and 6.5.2 are mainly parabolic; the problems considered in sections 6.5.3 and 6.5.4 are hyperbolic.

6.5.1. Burgers' equation

The propagating "shock", which is governed by Burgers' equation, was used in section 1.2.5 to demonstrate the traditional Galerkin method, in section 3.2.5 to demonstrate the finite-element method, and in section 5.2.2 to demonstrate the spectral method.

An indication of the accuracy per degree of freedom (i.e. per unknown in the trial solution) can be gained from the results presented in Table 6.1. The results are taken from sections 3.2.5 and 5.2.2. As can be seen, the spectral method is more accurate than the others. The error estimate given in section 5.1 suggests that for sufficiently smooth initial data the spectral method will have a much higher convergence rate than either of the local methods. From the data presented in section 3.4.4, both the linear finite-element method and the three-point finite-difference scheme have a second-order (N^{-2}) convergence rate.

No attempt has been made to incorporate into the spectral code (appendix 1), on which the results of section 5.2.2 were based, the efficient transform techniques described in section 5.3. Therefore no comparison of CPU time and computational efficiency will be made for the spectral method.

To obtain a comparison of the finite-element method and the finite-difference method it has been found convenient to consider a modified form of Burgers' equation,

$$\frac{\partial u}{\partial t} + 0.5\frac{\partial(u^2)}{\partial x} - \alpha\frac{\partial u}{\partial x} = -\frac{1}{Re}\frac{\partial^2 u}{\partial x^2} = 0, \qquad (6.5.1)$$

where α is a constant. Choosing $\alpha = 0.5$ "freezes" the convection of the shock. Thus the shock remains centered at $x = 0$ as $t \to \infty$. Consequently steady-state solutions can be obtained without requiring a semiinfinite spatial domain.

Table 6.2. Relative-CPU-time comparison for various schemes

Case	Relative CPU time per time step[a]	
	Explicit (2R–K)	Implicit (C–N)
Finite element:		
Linear	1.15	2.12
Quadratic	1.50	2.35
Cubic	1.65	2.62
Finite difference:		
3-point	1.00	2.27
5-point	1.31	2.65
7-point	1.62	3.27

[a] 2R–K = second-order Runge–Kutta; C–N = Crank–Nicolson.

For the various finite-element and finite-difference schemes considered in section 3.4.4, the relative CPU time per time step is shown in Table 6.2. The CPU times were obtained from a uniform grid of 121 points in the range $-1 \leq x \leq 1$ at Re = 100. The CPU times do not include the initial evaluation of the algebraic coefficients.

The time-integration schemes shown in Table 6.2 can be described as follows. After application of the finite-difference or finite-element methods to eq. (6.5.1), system of ordinary differential equations is obtained:

$$M\frac{\partial u}{\partial t} = S(u), \qquad (6.5.2)$$

where u is the vector of the nodal unknowns and $S(u)$ is the contribution from the spatial terms. For the finite-difference methods the mass matrix M is replaced by the identity matrix I. The second-order Runge–Kutta scheme can be written

$$u^* = u^n + \Delta t\, M^{-1} S(u^n),$$
$$u^{n+1} = u^n + 0.5\, \Delta t\, M^{-1}(S(u^n) + S(u^*)). \qquad (6.5.3)$$

The Crank–Nicolson scheme can be written

$$\left(M - 0.5\, \Delta t \frac{\partial S}{\partial u}\right)^n (u^{n+1} - u^n) = \Delta t\, S(u^n), \qquad (6.5.4)$$

where an element of $\partial S/\partial u$ is $\partial S_k/\partial u_j$. For the Crank–Nicolson scheme the nonlinear convective term has been linearized about the known solution at time level n. Thus the scheme is noniterative.

The results show that the second-order Runge–Kutta scheme is roughly twice as economical as the Crank–Nicolson scheme. For the finite-element methods used with the explicit scheme (2R–K), M can be factorized once and

Table 6.3. Computational-efficiency comparison, Burgers' equation, Re = 100

Case	Δt_{max}	Number of time steps	Relative CPU time	$\|u - u_{ex}\|_{rms}$
Finite element:				
Linear	0.32	32	0.95	0.94×10^{-3}
Quadratic	0.32	32	1.05	0.50×10^{-3}
Cubic	0.16	56	2.00	0.32×10^{-3}
Finite difference:				
3-point	0.32	32	1.00	0.93×10^{-3}
5-point	0.32	32	1.16	0.30×10^{-3}
7-point	0.32	32	1.42	0.32×10^{-3}

for all. For the Crank–Nicolson scheme a factorization of the left-hand-side matrix and subsequent multiplication of the right-hand side is required at all time steps. These two steps are carried out by BANFAC and BANSOL (see appendix 2), using a generalized Thomas algorithm which takes advantage of the narrower bandwidth associated with midside nodes of higher-order finite elements. Consequently higher-order implicit finite-element methods are more economical than the higher-order implicit finite-difference methods with the same maximum bandwidth.

The results shown in Table 6.2 indicate that the additional computational cost of using a higher-order scheme *in one dimension* is not very great. However, the results presented in section 3.1.2 indicated a significant increase in algebraic complexity, and hence CPU time, in using higher-order schemes in two and three dimensions, particularly for the finite-element method.

The computational efficiency of the finite-difference and finite-element methods can be compared by the considering the results shown in Table 6.3. The results were obtained on a 121-node variable grid; the mesh size varied between $\Delta x = 0.02$ at the shock and $\Delta x = 0.20$ adjacent to the boundaries. The error $\|u - u_{ex}\|_{rms}$ shown in Table 6.3 was evaluated after integrating the system of ordinary differential equations (6.5.2) from $t = 0.01$ to $t = 8.00$ with Re = 100 and a variable-time-step Crank–Nicolson scheme. Corresponding results using Runge–Kutta schemes are given by Fletcher (1982). However, it was found that explicit Runge–Kutta schemes had to be restricted to a very small time step to avoid instabilities. Consequently for the current problem explicit schemes are not competitive.

At $t = 8.00$ both the exact and the computational solutions have reached the steady state. Consequently the maximum time step can be much larger than suggested by an a priori error estimate, without degrading the accuracy of the solution. For the particular problem considered, higher-order schemes are generally more efficient than the three-point schemes, with both finite-element and finite-difference methods demonstrating comparable efficiency.

For a smoother exact solution (smaller value of Re) we would expect the cubic finite-element method and the seven-point finite-difference method to show a smaller relative (to the other schemes) error and a higher relative computational efficiency.

6.5.2. Model parabolic equations

Culham and Varga (1971) have considered a nonlinear parabolic equation in the form

$$\beta(p)\frac{\partial p}{\partial t} = \frac{\partial}{\partial x}\left(\alpha(p)\frac{\partial p}{\partial x}\right) + \lambda S(x,t), \qquad (6.5.5)$$

with initial condition

$$p(x,0) = p_0, \qquad 0 \leq x \leq L,$$

and boundary conditions

$$\frac{\partial p}{\partial x} = 0 \quad \text{at } x = 0, t > 0$$

and

$$p(L,t) = p_0(1 - \alpha_1 \gamma t e^{-\gamma t}).$$

In the above equations $\beta(p)$, $\alpha(p)$ are known functions of the pressure p, and S is a known function of x and t. Linear and cubic Galerkin finite-element spatial formulations were compared with a second-order centered finite-difference (three-point) formulation. The basis of comparison was the CPU time to achieve a predetermined accuracy. However, this depends on both the spatial representation and the temporal representation. For methods considered to be second-order accurate spatially (e.g. linear finite-element or three-point finite-difference), a 100-point uniform grid was used and the time-step adjusted until the desired error level was achieved. For methods considered to be fourth-order accurate spatially (e.g. cubic finite-element), a similar procedure was used with a 10-point uniform grid. The error was measured in the L_∞ norm, that is,

$$E(t) = \max_j |p(jh,t) - p_{\text{ex}}(jh,t)|, \qquad (6.5.6)$$

where $p(jh,t)$ is the computed solution at grid point (jh,t).

After spatial discretization of eq. (6.5.5), it can be written

$$B\frac{\partial \mathbf{P}}{\partial t} = A\mathbf{P} + \mathbf{S}, \qquad (6.5.7)$$

where $\mathbf{P}$ is the vector of unknown nodal values of p.

Culham and Varga introduced a first-order finite-difference representation for $\partial \mathbf{P}/\partial t$ and considered various temporal schemes depending on the time level n at which B, A, and S were evaluated. The following noniterative scheme was generally the most effective:

$$[B^{(n)} - \Delta t\, A^{(n)}]\mathbf{P}^{(n+1)} = B^{(n)}\mathbf{P}^{(n)} + \Delta t\, \mathbf{S}^{(n+1)}. \tag{6.5.8}$$

It was found that for error levels in the range 1 to 10%, a direct application of Galerkin methods was computationally less efficient than the three-point finite-difference method because of the repeated numerical quadratures associated with the nonlinear terms A, B, and S. However by interpolating A, B, and S in a manner similar to the term f in section 3.2.1, and evaluating the integrals once and for all, the Galerkin formulation was more competitive. For a 1% error level, a Hermitian cubic Galerkin formulation was slightly more efficient than the three-point finite-difference scheme.

We note, in passing, that the errors considered for this problem are of about the same order as for Burgers' equation, but the solution is considerably smoother.

Hopkins and Wait (1978) undertook an investigation of similar scope for a number of model parabolic equations of the form

$$\frac{\partial u}{\partial t} = \frac{\partial^2 u}{\partial x^2} + f(u). \tag{6.5.9}$$

Typical initial conditions, for $f(u) = e^{-u} + e^{-2u}$, were

$$u(x, 0) = \ln(x + 2) \tag{6.5.10}$$

and boundary conditions

$$u(0, t) = \ln(t + 2), \qquad u(1, t) = \ln(t + 3). \tag{6.5.11}$$

Various Lagrange Galerkin formulations up to cubic were compared with collocation and finite-difference formulations. The time-integration scheme was comparable to eq. (6.5.8) except that terms like B and A were evaluated at time level $n + 1$, which necessitated an iteration at each time level.

A three-point finite-difference method was found to be the most efficient method. Galerkin formulations were about two times slower for a given error level. However, this poor performance is attributed by Hopkins and Wait to the repeated numerical reevaluation of the source term f. The strategy of Culham and Varga (1971) would have improved the situation.

Swartz (1974) has considered the heat-conduction equation

$$\frac{\partial u}{\partial t} = \frac{\partial^2 u}{\partial x^2} \tag{6.5.12}$$

with initial conditions

$$u(x, 0) = \exp(i\, 2\pi w x) \tag{6.5.13}$$

and boundary conditions

$$u(0, t) = u(1, t) \quad \text{and} \quad \frac{\partial u}{\partial x}(0, t) = \frac{\partial u}{\partial x}(1, t). \tag{6.5.14}$$

Swartz evaluated the computational efficiency of various schemes by assuming that the truncation error e can be written

$$e = C_p(\Delta x)^p + C_q(\Delta t)^q \tag{6.5.15}$$

and the computation work w can be written

$$w = \frac{C_w}{\Delta x \, \Delta t}. \tag{6.5.16}$$

For various spatial and temporal schemes, Swartz chose Δx and Δt so that w was a minimum for a given e, and then evaluated the corresponding minimum work, $w_{\min}$. This process was undertaken for spline–Galerkin formulations, explicit centered-difference formulations, and implicit finite-difference formulations. All three groups were treated as being of general order m. The implicit finite-difference formulations were conceptually similar to the compact implicit method of Adam (1977) (section 3.3).

Various time-integration schemes of finite-difference form were considered. For a relative error $|E| = 10^{-2}$, the trapezoidal (Crank–Nicolson) time-integration scheme was as accurate as any. Because of the smoothness of the exact solution the optimum order was rather high, typically $m = 4$–8. The spline–Galerkin formulation gave a lower minimum-work estimate than the best explicit centered-difference scheme, but a higher minimum-work estimate than the best implicit finite-difference scheme.

Popinski and Baker (1976) and Soliman and Baker (1981a, b) have investigated the nonlinear parabolic equations governing two-dimensional laminar and turbulent boundary-layer flow. Linear and quadratic Galerkin formulations across the boundary layer were combined with an implicit finite-difference marching scheme in the flow direction. It was found that the accuracy of the finite-element solutions compared favorably with that of a second-order finite-difference scheme for a range of step sizes in each direction.

Swartz and Wendroff (1974) have applied the Swartz method of minimizing eq. (6.5.16) for a given e to the model hyperbolic problem

$$\frac{\partial u}{\partial t} = c \frac{\partial u}{\partial x} \tag{6.5.17}$$

with the initial data given by eq. (6.5.13). As with the parabolic problem (6.5.12), spline–Galerkin methods were found to be reasonably competitive with higher-order finite-difference schemes.

Orszag and Israeli (1974) report the application of a Chebyshev Galerkin spectral formulation to eq. (6.5.17) with $c = -1$ and initial and boundary conditions

$$u(x, 0) = 0, \qquad u(-1, t) = \sin w\pi t, \qquad (6.5.18)$$

where w is the number of full wavelengths in the domain $-1 \leq x \leq 1$. The accuracy of the solution at $t = 5$ was compared with that of second-order and fourth-order finite-difference schemes. To reduce the L_2 error to about 1%, the second-order finite-difference scheme required approximately 40 degrees of freedom per wavelength, the fourth-order scheme about 15, and the spectral method between 3 and 4. According to data presented by Swartz and Wendroff (1974) a twelfth-order finite-difference scheme would be required to achieve comparable accuracy for the spectral results. However, no comparison of computational efficiency was presented by Orszag and Israeli.

6.5.3. Passive scalar convection

Orszag (1971) has applied a Galerkin spectral formulation, with a Fourier-series trial solution, to the equations governing the inviscid convection of a passive scalar. In two-dimensional Cartesian coordinates, these equations are

$$\frac{\partial u}{\partial x} + \frac{\partial v}{\partial y} = 0 \qquad (6.5.19)$$

and

$$\frac{\partial A}{\partial t} + u\frac{\partial A}{\partial x} + v\frac{\partial A}{\partial y} = 0 \qquad (6.5.20)$$

with initial conditions

$$A(\mathbf{x}, 0) = \begin{cases} 1 - z/r & \text{when } z \leq r, \\ 0 & \text{when } z > r, \end{cases} \qquad (6.5.21)$$

where $z^2 = (x - x_0)^2 + y^2$. In a three-dimensional plot (Fig. 6.5), A appears as a cone. At $t = 0$ the cone is centered at $(x_0, 0)$ with base radius r. The velocity components u and v are assumed known and given by

$$u = \frac{\partial \psi}{\partial y}, \qquad v = \frac{-\partial \psi}{\partial x}, \quad \text{and} \quad \psi = -0.5\Omega(x^2 + y^2). \qquad (6.5.22)$$

Eqs. (6.5.19) and (6.5.20) constitute a hyperbolic system, and the exact solution is the rotation of the "cone" (eq. (6.5.21)) about the origin on a circular path with no distortion of its shape. The distortion of the shape of the cone is an easily observable qualitative measure of the accuracy of a particular computational scheme.

Orszag has applied a Galerkin spectral method to this problem. Periodic boundary conditions are applied on $x = \pm 1$, $y = \pm 1$ for $u, v,$ and A. Complex Fourier trial solutions are introduced for A and ψ, and the transform method (section 5.3.2) is used to make the method efficient.

Results obtained with the spectral method have been compared with those

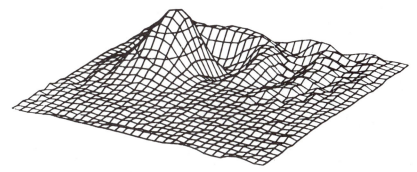

Figure 6.3. Convection of a passive scalar: second-order finite-difference method, 32 × 32 grid (after Orszag, 1971; reprinted with permission of Cambridge University Press)

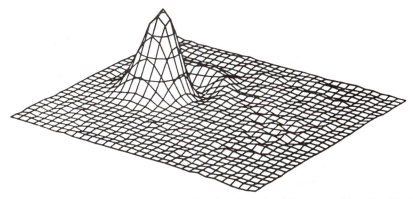

Figure 6.4. Convection of a passive scalar: fourth-order finite-difference method, 32 × 32 grid (after Orszag, 1971; reprinted with permission of Cambridge University Press)

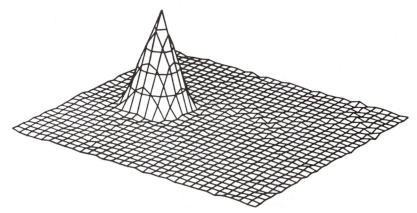

Figure 6.5. Convection of a passive scalar: Galerkin spectral method, 32 × 32 degrees of freedom (after Orszag, 1971; reprinted with permission of Cambridge University Press)

of second-order and fourth-order Arakawa finite-difference schemes. The Arakawa schemes have the important property of conserving

$$\sum_{i,j} (A_{i,j}^n)^2$$

in the limit $\Delta t \to 0$. The above expression represents a summation over the spatial grid at time level n. As indicated in section 3.3, Jespersen (1974) showed that the second-order Arakawa scheme is a linear finite-element formulation of the spatial terms in eq. (6.5.20) on rectangular elements. The ordinary differential equations that result from introducing the spectral or Arakawa schemes have been integrated using a leapfrog scheme,

$$A_{i,j}^{n+1} = A_{i,j}^{n-1} - 2\Delta t \left\{ u \frac{\partial A}{\partial x} + v \frac{\partial A}{\partial y} \right\}_{\text{fd}}^{n}. \tag{6.5.23}$$

Typical results for the second-order Arakawa scheme are shown in Fig. 6.3. These results were obtained on a 32×32 grid and show the distribution of A after one counterclockwise rotation of the "cone". The cone base radius is $r = 0.25$, and $x_0 = -0.5$. It is clear that the shape of the cone is no longer sharply defined, and an unrealistic wake is developing. Corresponding results for the fourth-order Arakawa scheme, also on a 32×32 grid, are shown in Fig. 6.4. It is apparent that after one revolution the shape of the cone is better defined and the wake disturbances are smaller. A Galerkin spectral solution has been obtained with a wave-number cutoff of 16 in each direction, and the result is shown in Fig. 6.5. In contrast to the results produced by the finite-difference schemes, the shape of the cone is sharply defined and no discernible wake is present.

Orszag attributes the errors in the finite-difference solutions to *dispersion*. That is, the representation of the first-derivative terms in eq. (6.5.20) introduces a phase error. Orszag notes that to achieve equal accuracy, the second-order method requires four times as many degrees of freedom per spatial direction and the fourth-order method requires twice as many degrees of freedom per spatial direction as the spectral method. The relative accuracies and the CPU times per time step are shown in Table 6.4. For this problem it is clear that the spectral method is computationally more efficient than either of the Arakawa schemes.

Table 6.4. CPU times and accuracy for scalar convection problem

Method	CPU time (CDC 6600) per time step (sec)	Approximate relative accuracy
Second-order Arakawa f.d.	0.08	1
Fourth-order Arakawa f.d.	0.13	4
Galerkin spectral	0.30	16

Gresho et al. (1978) have applied the Galerkin finite method to the passive-scalar-convection problem. They found that linear elements on a 30 × 30 grid produced results that appeared superior to the fourth-order Arakawa results (Fig. 6.4). However, if the mass matrix was *lumped*, the results coincided with the second-order Arakawa results (Fig. 6.3), as expected. The role of the finite-element mass matrix in reducing dispersion has also been demonstrated by Baker and Soliman (1979).

6.5.4. Open ocean modeling

Haidvogel et al. (1980) have considered a number of test problems that are related to limited-area ocean forecasting, coastal modeling, current-system simulation, and so on. The underlying physical mechanisms are substantially hyperbolic. That is, problems are characterized by the propagation of waves with little or no damping.

As a representative system of equations, Haidvogel et al. consider the following:

$$\frac{\partial \zeta}{\partial t} - \varepsilon \left\{ \frac{\partial \psi}{\partial y} \frac{\partial \zeta}{\partial x} - \frac{\partial \psi}{\partial x} \frac{\partial \zeta}{\partial y} \right\} + \frac{\partial \psi}{\partial x} = F(x, y) \tag{6.5.24}$$

and

$$\frac{\partial^2 \psi}{\partial x^2} + \frac{\partial^2 \psi}{\partial y^2} = \zeta, \tag{6.5.25}$$

where ε is the Rossby number, ζ is the vorticity, and ψ is the stream function. The problem is to find $\zeta(x, y, t)$ and $\psi(x, y, t)$ that satisfy eqs. (6.5.24) and (6.5.25). ε is a parameter that controls the degree of nonlinearity in the system. Boundary conditions are provided by specifying the stream function ψ on all boundaries, and the vorticity ζ only on inflow boundaries.

Amongst other problems, Haidvogel et al. consider the propagation of Rossby waves. On an infinite domain the solution of eqs. (6.5.24) and (6.5.25) can be written

$$\psi_{\text{ex}}(x, y, t) = -\gamma y + \sin(kx + ly + wt), \tag{6.5.26}$$

where

$$w = k(1 - \varepsilon \gamma)$$

and

$$k^2 + l^2 = 1.$$

A value $\gamma = 0$ corresponds to no wave propagation.

Haidvogel et al. considered three computational methods: a finite-difference formulation (FDF), a finite-element formulation (FEF), and a pseudospectral formulation (PSF). With each method either a leapfrog or an Adams–Bashforth time-differencing scheme (Gear, 1971) was used.

The FDF was a second-order centered finite-difference representation that used the second-order scheme for the terms

$$\frac{\partial \psi}{\partial y} \frac{\partial \zeta}{\partial x} - \frac{\partial \psi}{\partial x} \frac{\partial \zeta}{\partial y}$$

in eq. (6.5.24). Thus eq. (6.5.24) is treated in the same way that Orszag (1971) treated the convecting scalar equation (6.5.20). For the FDF, eq. (6.5.25) is solved by a cyclic-reduction direct solution technique.

The FEF used bilinear trial functions for ψ and ζ on rectangular elements. This generates the same algebraic coefficients for eq. (6.5.24) as the second-order Arakawa finite-difference scheme, except that a mass matrix multiplies the time derivative $\partial \zeta / \partial t$. By analogy with the one-dimensional convection equation $\partial \zeta / \partial t + u \, \partial \zeta / \partial x = 0$, which is known to be spatially fourth-order accurate (Fix, 1975), Haidvogel et al. assume that the bilinear FEF of eq. (6.5.24) is also fourth-order accurate.

Consequently a fourth-order solution of eq. (6.5.25) is sought by the method of deferred corrections. First a cyclic reduction direct solution ψ_1 is obtained by solving

$$\nabla_{\mathrm{fd}}^2 \psi_1 = h^2 \zeta_1, \tag{6.5.27}$$

where ∇_{fd}^2 is the five-point finite-difference representation of eq. (6.5.25), h is the mesh size, and ζ_1 is the solution obtained from eq. (6.5.24). Then a fourth-order accurate solution is obtained by solving

$$\nabla_{\mathrm{fd}}^2 \psi_2 = h^2 \zeta_1 + \frac{h^4}{12} \left\{ \nabla^2 \zeta_1 - 2 \frac{\partial^4 \psi_1}{\partial x^2 \, \partial y^2} \right\}. \tag{6.5.28}$$

The PSF is introduced by providing Chebyshev trial solutions for ψ, ζ, and R. Here R is the collective evaluation of all the spatial terms in eq. (6.5.24) at the current time level. Terms like $(\partial \psi / \partial y)(\partial \zeta / \partial x)$ are evaluated in the physical plane using the transform technique (section 5.3.2).

For the Rossby-wave problem (6.5.26), an alternating-direction implicit treatment of the vorticity boundary conditions is necessary. This essentially relaxes the stability time-step limitation that arises from the closely spaced Chebyshev sample points adjacent to the boundary (section 5.1). Application of the PSF to eq. (6.5.25) produces a nonsparse matrix equation, but this can be readily diagonalized and efficiently solved (Haidvogel et al., 1980).

Haidvogel et al. present results for $0 < \varepsilon < 0.8$, $N = 17, 33, 43$, $v = 16/3.5$, $32/3.5$, and $\eta = 64, 128$, where N is the number of spatial degrees of freedom in each direction, v is the number of spatial degrees of freedom per half wavelength, and η is the number of time steps per period.

Typically, solutions were integrated for five periods and the normalized rms errors in the stream function and vorticity compared for the FDF, FEF, and PSF. Typical rms errors were in the range 0.002 to 0.2.

Haidvogel et al. found that all three techniques were effective if ε was not too large. The FEF and PSF were both more accurate and more efficient than the FDF. However, for $\varepsilon \neq 0$ it was found that for the PSF, a *catastrophic instability* would develop at less than five periods unless a periodic (dissipative) smoothing was undertaken. The smoothing did not affect the accuracy adversely.

A general conclusion of Haidvogel et al. was that, for comparable execution times, the FEF ($N = 33$) was 4 times more accurate and the PSF ($N = 17$) was 15 times more accurate than the FDF ($N = 43$). However, a higher-order FDF or FEF would have improved the accuracy (and efficiency) comparison.

6.6. Closure

In this chapter we have endeavored to view the finite-difference, finite-element, and spectral methods in a perspective sufficiently broad so as to be able to assess the relative strengths and weaknesses of the various methods. Some of the more important points are summarized in Table 6.5.

An important feature of Galerkin finite-element and spectral methods is that they conserve physically meaningful global properties such as energy and mass. For fluid-flow problems where shocks are expected to occur, satisfaction of the conservation laws is necessary to ensure that the correct solution at the shock is obtained without explicitly imposing the local shock solution (Morton, 1977).

For the hyperbolic systems of equations that characterize meteorological flows, the length of time over which solutions can be obtained is often limited by the onset of *nonlinear instabilities*. It is found that computational schemes which conserve energy and enstrophy (e.g. the Galerkin schemes or the Arakawa finite-difference schemes) significantly extend the length of time over which stable (and accurate) solutions can be obtained.

Table 6.5. Comparison of the finite-difference, finite-element, and spectral methods

Method / Attribute	Finite difference	Finite element	Spectral
Trial solution	Local	Local	Global
Ease of coding	Very good	Good	Fair
Flexibility	Good	Very good	Fair
Accuracy per unknown	Fair	Good	Very good
Computational efficiency	Good	Good	Very good
Main strength(s)	Economy Ease of coding	Flexibility	High accuracy
Main weakness	Extension to higher order (b.c.'s)	Economy	Inflexibility

Some of the features listed in Table 6.5 are well established, but others are more contentious. There seems little doubt that per unknown (degree of freedom), the accuracy of the methods can be ranked as shown in Table 6.5. However, the computational efficiency depends on the economy of the various methods, and it is more difficult to be dogmatic about ranking the methods in that regard. Certainly cases have been presented where the spectral method is more efficient than either the finite-difference or the finite-element method. The relative efficiency of the finite-difference and finite-element methods is probably problem dependent, although the use of higher-order finite elements in two or three dimensions appears rather uneconomical (section 3.1.2).

The material and references provided in this chapter should help the reader to make up his or her own mind about the relative strengths and weaknesses of the methods considered. However, the more extreme value judgements expressed in Table 6.5 should be treated as nothing more than an expression of the author's prejudice.

References

Adam, Y. *J. Comp. Phys.* **24**, 10–22 (1977).

Baker, A. J., and Soliman, M. O. *J. Comp. Phys.* **32**, 289–324 (1979).

Culham, W. E., and Varga, R. S. *Soc. Pet. Eng. J.* **11**, 374–388 (1971).

Deconinck, H., and Hirsch, C. *Lecture Notes in Physics*, Vol. 141, pp. 138–143, Springer-Verlag, Berlin (1981).

Fix, G. J. *SIAM J. App. Math.* **29**, 371–387 (1975).

Fletcher, C. A. J. "A Comparison of the Finite Element and Finite Difference Methods for Computational Fluid Dynamics", in *Finite Element Flow Analysis* (ed. T. Kawai), pp. 1003–1010, Univ. of Tokyo Press (1982).

Fletcher, C. A. J., and Fleet, R. W. "A Dorodnitsyn Finite Element Boundary Layer Formulation", *8th International Conference on Numerical Methods in Fluid Dynamics*, Aachen, 1982.

Fletcher, C. A. J., and Holt, M. *J. Comp. Phys.* **18**, 154–164 (1975).

Gear, C. W. *Numerical Initial Value Problems in Ordinary Differential Equations*, Prentice-Hall, Englewood Cliffs, NJ (1971).

Gottlieb, D., and Orszag, S. A. *Numerical Analysis of Spectral Methods: Theory and Applications*, SIAM, Philadelphia (1977).

Gresho, P. M., Lee, R. L., and Sani, R. L. In *Finite Elements in Fluids*, Vol. 3, pp. 335–350, Wiley, London (1978).

Haidvogel, D. B., Robinson, A. R., and Schulman, E. E. *J. Comp. Phys.* **34**, 1–53 (1980).

Hopkins, T. R., and Wait, R. *Int. J. Num. Meth. Eng.* **12**, 1081–1107 (1978).

Jameson, A. In *Numerical Methods in Fluid Dynamics* (eds. H. J. Wirz and J. J. Smolderen), pp. 1–87, Hemisphere, Washington (1978).

Jespersen, D. C. *J. Comp. Phys.* **16**, 383–390 (1974).

Kreiss, H. O. *Lecture Notes in Mathematics*, Vol. 363, pp. 64–74, Springer-Verlag, Berlin (1973).

Majda, A., Donoghue, J., and Osher, S. *Math. Comp.* **32**, 1041–1081 (1978).

McCrory, R. L., and Orszag, S. A. *J. Comp. Phys.* **37**, 93–112 (1980).

Morchoisne, Y. *La Recherche Aerospatiale* **1979-5**, 11–31 (1979).

Morton, K. W. In *The State of the Art in Numerical Analysis* (ed. D. Jacobs), pp. 699–756, Academic Press, London (1977).

Orszag, S. A. *J. Fluid Mech.* **49**, 75–112 (1971).

Orszag, S. A. *J. Comp. Phys.* **37**, 70–92 (1980).

Orszag, S. A., and Israeli, M. *Ann. Rev. Fluid Mech.* **6**, 281–318 (1974).

Patterson, G. S. *Ann. Rev. Fluid Mech.* **10**, 289–300 (1978).

Pinder, G. F., and Gray, W. G. *Finite Element Simulation in Surface and Subsurface Hydrology*, Academic Press, New York (1977).

Popinski, Z., and Baker, A. J. *J. Comp. Phys.* **21**, 55–84 (1976).

Soliman, M. O., and Baker, A. J. *Comp. Fluids* **9**, 43–62 (1981a).

Soliman, M. O., and Baker, A. J. *Comp. Meth. App. Mech. Eng.* **28**, 81–102 (1981b).

Swartz, B. In *Mathematical Aspects of Finite Elements in Partial Differential Equations* (ed. C. de Boor), pp. 279–312 Academic Press, New York (1974).

Swartz, B., and Wendroff, B. *SIAM J. Num. Anal.* **11**, 979–993 (1974).

Thompson, J. F., Thames, F. C. and Mastin, C. W. *J. Comp. Phys.* **15**, 299–319 (1974).

Wigton, L. B., and Holt, M. Viscous–Inviscid Interaction in Transonic Flow, *AIAA 5th Computational Fluid Dynamics Conference*, Palo Alto, paper 81–1003 (1981).

Zienkiewicz, O. C. *The Finite Element Method*, 3rd Ed., McGraw-Hill, London (1977).

CHAPTER 7

Generalized Galerkin Methods

When evaluated on a uniform grid, the Galerkin method produces *symmetric* formulae. Thus odd-order derivatives lead to zero coefficients associated with the node at which the function is centered. The steady, two-dimensional convection–diffusion equation is

$$u\frac{\partial T}{\partial x} + v\frac{\partial T}{\partial y} - \mathcal{D}\left(\frac{\partial^2 T}{\partial x^2} + \frac{\partial^2 T}{\partial y^2}\right) = 0,$$

where T is the property being convected and u and v are the velocity components. If the algebraic equations associated with each node are put into a matrix form, the entries on the diagonal only have contributions coming from the diffusive terms, $\mathcal{D}(\partial^2 T/\partial x^2 + \partial^2 T/\partial y^2)$. If the problem is dominated by convection (i.e., $\mathcal{D}$ is small), then the matrix is not diagonally dominant and the solution is often oscillatory, depending on the boundary conditions.

In this chapter we examine techniques for introducing asymmetric test functions and their ability to generate physically more realistic solutions. In section 7.1 we analyse the steady one-dimensional convection–diffusion equation. In section 7.2 the *Petrov–Galerkin* method is introduced, and we consider different strategies for choosing the test function. Section 7.3 is mainly concerned with steady two-dimensional convection–diffusion problems and the phenomenon of crosswind diffusion. In sections 7.4 and 7.5 we extend the Petrov–Galerkin formulation to time-dependent problems: parabolic equations in section 7.4 and hyperbolic equations in section 7.5.

7.1. A Motivation

Many problems involving fluid flow embody the processes of convection and diffusion. It has already been seen that Burgers' equation is an appropriate one-dimensional model when the dependent variable is velocity-like. If the dependent variable is a passive scalar such as temperature, and the problem is steady, an even simpler model equation is available:

$$u\frac{dT}{dx} - \mathcal{D}\frac{d^2 T}{dx^2} = 0. \tag{7.1.1}$$

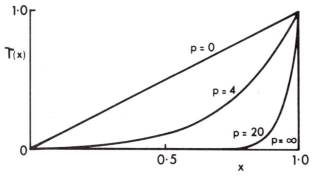

Figure 7.1. Exact solution of eqs. (7.1.2) and (7.1.3)

Here u is the convecting velocity and is assumed constant; $\mathscr{D}$ is the diffusivity. Combining u and $\mathscr{D}$ into a single parameter p gives

$$L(T) = p\frac{dT}{dx} - \frac{d^2T}{dx^2} = 0. \tag{7.1.2}$$

For the boundary conditions

$$T(0) = 0, \qquad T(1) = 1, \tag{7.1.3}$$

eq. (7.1.2) has the exact solution

$$T = \frac{e^{px} - 1}{e^p - 1}. \tag{7.1.4}$$

The exact solutions for various values of p are shown in Fig. 7.1. It can be seen that for large values of p (i.e. when the convection is much stronger than the diffusion), the solution is almost unchanged until close to the downstream boundary, where a rapid change occurs in order to satisfy the downstream boundary condition (at $x = 1$). Thus for large p a *boundary layer* occurs adjacent to $x = 1$.

Application of a Galerkin finite-element method to eq. (7.1.2), with linear trial functions on a uniform grid, produces the following algebraic equation at internal nodes:

$$0.5p\,\Delta x\,(\overline{T}_{k+1} - \overline{T}_{k-1}) - (\overline{T}_{k-1} - 2\overline{T}_k + \overline{T}_{k+1}) = 0 \quad \text{for } k = 1, \ldots, N-1, \tag{7.1.5}$$

with

$$\overline{T}_0 = 0 \quad \text{and} \quad \overline{T}_N = 1. \tag{7.1.6}$$

Equation (7.1.5) has the exact solution

$$T_k = A_0 + B_0 \left[\frac{1+\beta}{1-\beta}\right]^k, \tag{7.1.7}$$

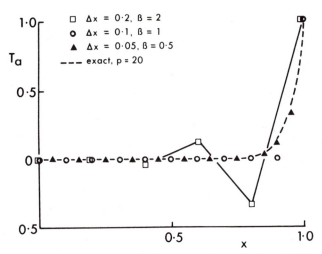

Figure 7.2. Conventional Galerkin solution of eqs. (7.1.2) and (7.1.3)

where A_0 and B_0 are determined to satisfy eq. (7.1.6) and $\beta = 0.5p\,\Delta x$. Solutions for $p = 20$ and $\Delta x = 0.05, 0.1$, and 0.2 are shown in Fig. 7.2. The solution on the finest grid, corresponding to $\beta = 0.5$, is in reasonable agreement with the exact solution. The solution corresponding to $\beta = 1$ is accurate except in the boundary layer. The solution corresponding to $\beta = 2$ is not only inaccurate but also oscillatory.

An examination of eq. (7.1.7) indicates that the requirement for nonoscillatory solution is

$$\beta \leq 1. \tag{7.1.8}$$

For a given value of p, this implies Δx must be restricted in size to satisfy eq. (7.1.8). The same equation, (7.1.5), arises in a three-point finite-difference formulation of eq. (7.1.2). Then the restriction on nonoscillatory solutions, eq. (7.1.8), is often referred to as a *cell Reynolds number* limitation (Roache, 1972). Thus if $p\,\Delta x$ is interpreted as a cell Reynolds number, eq. (7.1.8) is written

$$p\,\Delta x \leq 2. \tag{7.1.9}$$

A popular technique to avoid oscillatory solutions in finite-difference formulations of eq. (7.1.2), is to represent dT/dx with the backward-difference formula, $\{T_k - T_{k-1}\}/\Delta x$. This is referred to as an *upwinding* formulation, since only information upwind of node k is transmitted to node k by convection. An upwind solution for $\Delta x = 0.2$ and $p = 20$ is shown in Fig. 7.3. Although the upwind solution is free of oscillations, it is also inaccurate.

It is instructive to write the upwind convective formula in the following way:

$$\frac{p}{\Delta x}\{\bar{T}_k - \bar{T}_{k-1}\} = \frac{p}{2\,\Delta x}\{\bar{T}_{k+1} - \bar{T}_{k-1}\} - \frac{p}{2\,\Delta x}\{\bar{T}_{k-1} - 2\bar{T}_k + \bar{T}_{k+1}\}. \tag{7.1.10}$$

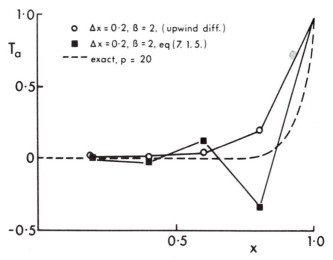

Figure 7.3. Comparison of Galerkin and upwind finite-difference solutions of eqs. (7.1.2) and (7.1.3)

Thus the upwind convective formula is equivalent to the Galerkin convective formula, as in eq. (7.1.5), with an additional *diffusive* term. Consequently, using the upwind convective formula is equivalent to solving eq. (7.1.2) with an effective diffusion coefficient of $1 + \beta$.

Therefore the challenge to the finite-element method is to find a way to modify the Galerkin method so that oscillatory solutions are avoided if possible, but without incurring the inaccuracies associated with the upwinding scheme.

It may be recalled that the conventional Galerkin finite-element formulation for eq. (7.1.2) proceeds by introducing a trial solution,

$$T_a = \sum_j \overline{T}_j \phi_j(x) \qquad (7.1.11)$$

and by evaluating the inner product

$$(L(T_a), \psi_k(x)) = 0, \qquad k = 1, \ldots, N, \qquad (7.1.12)$$

with $\psi_k(x) = \phi_k(x)$. For linear trial functions ϕ_j on a uniform grid, this generates the algebraic equations (7.1.5). The symmetric nature of the convective formula, $\beta\{\overline{T}_{k+1} - \overline{T}_{k-1}\}$, arises from evaluating (ϕ_j, ϕ_k) over two neighboring elements.

To obtain more control over the coefficients associated with $\overline{T}_{k-1}$, $\overline{T}_k$, and $\overline{T}_{k+1}$ in the algebraic equation, it is convenient to modify the test function ψ_k as follows. Let

$$\psi_k\{x\} = \phi_k(x) + \alpha\gamma_k(x), \qquad (7.1.13)$$

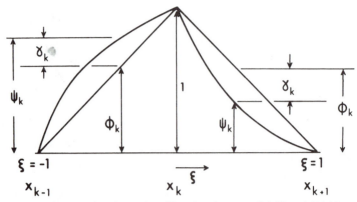

Figure 7.4. Test function and auxiliary function, eqs. (7.1.13) and (7.1.14)

where

$$\gamma_k(\xi) = \begin{cases} -3\xi(1 - \xi) & \text{in element } [k, k + 1], \\ 3\xi(1 + \xi) & \text{in element } [k - 1, k], \end{cases} \qquad (7.1.14)$$

and ξ is a local coordinate (Fig. 7.4). In eq. (7.1.13) α is a parameter that controls the influence of the antisymmetric function γ_k. The functions ψ_k, ϕ_k, and γ_k are shown schematically in Fig. (7.4). As α increases from zero the contribution from element $[k - 1, k]$ increases and the contribution from element $[k, k + 1]$ decreases. Because γ_k is antisymmetric about node k, no contribution arises from the diffusive term $d^2 T/dx^2$.

Evaluation of the modified Galerkin formulation (7.1.12), (7.1.13) on a uniform grid gives the following algebraic equations in place of eq. (7.1.5):

$$\beta\{\overline{T}_{k+1} - \overline{T}_{k-1}\} - \alpha\beta\{\overline{T}_{k-1} - 2\overline{T}_k + \overline{T}_{k+1}\} - \{\overline{T}_{k-1} - 2\overline{T}_k + \overline{T}_{k+1}\} = 0,$$

$$k = 1 \dots N - 1. \qquad (7.1.15)$$

Like the use of the upwind convective formula (7.1.10), the additional function γ_k has introduced an additional diffusive term. However, through the use of the parameter α, this additional diffusion can be controlled. It follows directly that $\alpha = 0$ gives the conventional Galerkin result (7.1.5), and $\alpha = 1$ gives the upwinded form (7.1.10).

Eq. (7.1.15) has an exact solution (Christie et al., 1976):

$$T_k = A_0 + B_0 \left[\frac{1 + \beta(\alpha + 1)}{1 + \beta(\alpha - 1)}\right]^k, \qquad (7.1.16)$$

where A_0 and B_0 are chosen to satisfy the boundary conditions at $x = 0$ and 1. It is clear that the solution (7.1.16) is free of oscillations if

$$\alpha \geq 1 \quad \text{or} \quad -\infty < \alpha < 1, \ \beta < \frac{1}{1 - \alpha}. \qquad (7.1.17)$$

It turns out (Christie et al., 1976) that an optimum value of α can be chosen by equating the solution of the exact equation, (7.1.2), with the solution of the algebraic equation (7.1.15). The *optimum* value of α is

$$\alpha_{opt} = \coth \beta - \frac{1}{\beta}. \tag{7.1.18}$$

For this particular choice, the solution (7.1.16) of the algebraic equation (7.1.15) is also the solution of the differential equation (7.1.2). For the more general two- or three-dimensional problem it is unlikely that the solution of the algebraic equation will also be the exact solution. On the other hand, it is likely that an appropriate choice of α will generate a more accurate solution than the conventional Galerkin solution. The difficulty is in devising a strategy for choosing α.

For large values of β eq. (7.1.18) has the asymptotic form

$$\alpha_{opt} = 1 - \frac{1}{\beta}. \tag{7.1.19}$$

Substituting this expression into eq. (7.1.15) causes the coefficient multiplying $\bar{T}_{k+1}$ to be zero. Consequently the downstream boundary condition in a Dirichlet form, eq. (7.1.3), then has no influence on the upstream solution. The solution appears as the $\beta = 1$ case in Fig. 7.2 for any value of β. It may also be noted that α given by eq. (7.1.19) coincides with the boundary between oscillatory and nonoscillatory solutions. In the limit of $\beta \to \infty$ (i.e. a problem with no diffusion), the use of eq. (7.1.19) correctly excludes the downstream boundary condition and gives a fully upwinded formulation.

It may be noted that the form of the algebraic equation (7.1.15) can also be obtained by assuming that γ_k in eq. (7.1.14) is given by three contiguous linear functions in elements $[k-1, k+1]$. Also eq. (7.1.15) may be obtained by a specially chosen quadrature point (Hughes, 1978).

7.2. Theoretical Background

Here we treat the modified Galerkin formulation (7.1.12), (7.1.13) as an example of a *Petrov–Galerkin* formulation. In addition, we examine the various approaches that have been made to choosing the form of the test function in eq. (7.2.5) to optimize the resulting algebraic equation.

7.2.1. Petrov–Galerkin formulation

The Galerkin method is an example of a projection method in the sense that the Galerkin solution is obtained by making an orthogonal projection of the exact solution u into the subspace of trial functions $\boldsymbol{\phi}_a \equiv \{\phi_1, \phi_2, \ldots, \phi_N\}$.

A solution of the linear differential equation

$$L(u) = f \tag{7.2.1}$$

is sought by the projection method. Here $u \in B_1$ and $f \in B_2$, where B_1, B_2 are Banach spaces. A trial solution in the N-dimensional subspace, $\phi_a \subset B_1$, is introduced by

$$u_a = \sum_{j=1}^{N} a_j \phi_j. \tag{7.2.2}$$

A second N-dimensional subspace, $\psi_a \subset B_2$ is introduced, where $\psi_a \equiv \{\psi_1, \psi_2, \ldots, \psi_N\}$. If Pa is the projection operator from space B_2 to subspace ψ_a, the projection method can be written

$$Pa(L(u_a)) = Pa(f). \tag{7.2.3}$$

For the Petrov–Galerkin method we have $B_1 = B_2 = H$, a real Hilbert space, and the projection operator is

$$Pa(v) = (v, \psi_k) = \int_R v \psi_k \, d\mathbf{x} \tag{7.2.4}$$

so that eq. (7.2.3) gives the following system of equations for the unknown coefficients a_j:

$$\sum_{j=1}^{N} (\phi_j, \psi_k) a_j = (f, \psi_k), \qquad k = 1, \ldots, N. \tag{7.2.5}$$

If the subspaces ϕ_a and ψ_a coincide, then eq. (7.2.5) is a conventional Galerkin formulation.

The nature of eq. (7.2.3) is such that all the methods of weighted residuals, described in section 1.3, are projection methods. We note in passing that Anderssen and Mitchell (1979) call the choice $\psi_j = K(\phi_j)$, where K is a particular operator, a *generalized moment method*. The definition of the method of moments, in Section 1.3.4, is consistent with this if the trial functions ϕ_j are polynomials.

7.2.2. Construction of the test function, ψ_k

Griffiths and Lorenz (1978) study the application of the Petrov–Galerkin formulation to the model equation

$$L(u) = -\frac{d^2 u}{dx^2} + \frac{p \, du}{dx} = f(x) \tag{7.2.6}$$

in the interval $0 < x < 1$, with boundary conditions

$$u(0) = u(1) = 0. \tag{7.2.7}$$

The above equations are similar to eqs. (7.1.2) and (7.1.3). Here the source term $f(x)$ has been substituted for an inhomogenous boundary condition.

By integrating by parts, a weak form of eqs. (7.2.6) and (7.2.7) can be manipulated into the following form:

$$a(u,v) = (f,v), \qquad (7.2.8)$$

where

$$a(u,v) \equiv \left(\frac{du}{dx}, \frac{dv}{dx} + pv \right). \qquad (7.2.9)$$

The corresponding Petrov–Galerkin formulation is

$$a(u_a, v_a) = (f, v_a), \qquad (7.2.10)$$

where the trial solution $u_a \in \phi_a$ and the test function $v_a \in \psi_a$. Griffiths and Lorenz assume that v_a depends on the trial functions $\phi_k(x)$, a controlling parameter α, and auxiliary functions $\gamma_k(x)$, as in eq. (7.1.13).

Griffiths and Lorenz establish the important result

$$|u - u_a|_1 \leq 1 + \frac{C_1}{C_2(\Delta x)} \min_{\tilde{u}_a \in \phi_a} |u - \tilde{u}_a|_1. \qquad (7.2.11)$$

In eq. (7.2.11), $|\ |_1$ is the seminorm defined by

$$|u|_1 = \left[\int_0^1 \left\{ \frac{du}{dx} \right\}^2 dx \right]^{1/2}. \qquad (7.2.12)$$

For the particular problem considered, eqs. (7.2.6) and (7.2.7),

$$|u|_1 = \|u\|_A \equiv (L(u), u). \qquad (7.2.13)$$

That is, the seminorm coincides with the energy norm. In eq. (7.2.11) C_1 is the continuity constant for $a(\cdot, \cdot)$.

Griffiths and Lorenz seek to optimize the Petrov–Galerkin formulation by choosing α in eq. (7.1.13) so that $C_2(\Delta x)$ in eq. (7.2.11) is a maximum. Then eq. (7.2.11) indicates that the Petrov–Galerkin solution u_a is the best of any possible trial solutions $\tilde{u}_a$ in the seminorm (7.2.12). The dependence of the seminorm on the derivative suggests that the optimum solution will not be oscillatory.

After introducing linear trial functions and the test functions defined by eqs. (7.1.13) and (7.1.14), the following bounds on $C_2(\Delta x)$ can be obtained:

$$C < C_2(\Delta x) < C(1 + \lambda^2 \Delta x)^{1/2}, \qquad (7.2.14)$$

where

$$C = (1 + \alpha\beta)(1 + 3\alpha^2)^{-1/2} \qquad (7.2.15)$$

and

$$\lambda = \frac{\beta}{1 + \alpha\beta}. \qquad (7.2.16)$$

Here, as before, $\beta = 0.5p\,\Delta x$. For typical values of the various parameters, $(1 + \lambda^2\,\Delta x)^{1/2} \approx 1$, so that $C_2(\Delta x) \approx C$. It is then straightforward to choose the optimum α to maximize $C_2(\Delta x)$. This gives

$$\alpha_{\text{opt}} = \frac{\beta}{3} \text{ or } \frac{p\,\Delta x}{6}. \tag{7.2.17}$$

For the particular choice of trial and test functions indicated above, Griffiths and Lorenz obtained the following L_2 error estimate:

$$\|u - u_a\|_2 \le C\left(1 + \frac{C_1}{C_2(\Delta x)}\right)((1 + \alpha)\,\Delta x + \alpha)\,\Delta x\,|u|_2. \tag{7.2.18}$$

Here C is a constant which is independent of Δx, α, and u. Examination of eq. (7.2.18) indicates that the L_2 error diminishes like Δx^2 if $\alpha = \alpha_{\text{opt}}$ (eq. (7.2.17)) or $\alpha = 0$ (conventional Galerkin). Otherwise the L_2 error diminishes like Δx.

If the inner products in eq. (7.2.5) are evaluated for a uniform grid, eq. (7.1.15) is obtained with an additional term arising from (f, ψ_k). A Taylor expansion about the jth node indicates an order of convergence consistent with eq. (7.2.18), except when $\alpha = \alpha_{\text{opt}} = \beta/3$. Then the truncation error is $O(\Delta x^4)$. This *superconvergence* at the nodes is confirmed by numerical experiments (Griffiths and Lorenz, 1978).

Some typical results are shown in Table 7.1 for the case $f(x) = p$. Then eqs. (7.2.6) and (7.2.7) have the exact solution

$$u(x) = x - \frac{e^{px} - 1}{e^p - 1}. \tag{7.2.19}$$

All solutions were obtained with $p = 60$. For Δx sufficiently small the optimal Petrov–Galerkin solution is better, as measured by the maximum error $\|u - u_a\|_\infty$, than either the conventional Galerkin ($\alpha = 0$) or the fully upwinded solution ($\alpha = 1$). However, on a coarse grid the optimal Petrov–Galerkin

Table 7.1. Maximum error and error in energy variation with grid size (after Griffiths and Lorenz, 1978).

Δx	$\alpha = p\,\Delta x = 10\,\Delta x$		$\alpha = 0$ (conv. Galerkin)		$\alpha = 1$ (full upwinding)	
	$\|u - u_a\|_\infty$	$\|u - u_a\|_A$	$\|u - u_a\|_\infty$	$\|u - u_a\|_A$	$\|u - u_a\|_\infty$	$\|u - u_a\|_A$
$\frac{1}{5}$	0.364	5.10	0.587[a]	6.01	0.0769	5.01
$\frac{1}{10}$	0.140	4.52	0.504[a]	5.49	0.140	4.52
$\frac{1}{20}$	0.0271	3.45	0.250[a]	3.85	0.200	3.62
$\frac{1}{40}$	0.00268	2.14	0.0803	2.23	0.179	2.49
$\frac{1}{80}$	0.000161	1.15	0.0178	1.17	0.103	1.50

[a]Solution oscillatory; $\|u - u_a\|_\infty$ is the maximum error.

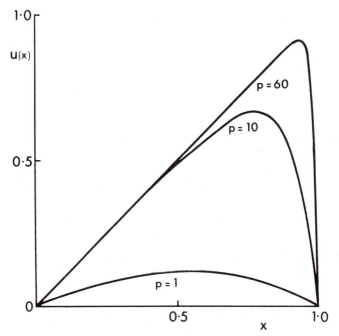

Figure 7.5. Exact solution of eqs. (7.2.6) and (7.2.7)

solution is inferior to the fully upwinded solution, and the conventional Galerkin solution is oscillatory.

It may be recalled that the optimum value of α was chosen to minimize the error in energy, $\|u - u_a\|_A$. Even on the finest mesh the error in energy does not appear to be overly sensitive to the choice of α. However, this problem is characterized by a uniform growth in u terminated by a boundary layer adjacent to $x = 1$ (Fig. 7.5). For a value of $p = 60$ the error in energy is dominated by the solution in the boundary layer.

Griffiths and Lorenz found that the value

$$\alpha_{opt} = \coth \beta - \frac{1}{\beta} \qquad (7.2.20)$$

is also the best choice for the current problem. For small values of β, α_{opt} has the following asymptotic value:

$$\alpha_{opt} = \frac{\beta}{3},$$

which agrees with eq. (7.2.17). This lends support to the above analysis.

An alternative technique for constructing an *optimal* Petrov–Galerkin formulation for solving convection-dominated problems has been studied by

Barrett and Morton (1980, 1981). The technique can be illustrated with the model equation

$$L(u) = -\frac{d^2u}{dx^2} + p\frac{du}{dx} = 0 \tag{7.2.21}$$

and boundary conditions

$$u(0) = 0 \quad \text{and} \quad u(1) = 1. \tag{7.2.22}$$

This is the same problem as considered previously (eqs. (7.1.2) and (7.1.3)). It may be recalled that for large values of p we expect a boundary layer to develop adjacent to $x = 1$ (Fig. 7.1).

A weak form of the above equations can be written

$$\left(\frac{du}{dx} - pu, \frac{dv}{dx}\right) = 0. \tag{7.2.23}$$

This form may be contrasted with the weak form considered by Griffiths and Lorenz, eq. (7.2.9) above. A Petrov–Galerkin formulation equivalent to eq. (7.2.23) is

$$\left(\frac{du_a}{dx} - pu_a, \frac{d\psi_k}{dx}\right) = 0, \qquad k = 1, \ldots, N, \tag{7.2.24}$$

where ψ_k is the test function. The trial solution u_a is defined by eq. (7.2.2).

It is well known that for a general diffusion problem (with no first-derivative terms) the Galerkin formulation produces the best approximation in the energy norm, and that the weak form equivalent to eq. (7.2.24) is symmetric. Barrett and Morton argue that by choosing $\psi_k = M(\phi_k)$, where M is a linear operator constructed to make eq. (7.2.24) symmetric, the Petrov–Galerkin formulation will be the best approximation in the corresponding norm.

One possibility is to set $\psi_k = L(\phi_k)$. However, this is just the least-squares method, which is rejected on the basis that it requires the use of elements with higher continuity.

To make eq. (7.2.24) symmetric requires that ψ_k satisfy the following equations:

$$\frac{d\psi_k}{dx} = \rho(x)\left(\frac{d}{dx} - p\right)\phi_k(x) \tag{7.2.25}$$

with boundary contitions

$$\psi_k(0) = \psi_k(1) = 0. \tag{7.2.26}$$

In eq. (7.2.25), $\rho(x)$ is a positive weight function that can be used to balance the relative importance of different parts of the domain.

The exact solutions of eqs. (7.2.25) and (7.2.26), which behave properly as the grid is refined, are of the form

$$\psi_k(x) = e^{-px}\phi_k(x). \tag{7.2.27}$$

This choice is optimal for eq. (7.2.21), but it is inappropriate for more general equations, due to inaccuracies associated with ill-conditioning.

Consequently, to keep the approach sufficiently general, Barrett and Morton sought only an approximately symmetric form for eq. (7.2.24). To this end eq. (7.2.25) is replaced by

$$\frac{d\psi_k^\varepsilon}{dx} = \rho(x)\left(\frac{d}{dx} - p\right)\phi_k(x) + \varepsilon_k. \tag{7.2.28}$$

The solution can be written

$$\psi_k(x) = \rho(x)\phi_k(x) - \int_0^x \left\{\left(\frac{d\rho}{ds} + p\rho\right)\psi_k - \varepsilon_k\right\}ds. \tag{7.2.29}$$

To satisfy the boundary condition, $\psi_k^\varepsilon(1) = 0$, requires

$$\int_0^1 \varepsilon_k \, ds = \int_0^1 \left(\frac{d\rho}{ds} + p\rho\right)\phi_k \, ds = \alpha_k. \tag{7.2.30}$$

In order to generate a sparse system of equations it is desirable for ψ_k^ε to span the same number of nodes x_k as ϕ_k. This can be achieved for one node n by setting

$$\varepsilon_n(x) = \alpha_n C_n(x), \tag{7.2.31}$$

where

$$C_n(x) = \begin{cases} 1/(x_{n+1} - x_n) & \text{when } x_n \leq x \leq x_{n+1}, \\ 0 & \text{otherwise.} \end{cases} \tag{7.2.32}$$

Then $\psi_n = \psi_n^\varepsilon$ from eq. (7.2.29), and $\psi_{n+1} = \psi_{n+1}^\varepsilon$ with $\varepsilon_{n+1} = \alpha_{n+1}C_n(x)$ in eq. (7.2.29). For nodes $1 < k < n$,

$$\psi_k(x) = \alpha_{k+1}\psi_k^\varepsilon(x) - \alpha_k\psi_{k+1}^\varepsilon(x) \tag{7.2.33}$$

For nodes $n + 1 < k < N$,

$$\psi_k(x) = \alpha_{k-1}\psi_k^\varepsilon(x) - \alpha_k\psi_{k-1}^\varepsilon(x). \tag{7.2.34}$$

In obtaining eqs. (7.2.33) and (7.2.34), the following relationship has been utilized:

$$\alpha_{k-1}\psi_k^\varepsilon - \alpha_k\psi_{k-1}^\varepsilon = \alpha_{k-1}\psi_k - \alpha_k\psi_{k-1}. \tag{7.2.35}$$

Using the test functions defined by eqs. (7.2.31) to (7.2.34) in eq. (7.2.24) produces a solution u_a which is approximately the best fit in the norm

$$\left\|\rho^{1/2}\frac{dv}{dx}\right\|_2^2 + \left\|\left(p^2\rho + \frac{pd\rho}{dx}\right)^{1/2}v\right\|_2^2. \tag{7.2.36}$$

For the problem governed by eqs. (7.2.21) and (7.2.22), which has a boundary layer adjacent to $x = 1$, it is desirable to set the perturbation $\varepsilon_n(x)$ at

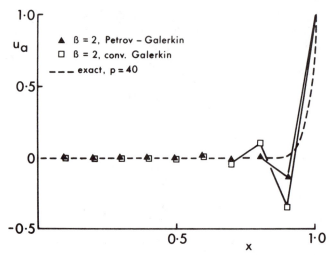

Figure 7.6. Comparison of Galerkin and Petrov–Galerkin solutions of eqs. (7.2.21) and (7.2.22)

$x = 0$. For a linear trial function $\phi_k(x)$ and $\rho(x) \equiv 1$, Barrett and Morton find that a typical test function is similar to the one shown in Fig. 7.4.

For an 11-point equally spaced grid the Petrov–Galerkin solution for $p = 40$ is shown in Fig. 7.6. Also shown are the Galerkin solution and the exact solution. In contrast to the Petrov–Galerkin solution defined by eq. (7.1.16), the present Petrov–Galerkin formulation produces oscillatory solutions with linear trial functions ϕ_k if $p\,\Delta x > 6^{1/2}$.

Because of the nature of the present Petrov–Galerkin formulation, it is possible to recover a solution u_R of higher accuracy. This will be illustrated for the boundary-layer region (near $x = 1$) of the solution shown in Fig. 7.6.

The following trial solution is assumed for u_R:

$$u_R = a_0 \exp(a_1(x - 1)) + a_2. \tag{7.2.37}$$

Equations connecting the three coefficients a_0, a_1, and a_2 are obtained by requiring that $u_R(1) = 1$ and evaluating

$$\left(\left(\frac{d}{dx} - p\right) u_R, \left(\frac{d}{dx} - p\right) \phi_k\right) = \left(\left(\frac{d}{dx} - p\right) u_{PG}, \left(\frac{d}{dx} - p\right) \phi_k\right),$$

$$k = N - 2, N - 1. \tag{7.2.38}$$

Here u_{PG} is the Petrov–Galerkin solution. For this problem the recovered solution was indistinguishable from the exact solution. This is partly due to the form of the trial solution for u_R, eq. (7.2.37). For the more general situation where the form of the trial solution is not obvious, one would still expect u_R to be more accurate than u_{PG}.

The ability to recover a more accurate solution is a feature of the present formulation that follows from the approximate best-fit property (7.2.36). Barrett and Morton (1980) demonstrate the superior accuracy of the recovered solution for a number of more complex one-dimentional problems.

Barrett and Morton (1981) discuss the extension of the above formulation to more than one dimension but find that it is not possible to keep the equivalent perturbation $\varepsilon_k(x, y)$ small. Even when this aspect was ignored, it was found that the resulting algebraic equations (after evaluation of eq. (7.2.24)) had a large number of contributing nodes. Consequently the procedure is not very economical.

An alternative treatment of convection-dominated problems is suggested by the form of eq. (7.1.15). That is to deliberately introduce an *artificial* dissipative term, $(1/p_e) \, d^2u/dx^2$, and to apply the conventional Galerkin method. This is the approach of Hughes and Brooks (1979) and Kelly et al. (1980). Although the proper choice of p_e is obvious in the model problems considered above, it is not obvious for more general multidimensional problems. The same technique of adding an artifical dissipative term has been used for many years for stabilizing finite-difference computations of inviscid compressible flow in which shocks occur (Richtmyer and Morton, 1967).

7.3. Steady Convection—Diffusion Problems

Steady convection–diffusion problems in one dimension have already been studied in sections 7.1 and 7.2. Here we briefly examine the generation of higher-order schemes in one dimension and then look at two-dimensional steady convection–diffusion problems in more depth.

7.3.1. Higher-order one-dimensional formulations

For a uniform grid a three-point algebraic equation that results from a finite-difference or finite-element formulation of eq. (7.1.2) is said to be *consistent* if it can be written as

$$c\overline{T}_{k-1} - (a + c)\overline{T}_k + a\overline{T}_{k+1} = 0. \tag{7.3.1}$$

Clearly eq. (7.1.15) is consistent. Christie and Mitchell (1978) note that eq. (7.3.1) will produce the exact solution of eq. (7.1.2) if

$$c/a = e^{2\beta}. \tag{7.3.2}$$

For linear, quadratic, and cubic Lagrangian trial functions Christie and Mitchell generate algebraic formulae, in the form of eq. (7.3.1), using a Petrov–Galerkin formulation equivalent to eqs. (7.1.12) and (7.1.13). We will illustrate the technique for a quadratic trial function.

In this case the test function $\psi_k(x)$ is given by

$$\psi_k(x) = \phi_k(x) + \alpha_1 \gamma_k(\xi) + \alpha_2 \gamma_{k+1}(\xi) \qquad (7.3.3)$$

for corner nodes $k - 1, k, k + 1, \ldots$. For midside nodes $k - \frac{1}{2}, k + \frac{1}{2}, \ldots,$ $\psi_{k-1/2}(x)$ is given by

$$\psi_{k-1/2}(x) = \phi_{k-1/2}(x) + 4\alpha_3 \gamma_k(\xi), \qquad (7.3.4)$$

where

$$\gamma(\xi) = \begin{cases} -5\xi(\xi + 0.5)(\xi + 1.0), & -1 < \xi < 0, \\ 0 & \text{otherwise,} \end{cases} \qquad (7.3.5)$$

and ξ is the element-based coordinate indicated in Fig. 7.4. In eqs. (7.3.3) and (7.3.4), α_1, α_2, and α_3 are adjustable constants equivalent to α in eq. (7.1.13). It may be noted that ψ_k is cubic, that is, one order higher than the trial function ϕ_k.

Evaluation of the Petrov–Galerkin inner product (7.1.12) and elimination of the midside nodal values $T_{k-1/2}, \ldots$ gives the following expressions for a and c in eq. (7.3.1):

$$a = 1 - (1 - 0.5\alpha_3)\beta + (1 + 0.5(\alpha_3 - \alpha_2))\beta^2/3,$$
$$c = 1 + (1 - 0.5\alpha_3)\beta + (1 - 0.5(\alpha_3 - \alpha_1))\beta^2/3. \qquad (7.3.6)$$

The values $\alpha_1 = \alpha_2 = \alpha_3 = 0$ produce the conventional Galerkin formulation, and $\alpha_1 = 8$, $\alpha_2 = 0$, $\alpha_3 = -2$ produce a quadratic upwind scheme. A similar result to eq. (7.3.6) has been generated by Heinrich and Zienkiewicz (1977), but with $\alpha_1 = \alpha_2$.

Christie and Mitchell note that the conventional quadratic Galerkin scheme is fourth-order accurate for small β and is free of oscillation for all values of β. Unfortunately, in the limit $\beta \to \infty$ it becomes consistent with $d^2 T/dx^2 = 0$ rather than the required $dT/dx = 0$.

For a problem similar to that defined by eqs. (7.1.2) and (7.1.3) Christie and Mitchell have compared solutions from linear, quadratic, and cubic trial functions for various values of a and c with $\Delta x = 0.1$ and values of $\beta = 2, 5,$ and 10. For such a coarse grid the boundary layer at $x = 1$ is only just resolved at $\beta = 2$ and is lost between nodes $N - 1$ and N for $\beta = 5$ and 10. For higher-order schemes the oscillation-free limit on β rises slightly. All fully upwinded schemes are oscillation free, and the quadratic and cubic upwind schemes are relatively accurate, particularly for large β.

7.3.2. Two-dimensional formulations

The temperature distribution in the entrance of a two-dimensional duct is governed by the following equation:

$$L(T) = u\frac{\partial T}{\partial x} + v\frac{\partial T}{\partial y} - \mathscr{D}\left(\frac{\partial^2 T}{\partial x^2} + \frac{\partial^2 T}{\partial y^2}\right) = 0, \qquad (7.3.7)$$

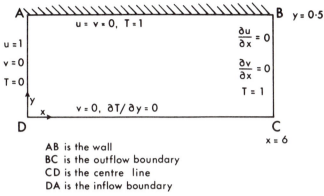

AB is the wall
BC is the outflow boundary
CD is the centre line
DA is the inflow boundary

Figure 7.7. Two-dimensional thermal-entry problem

where u and v are nondimensional velocity components and are assumed known. It can be seen that this equation reduces to eq. (7.1.1) in one dimension.

A typical set of boundary conditions are shown in Fig. 7.7. The boundary condition $T = 1$ on the outflow boundary BC in Fig. 7.7, may be compared with the use of $\partial T/\partial x = 0$ for the same problem considered in section 4.1.1 (Fig. 4.2). From physical considerations $\partial T/\partial x = 0$ is the correct boundary condition. However, setting $T = 1$ causes a boundary layer to develop adjacent to the outflow boundary if $\mathscr{D}$ is small, and this situation contributes to the oscillatory solution produced by the conventional Galerkin method. Consequently the impact of the Petrov–Galerkin formulation on the problem is then accentuated.

The Petrov–Galerkin formulation proceeds in the same manner as for the one-dimensional problem. A trial solution T_a is introduced by

$$T_\alpha = \sum_{j=1}^N \overline{T}_j \phi_j(x, y), \tag{7.3.8}$$

and the Petrov–Galerkin formulation requires that

$$(L(T_a), \psi_k(x, y)) = 0, \qquad k = 1, \ldots, N. \tag{7.3.9}$$

Application of Green's theorem, to reduce the order of the highest derivative appearing in eq. (7.3.9), gives

$$\left(u\frac{\partial T_a}{\partial x}, \psi_k \right) + \left(v\frac{\partial T_a}{\partial y}, \psi_k \right) + \mathscr{D}\left(\frac{\partial T_a}{\partial x}, \frac{\partial \psi_k}{\partial x} \right) + \mathscr{D}\left(\frac{\partial T_a}{\partial y}, \frac{\partial \psi_k}{\partial y} \right) = 0,$$

$$k = 1, \ldots, N. \tag{7.3.10}$$

Because of the boundary conditions for this problem, all the inner products in eq. (7.3.10) are area integrals; for example

$$\left(\frac{\partial T_a}{\partial x}, \frac{\partial \psi_k}{\partial x} \right) = \iint \frac{\partial T_a}{\partial x} \frac{\partial \psi_k}{\partial x} dx \, dy. \tag{7.3.11}$$

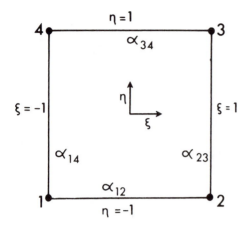

Figure 7.8. Upwinding parameters for a
linear rectangular element

As with the one-dimensional formuation, *the major difficulty is to arrive at
a suitable form for* $\psi_k(x,y)$. For Lagrangian rectangular elements the two-
dimensional trial functions can be expressed as the product of the one-
dimensional trial functions. For the linear element shown in Fig. 7.8, the
trial function ϕ_1 can be written

$$\phi_1 = \phi_1^x(\xi)\phi_1^y(\eta), \qquad (7.3.12)$$

where

$$\phi_1^x(\xi) = 0.5(1 - \xi)$$

and

$$\phi_1^y(\eta) = 0.5(1 - \eta).$$

Consequently eq. (7.1.13) can be generalized to

$$\psi_k(x,y) = (\phi_k^x(x) + \alpha^x\gamma_k^x(x))(\phi_k^y(y) + \alpha^y\gamma_k^y(y)). \qquad (7.3.13)$$

Thus, in the notation of Fig. 7.8,

$$\psi_1(x,y) = \psi_1^x(\xi)\psi_1^y(\eta),$$

where

$$\psi_1^x(\xi) = \phi_1^x(\xi) + \alpha_{12}\gamma_1^x(\xi) = 0.5(1 - \xi)(1 + 1.5\alpha_{12}(1 + \xi))$$

and

$$\psi_1^y(\eta) = \phi_1^y(\eta) + \alpha_{14}\gamma_1^y(\eta) = 0.5(1 - \eta)(1 + 1.5\alpha_{14}(1 + \xi)).$$

Similarly

$$\psi_2^x(\xi) = 0.5(1 + \xi)(1 - 1.5\alpha_{12}(1 - \xi))$$

and

$$\psi_2^y(\eta) = 0.5(1 - \eta)(1 + 1.5\alpha_{23}(1 + \eta)).$$

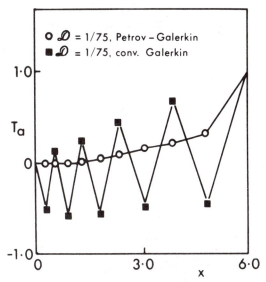

Figure 7.9. Centerline temperature distribution

The above approach has been used by Huyakorn (1977) and by Heinrich et al. (1977). It is clear that the degree of upwinding is now governed by the four α parameters in each element for linear trial functions.

Huyakorn (1977) makes use of the optimum one-dimensional α value (7.1.18) to determine α_{ij} on an element edge. Thus, if $\mathbf{l}_{ij}$ is the direction vector of the element edge and $\mathbf{v}_i$ and $\mathbf{v}_j$ are the velocity vectors at nodes i and j, an average velocity along ij can be defined as follows:

$$v_{\mathrm{av}} = 0.5(\mathbf{v}_i + \mathbf{v}_j) \cdot \mathbf{l}_{ij}. \tag{7.3.14}$$

If h is the length of the edge ij, the following expression is used for α_{ij}:

$$\alpha_{ij} = \coth \beta - \frac{1}{\beta}, \tag{7.3.15}$$

where $\beta = 0.5 h v_{\mathrm{av}}/\mathscr{D}$.

A typical result for the centerline temperature distribution (Fig. 7.7) is shown in Fig. 7.9. The value α_{ij} given by eq. (7.3.15) is used to compute the Petrov–Galerkin solution. Huyakorn (1977) reports that a conventional Galerkin solution on a refined grid gives good agreement with the Petrov–Galerkin solution.

Huyakorn (1977) has also used the Petrov–Galerkin formulation with linear triangular elements, and Heinrich and Zienkiewicz (1977) have extended the above formulation to quadratic elements.

Griffiths and Mitchell (1979) have considered a Petrov–Galerkin formulation, similar to that described above, with linear trial functions on a uniform rectangular mesh ($\Delta x = \Delta y = h$). As well as the product form for the test function ψ_k given by eq. (7.3.13), Griffiths and Mitchell consider the following:

$$\psi_k(\xi, \eta) = \phi_k^x(\xi)\phi_k^y(\eta) + \alpha^x\gamma_k^x(\xi)\phi_k^y(\eta) + \alpha^y\gamma_k^y(\eta)\phi_k^x(\xi). \qquad (7.3.16)$$

The coordinates (ξ, η) are defined in Fig. 7.8. After evaluation of eq. (7.3.10), only the two parameters α^x, α^y are available to control the degree of up-winding, that is, these two parameters are associated with the kth node.

The resultant algebraic equations are lengthy, but can be expressed conveniently in terms of the following *grid operators* (see section 3.3):

$$\delta \equiv 0.5\{-1, 0, 1\},$$

$$\delta^2 \equiv \{1, -2, 1\}, \qquad (7.3.17)$$

$$m_x = m_y^t \equiv \{\tfrac{1}{6}, \tfrac{2}{3}, \tfrac{1}{6}\}.$$

It may be noted that δ and δ^2 are also three-point finite-difference operators. The operators m_x, m_y were used in section 4.1.

It is helpful to consider term in eq. (7.3.7) consecutively. Thus

$$h\frac{\partial \bar{T}}{\partial x} \Rightarrow m_y \otimes \delta_x \bar{T}_k - 0.5\alpha^x m_y \otimes \delta_x^2 \bar{T}_k - 0.5\alpha^y \delta_x \otimes \delta_y \bar{T}_k, \qquad (7.3.18)$$

$$h\frac{\partial \bar{T}}{\partial y} \Rightarrow m_x \otimes \delta_y \bar{T}_k - 0.5\alpha^x \delta_x \otimes \delta_y \bar{T}_k - 0.5\alpha^y m_x \otimes \delta_y^2 \bar{T}_k, \qquad (7.3.19)$$

$$h^2\left(\frac{\partial^2 \bar{T}}{\partial x^2} + \frac{\partial^2 \bar{T}}{\partial y^2}\right) \Rightarrow \begin{aligned} & m_y \otimes \delta_x^2 \bar{T} + m_x \otimes \delta_y^2 \bar{T} \\ & - 0.5\alpha^x \delta_x \otimes \delta_y^2 \bar{T}_k - 0.5\alpha^y \delta_x^2 \otimes \delta_y^2 \bar{T}_k. \end{aligned} \qquad (7.3.20)$$

An advantage of the above notation is that the effect of a Petrov–Galerkin formulation can be interpreted as introducing higher order derivatives. This is equivalent to the interpretation of the one-dimensional algebraic equation (7.1.15). Consideration of eqs. (7.3.18) to (7.3.20) indicates that the Petrov–Galerkin formulation introduces higher-order contributions associated with the diffusive terms $\partial^2 T/\partial x^2 + \partial^2 T/\partial y^2$. This was not the case in one dimension. For the first derivatives, diffusive terms are introduced as in one dimension, and also cross-derivatives are introduced. It turns out that the cross-derivatives have an important influence on crosswind diffusion.

The standard upwind three-point finite-difference representation of eq. (7.3.7), equivalent to eqs. (7.3.18) to (7.3.20), is

$$u\frac{\delta_x \bar{T}_k - 0.5\alpha^x \delta_x^2 \bar{T}}{h} + v\frac{\delta_y \bar{T}_k - 0.5\alpha^y \delta_y^2 \bar{T}_k}{h} - \frac{\delta_x^2 \bar{T}_k + \delta_y^2 \bar{T}_k}{h^2} = 0. \qquad (7.3.21)$$

In two dimensions the preferred strategy with upwind differencing is to add dissipation in the local flow direction, but without introducing dissipation normal to the local flow direction. Then oscillations induced by the symmetric treatment of the convective terms can be suppressed without causing the side effect of spurious crosswind diffusion.

For a uniform grid it is straightforward to obtain the truncation error of the different schemes expressed in natural coordinates. We let ∂_s denote differentiation in the local flow direction, and ∂_n denote differentiation normal to the local flow direction. Also, if θ is the local flow direction relative to the x-axis,

$$u = q\cos\theta \quad \text{and} \quad v = q\sin\theta. \tag{7.3.22}$$

For the situation where u and v are both positive, the leading term in the truncation error of the finite-difference scheme (7.3.21) is

$$-0.5h\big[(\alpha^x u^3 + \alpha^y v^3)\partial_s^2 - 2uv(\alpha^x u - \alpha^y v)\partial_s\partial_n + uv(\alpha^x v + \alpha^y u)\partial_n^2\big]\frac{T}{q^2}. \tag{7.3.23}$$

Thus no choice of α^x and α^y will eliminate crosswind diffusion, unless the flow is aligned with a coordinate direction, that is, $u \approx 0$ or $v \approx 0$. The leading term in the truncation error for the Petrov–Galerkin scheme (7.3.18)–(7.3.20) is

$$-0.5h\big[(\alpha^x u + \alpha^y v)\partial_s^2 + (\alpha^y u - \alpha^x v)\partial_s\partial_n\big]T. \tag{7.3.24}$$

Thus if α^x and α^y are chosen so that

$$\frac{\alpha^y}{\alpha^x} = \frac{v}{u} = \tan\theta, \tag{7.3.25}$$

crosswind diffusion can be minimized. However, there may well be a minor contribution from higher-order terms in the truncation error.

Griffiths and Mitchell have compared the above schemes for the inclined-thermal-shear-layer problem represented schematically in Fig. 7.10. The

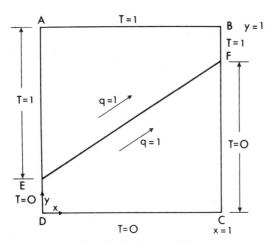

Figure 7.10. Inclined thermal shear layer

Table 7.2. Nondimensional temperature variation with y at $x = 0.5$

		Petrov–Galerkin scheme (7.3.18)–(7.3.20)		U.F.D. (7.3.21)
y	"Exact" solution	$\alpha^x = \alpha^y = 1$	$\alpha^x = 1,$ $\alpha^y = \tan\theta$	$\alpha^x = \alpha^y = 1$
0	0	0	0	0
0.1	0	0	0	0.03
0.2	0	-0.02	-0.01	0.06
0.3	0.016	-0.03	-0.02	0.13
0.4	0.155	0.14	0.12	0.25
0.5	0.500	0.52	0.49	0.44
0.6	0.854	0.87	0.89	0.67
0.7	0.992	1.01	1.03	0.89
0.8	1.000	1.01	1.00	0.99
0.9	1.000	1.00	0.99	1.00
1.0	1.000	1.00	1.00	1.00

velocity q is assumed constant and parallel to the line EF throughout the domain. The line EF is constrained to pass through the center point of the region $(0.5, 0.5)$, but the angle of the line, θ, is treated as a controlling parameter for the problem. On the boundary $EABF$, the nondimensional temperature $T = 1$. On the boundary $FCDE$, $T = 0$. For the outflow boundaries ABC for θ positive, this is not a physically realistic boundary condition for those regions adjacent to F that are influenced by the diffusion normal to EF. At points E and F, T is set to 0.5.

Solutions have been obtained on a uniform grid ($\Delta x = \Delta y = 0.1$) for a value $q/\mathscr{D} = 500$. The nondimensional temperature distribution for $x = 0.5$ and $\theta = 21.8°$ is shown in Table 7.2. The results indicate that the Petrov–Galerkin scheme produces solutions with far less crosswind diffusion than the corresponding upwind finite-difference scheme. Griffiths and Mitchell found that the lack of terms $\delta_x \delta_y T$ in eq. (7.3.21) was responsible for the poor performance. Using a tensor-product formulation for ψ_k in eq. (7.3.16)—that is, as in eq. (7.3.13)—added extra higher-order terms to eqs. (7.3.18) and (7.3.19) but made little difference to the results. It is interesting that for this problem the use of eq. (7.3.25) gives no improvement over the use of full upwinding ($\alpha^x = \alpha^y = 1$). However, this might not be the case for larger values of the diffusivity.

Higher-order upwind finite-difference schemes are available, and two have been tested by Leschziner (1980) and found to give accurate results for this problem.

Hughes and Brooks (1979) have solved the problem posed by eq. (7.3.7) by introducing an artificial diffusion aligned with the local flow direction and of magnitude given by the one-dimensional result (7.2.20). For the inclined-thermal-shear-layer problem with $\mathscr{D} = 10^{-6}$ the scheme is effectively a fully upwinded scheme. Results presented by Hughes and Brooks indicate that the method is smoother than a conventional Galerkin formulation and only slightly more dissipative.

7.4. Parabolic Problems

Here we restrict our attention to two related one-dimensional problems:

(i) the transient convection–diffusion equation,

$$\frac{\partial T}{\partial t} + u\frac{\partial T}{\partial x} - \mathscr{D}\frac{\partial^2 T}{\partial x^2} = 0. \tag{7.4.1}$$

(ii) Burgers' equation

$$\frac{\partial u}{\partial t} + u\frac{\partial u}{\partial x} - \frac{1}{\text{Re}}\frac{\partial^2 u}{\partial x^2} = 0. \tag{7.4.2}$$

7.4.1. Transient convection–diffusion equation

Equation (7.4.1) is the transient equivalent to eq. (7.1.1). For the linear trial function ϕ_k and quadratic test function ψ_k given by eq. (7.1.13), the Petrov–Galerkin formulation (7.1.12), applied on a uniform grid, gives

$$\left(\frac{1}{6} + \frac{\alpha}{4}\right)\frac{d\bar{T}_{k-1}}{dt} + \frac{2}{3}\frac{d\bar{T}_k}{dt} + \left(\frac{1}{6} - \frac{\alpha}{4}\right)\frac{d\bar{T}_{k+1}}{dt} + \frac{u(\bar{T}_{k+1} - \bar{T}_{k-1})}{2\,\Delta x}$$

$$- \frac{\alpha u}{2\,\Delta x}(\bar{T}_{k-1} - 2\bar{T}_k + \bar{T}_{k+1}) - \frac{\mathscr{D}}{\Delta x^2}(\bar{T}_{k-1} - 2\bar{T}_k + \bar{T}_{k+1}) = 0. \tag{7.4.3}$$

It may be noted that the upwinding parameter α now affects the time-derivative terms.

For time-dependent problems Griffiths and Mitchell (1979) have generalized the above approach by using three parameters α_1, α_2, and α_3 defined by

$$\alpha_1 = \alpha = \int_0^1 (\psi_k(-\xi) - \psi_k(\xi))\,d\xi,$$

$$\alpha_2 = -\int_{-1}^1 \xi\psi_k(\xi)\,d\xi, \tag{7.4.4}$$

$$\alpha_3 = \int_{-1}^1 |\xi|\psi_k(\xi)\,d\xi.$$

As in section 7.3.2, it is possible to interpret the Petrov–Galerkin formulation as a conventional Galerkin formulation applied to a *perturbation* of the original equation. Thus, with the three-parameter approach of Griffiths and Mitchell, eq. (7.4.3) is replaced by

$$(1 - \alpha_2 \delta + 0.5\alpha_3 \delta^2)\frac{d\bar{T}_k}{dt} + \frac{u}{\Delta x}(\delta - 0.5\alpha_1 \delta^2)\bar{T}_k - \frac{\mathscr{D}\delta^2 \bar{T}_k}{\Delta x^2} = 0, \qquad (7.4.5)$$

where δ and δ^2 are operators defined by eq. (7.3.17), with $\Delta x \equiv h$. If $\alpha_2 = 0.5\alpha_1$ and $\alpha_3 = \frac{1}{3}$, eq. (7.4.5) reverts to eq. (7.4.3). If $\alpha_1 = \alpha_2 = 0$ and $\alpha_3 = \frac{1}{3}$, the conventional Galerkin formulation is obtained.

If eq. (7.4.5) is discretized in time using a Crank–Nicolson scheme, the result is second-order accurate for all values of α_3 when $\alpha_2 = 0.5\alpha_1$. If, in addition, $\alpha_3 = \frac{1}{3} + \frac{1}{6}(u\,\Delta t/\Delta x)^2$, then the truncation error is proportional to $\mathscr{D}$, which suggests that this scheme will be effective for convection-dominated flows.

Although no results were presented, Gresho and Lee (1981) note that the steady-state optimum value for α_1, given by eq. (7.1.18), introduces unacceptable dissipation in transient problems.

7.4.2. Burgers' equation

The application of the Petrov–Galerkin formulation to eq. (7.4.2), with linear trial functions and quadratic test functions given by eq. (7.1.13), produces the same algebraic formula on a uniform grid as for eq. (7.4.1), except for the treatment of the nonlinear convective term $u\,\partial u/\partial x$. With a trial solution u_a in the form of eq. (7.1.11),

$$u\frac{\partial u}{\partial x} \Rightarrow \frac{\bar{u}_{k-1} + \bar{u}_k + \bar{u}_{k+1}}{3}\frac{\bar{u}_{k+1} - \bar{u}_{k-1}}{2\,\Delta x} - 0.25\alpha\,\Delta x\frac{\bar{u}_{k-1}^2 - 2\bar{u}_k^2 + \bar{u}_{k+1}^2}{\Delta x^2}. \qquad (7.4.6)$$

The first part of this expression coincides with the result for the conventional Galerkin formulation, eq. (3.2.63). The second part, associated with the Petrov–Galerkin formulation, indicates that the additional dissipation is *in terms of u^2 rather than u*.

If the convective term is written in the form

$$0.5\partial(u^2)/\partial x$$

and an auxiliary trial solution introduced for u_a^2, namely

$$u_a^2 = \sum_{j=1}^{N} \bar{u}_j^2 \phi_j(x), \qquad (7.4.7)$$

then the Petrov–Galerkin treatment is as follows:

$$0.5\frac{\partial(u^2)}{\partial x} \Rightarrow \frac{\bar{u}_{k+1}^2 - \bar{u}_{k-1}^2}{4\,\Delta x} - 0.25\alpha\,\Delta x\frac{\bar{u}_{k-1}^2 - 2\bar{u}_k^2 + \bar{u}_{k+1}^2}{\Delta x^2}. \qquad (7.4.8)$$

The first part of this expression coincides with the conventional Galerkin result (3.2.69). The second part is identical to that in eq. (7.4.6). If $\alpha = 1$ in eq. (7.4.8), it becomes identical with a three-point upwind finite-difference formulation of $0.5\partial(u^2)/\partial x$.

It is possible for both the above treatments of the convective terms to be combined with the higher-order Petrov–Galerkin schemes described in section 7.3.1.

Mitchell and Griffiths (1980) and Christie et al. (1981) have applied Petrov–Galerkin schemes to Burgers' equation (7.4.2) with the initial condition

$$u(x, 0) = u_0(x) = \sin \pi x, \qquad 0 < x < 1, \tag{7.4.9}$$

and boundary conditions

$$u(0, t) = u(1, t) = 0. \tag{7.4.10}$$

For the propagating sine wave the exact solution is given by Cole (1951). The behavior with increasing time is qualitatively similar to that of the propagating shock described in section 1.2.5. As the sine wave moves downstream (to the right), the nonlinear convective term causes the leading face to steepen, and the diffusive term $(\partial^2 u/\partial x^2)/\text{Re}$ causes the amplitude of the sine wave to diminish. Due to the downstream boundary condition $u(1, t) = 0$, a boundary layer develops adjacent to $x = 1$ at intermediate time. The boundary layer becomes more severe as Re increases. At large time the distorted sine wave is diffused away.

For a coarse grid $(\Delta x = \frac{1}{18})$ and a value $\text{Re} = 10^4$, Mitchell and Griffiths found that a quadratic upwind Petrov–Galerkin formulation, as in section 7.3.1, of eq. (7.4.2) gave more accurate results than the linear upwind formulation (7.4.2) with (7.4.6). As might be expected for such a large value of Re

Table 7.3. Quadratic Petrov–Galerkin solutions for propagating-sine-wave problem

	Petrov–Galerkin		
x ($\Delta x = \frac{1}{18}$)	$u\,\partial u/\partial x$	$0.5\,\partial(u^2)/\partial x$ + (7.4.7)	Exact solution
0.500	0.594	0.595	0.595
0.556	0.656	0.656	0.656
0.611	0.715	0.715	0.715
0.667	0.772	0.772	0.772
0.722	0.826	0.826	0.826
0.778	0.876	0.876	0.876
0.833	0.906	0.921	0.921
0.889	0.902	0.960	0.959
0.944	0.835	0.854	0.959
1.000	0	0	0

on a coarse grid, the conventional linear and quadratic Galerkin solutions were highly oscillatory.

For the same problem, Christie et al. (1981) have compared the alternative treatments of the convective terms, $u\,\partial u/\partial x$ and $0.5\,\partial(u^2)/\partial x$ using eq. (7.4.7), with a quadratic upwind Petrov–Galerkin formulation. It may be recalled that eqs. (7.4.6) and (7.4.8) were obtained with a linear (trial function) Petrov–Galerkin formulation. For $Re = 10^4$, $x = \frac{1}{18}$, and $t = 0.5$, results are shown in Table 7.3. The results were obtained with a trapezoidal time-integration scheme and $\Delta t = 0.001$. Thus errors in the solution presented in Table 7.3 are due to the spatial approximation only. Both solutions are free of oscillations and agree with the exact solution up to the start of the boundary-layer region. In the boundary-layer region the alternative treatment of the convective term, $0.5\,\partial(u^2)/\partial x$ with eq. (7.4.7), produces a more accurate solution.

7.5. Hyperbolic Problems

Here we will restrict our attention, for the most part, to the one-dimensional model equation

$$L(u) = \frac{\partial u}{\partial t} - w\frac{\partial u}{\partial x} = 0. \tag{7.5.1}$$

From the theory of characteristics, it is known that the solution u_k^{n+1} at node k at time level $n + 1$ (A in Fig. 7.11) is equal to the solution u^n at point O. Line OA is straight if w is a constant. Therefore eq. (7.5.1) represents the uniform translation of u to the left at an angle $\tan^{-1}(1/w)$ to the x axis in the x–t plane (Fig. 7.11).

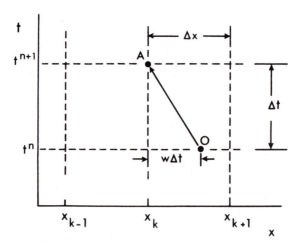

Figure 7.11. Characteristic direction for eq. (7.5.1)

For the special case $\Delta x = w \, \Delta t$, we see that $u_k^{n+1} = u_{k+1}^n$. The ratio

$$c = w \frac{\Delta t}{\Delta x}$$

is called the *Courant–Friedrichs–Lewy* (CFL) number, and the property that $u_k^{n+1} = u_{k+1}^n$ we will refer to as the unit-CFL property. Almost all practical finite-difference schemes satisfy the unit-CFL property. Unfortunately, the conventional Galerkin formulation, combined with typical finite-difference time discretization, does not.

Since hyperbolic problems in time involve no dissipative mechanism, the discretization of the time derivative is more critical than for parabolic problems. Morton and Parrott (1980) have introduced a Petrov–Galerkin formulation of eq. (7.5.1) in space *after* choosing an appropriate finite-difference time discretization.

The Petrov–Galerkin formulation for hyperbolic equations proceeds as in the previous sections. Firstly a trial solution u_a is chosen:

$$u_a = \sum_j \bar{u}_j \phi_j(x). \qquad (7.5.2)$$

For simplicity we will assume that $\phi_j(x)$ are linear trial functions on a uniform grid. Secondly inner products

$$(L(u), \psi_k(x)) = 0, \qquad k = 1, \ldots, N, \qquad (7.5.3)$$

are evaluated to provide algebraic equations similar to (7.1.15).

A unique feature of Morton and Parrott's approach is the way in which $\psi_j(x)$ is chosen. For $c = 1$ (i.e. $\Delta t = \Delta x / w$), it is required that eq. (7.5.3) satisfy the unit-CFL property. In this limit

$$\psi_k(x) = \chi_k(x), \qquad (7.5.4)$$

where $\chi_k(x)$ is to be determined. For $\Delta t = 0$ the Petrov–Galerkin method should revert to the conventional Galerkin method. In this limit $\psi_k(x) = \phi_k(x)$. An interpolation between these two conditions gives $\psi_k(x)$.

For the Euler time-differencing scheme, eq. (7.5.3) becomes

$$(u_a^{n+1} - u_a^n - w \, \Delta t \, u_{ax}^n, \psi_k(x)) = 0, \qquad k = 1, \ldots, N. \qquad (7.5.5)$$

Here $u_{ax} \equiv \partial u_a / \partial x$. The interpolation of ψ_k between ϕ_k and χ_k is controlled by a parameter α. Thus

$$\psi_k = \phi_k + \alpha(\chi_k - \phi_k). \qquad (7.5.6)$$

This equation has a structure similar to eq. (7.1.13). Morton and Parrott choose the following form for χ_k:

$$\chi_k(\xi) = \begin{cases} 4 - 6\xi & \text{if } 0 < \xi < 1, \\ 0 & \text{otherwise.} \end{cases} \qquad (7.5.7)$$

Table 7.4. Comparison of phase errors Δ

$c = w\dfrac{\Delta t}{\Delta x}$	$\xi = k\,\Delta x$	Lax– Wendroff	Euler Petrov– Galerkin	Crank– Nicolson Petrov– Galerkin	Crank– Nicolson Galerkin
0.2	$\pi/4$	-0.095	-0.018	-0.002	-0.004
	$\pi/2$	-0.346	-0.108	-0.042	-0.052
	$3\pi/4$	-0.680	-0.388	-0.290	-0.310
0.55	$\pi/4$	-0.066	-0.024	-0.001	-0.017
	$\pi/2$	-0.227	-0.088	-0.017	-0.094
	$3\pi/4$	-0.477	-0.196	-0.108	-0.346
0.90	$\pi/4$	-0.016	-0.007	0.001	-0.041
	$\pi/2$	-0.036	-0.011	0.010	-0.160
	$3\pi/4$	-0.004	$+0.026$	0.046	-0.400

The coordinate ξ is defined in Fig. 7.4. It may be noted that ψ_k is linear in ξ. This may be contrasted with the quadratic form assumed in eqs. (7.1.13) and (7.1.14).

Evaluation of eq. (7.5.5) gives the following algebraic equation, the Euler Petrov–Galerkin scheme, on a uniform grid:

$$\{1 + \tfrac{1}{6}(1 - \alpha)\delta^2\}(\bar{u}_k^{n+1} - \bar{u}_k^n) = c\,\delta\bar{u}_k^n + 0.5\alpha c\,\delta^2\bar{u}_k^n. \tag{7.5.8}$$

The structure of this equation is similar to that of the Lax–Wendroff finite-difference scheme. To obtain the Lax–Wendroff scheme would require simultaneously $\alpha = c$ on the right-hand side and $\alpha = 1$ on the left-hand side.

Morton and Parrott show that eq. (7.5.8) is stable if

$$0 \le w\frac{\Delta t}{\Delta x} \le \alpha \le 1. \tag{7.5.9}$$

In solving hyperbolic problems it is important that any computational scheme introduce as little dissipation and dispersion as possible. Both effects can be checked for a linear equation like eq. (7.5.1) by introducing a Fourier mode e^{ikx} as an initial condition and comparing the subsequent behavior predicted by the algebraic equation (7.5.8) with the behavior predicted by the exact equation (7.5.1). In particular, the amplitude will be reduced by positive dissipation, and the speed of propagation and hence subsequent position (phase) will be altered by dispersion.

Here the damping factor κ is defined as the relative amplitude after passage of one wave, that is, $\kappa = 1$ implies no damping. If we set $\xi = k\,\Delta x$, then n ($= 2\pi/c\xi$) time steps will be required for one wave to pass. During this time a nondimensional phase error of Δ (units of 2π) will have occurred.

The Euler Petrov–Galerkin (EPG) scheme is compared with the Lax–Wendroff scheme in Tables 7.4 and 7.5. The phase errors are generally less,

Table 7.5. Comparison of damping factor κ

$c = w\dfrac{\Delta t}{\Delta x}$	$\xi = k\,\Delta x$	Lax–Wendroff	Euler Petrov–Galerkin	Crank–Nicolson Petrov–Galerkin	Crank Nicolson Galerkin
0.2	$\pi/4$	0.936	0.966	0.995	1.000
	$\pi/2$	0.675	0.724	0.944	1.000
	$3\pi/4$	0.462	0.303	0.765	1.000
0.55	$\pi/4$	0.879	0.923	0.973	1.000
	$\pi/2$	0.436	0.528	0.756	1.000
	$3\pi/4$	0.148	0.134	0.291	1.000
0.90	$\pi/4$	0.948	0.965	0.983	1.000
	$\pi/2$	0.715	0.797	0.888	1.000
	$3\pi/4$	0.551	0.672	0.802	1.000

particularly for small values of ξ and c. The EPG damping factors are generally closer to one, except for small values of c and large values of ξ (short wavelength). Since dispersion is often a more serious problem than dissipation, it can be argued that waves of short wavelength, which are propagated at the wrong speed, should be dissipated as much as possible (small κ).

The results shown in Table 7.4 and 7.5 have been obtained with $\alpha = c$ for which the spatial truncation error is $O(\Delta x^2)$. For $\alpha > c$ the spatial truncation error is $O(\Delta x)$.

The Euler time-differencing scheme can be generalized by evaluating the spatial terms as a weighted average of their values at the nth and $(n + 1)$th time steps. Thus eq. (7.5.5) is replaced by

$$((u_a^{n+1} - w\,\Delta t\,\theta u_{ax}^{n+1} - u_a^n - w\,\Delta t\,(1 - \theta)u_{ax}^n), \psi_k(x)) = 0, \qquad k - 1, \ldots, N, \tag{7.5.10}$$

where $\theta = 0$ gives the Euler Petrov–Galerkin formulation. For this formulation the test function that satisfies the unit-CFL condition is

$$\chi_k(\xi) = \begin{cases} 2(2 - 3\theta) - (6(1 - 2\theta)\xi) & \text{if } 0 < \xi < 1, \\ 0 & \text{otherwise.} \end{cases} \tag{7.5.11}$$

Making use of eq. (7.5.6) and evaluating eq. (7.5.10) with linear trial functions on a uniform grid gives the following algebraic equation for $\theta = 0.5$:

$$[1 + 0.5\alpha\delta + \tfrac{1}{6}(1 + 0.5\alpha)\delta^2](\bar{u}_k^{n+1} - \bar{u}_k^n) = 0.5c[\delta + 0.5\alpha\delta^2](\bar{u}_k^{n+1} + \bar{u}_k^n). \tag{7.5.12}$$

This scheme is unconditionally stable and second-order accurate for $\theta < \alpha < 1$. For the special case $\alpha = c^2$, the scheme is third-order accurate. This is the scheme referred to as Crank–Nicolson Petrov–Galerkin in Tables 7.4 and 7.5.

The results indicate that this scheme has less phase error than either the Euler Petrov–Galerkin scheme or the conventional Crank–Nicolson Galerkin scheme ($\alpha = 0$). Although it is less dissipative than the Euler Petrov–Galerkin scheme, it is more dissipative than the Crank–Nicolson Galerkin scheme, which is nondissipative. Like the conventional Crank–Nicolson Galerkin scheme, the Crank–Nicolson Petrov–Galerkin scheme is conservative and consequently suitable for solving nonlinear hyperbolic equations.

Morton and Stokes (1981) have found that the above techniques can be extended to two dimensions without difficulty.

The notation used in eq. (7.4.5) has also been applied to eq. (7.5.1) by Griffiths and Mitchell (1979). More recently Griffiths (1981) has shown that the schemes of Morton and Parrott (e.g. eqs. (7.5.8) and (7.5.12)) can be interpreted within the framework of eq. (7.4.5). In particular Griffiths shows that the unit-CFL condition implies that

$$\alpha_1 = 2\alpha_2 + 1 - 2\theta,$$
$$\alpha_3 = 0.5 + \alpha_1(\theta - 0.5), \tag{7.5.13}$$

where α_1, α_2, and α_3 are the same as the parameters considered in section 7.4 and θ is as given in eq. (7.5.10).

Griffiths offers the following alternative interpretation of the Crank–Nicolson Petrov–Galerkin formulation applied to eq. (7.5.1): that it is equivalent to a conventional Galerkin method applied to the modified equation,

$$\left[1 - 0.5\alpha_1 w\,\Delta x\,\frac{\partial}{\partial x}\right]\left\{\frac{\partial u}{\partial t} + w\frac{\partial u}{\partial x}\right\}$$
$$- \Delta x(\alpha_2 - 0.5\alpha_1)\frac{\partial^2 u}{\partial x\,\partial t} + 0.5(\alpha_3 - \tfrac{1}{3})\,\Delta x^2\frac{\partial^3 u}{\partial x^2\,\partial t} = 0. \tag{7.5.14}$$

This approach is conceptually similar to that of Hughes and Brooks (1979) for adding dissipation to the convection–diffusion equation (section 7.3). Brooks and Hughes (1980) have considered the two-dimensional inviscid convection equation which governs the rotating-cone problem (Figs. 6.3 to 6.5). As in the treatment of the convection–diffusion equation, Brooks and Hughes add artificial diffusion aligned with the local flow direction to avoid the introduction of crosswind diffusion. They chose the streamline diffusion parameter to minimize phase error.

7.6. Closure

In this chapter we have looked at convection–dominated flows and the use of the Petrov–Galerkin method to avoid the spatial oscillations associated with the conventional Galerkin method.

For steady convection–diffusion problems in one dimension, the free parameter in the simplest Petrov–Galerkin formulation can be chosen to give the exact answer. In more than one dimension the Petrov–Galerkin formulation must be constrained to avoid the introduction of spurious crosswind diffusion. For transient problems more free parameters are required to cope with the additional problem of dispersion, namely, incorrect wave propagation speed.

Perhaps the most effective way of using the Petrov–Galerkin formulation is to indicate what additional terms should be added to the original equation so that, after application of the conventional Galerkin method, the subsequent approximate solution will have minimized the nonphysical behavior.

References

Anderssen, R. S., and Mitchell, A. R. *Math. Meth. Appl. Sci.* **1**, 3–15 (1979).

Barrett, J. W., and Morton, K. W. *Int. J. Num. Meth. Eng.* **15**, 1457–1474 (1980).

Barrett, J. W., and Morton, K. W. Optimal Finite Element Approximation for Diffusion–Convection Problems, *Conference on Mathematics of Finite Elements and Their Applications*, Brunel University, May 1981.

Brooks, A., and Hughes, T. J. R. *Proceedings 3rd International Conference on Finite Element Methods in Flow Problems*, Banff, Canada, pp. 283–292 (1980).

Christie, I., Griffiths, D. F., Mitchell, A. R., and Zienkiewicz, O. C. *Int. J. Num. Meth. Eng.* **10**, 1389–1396 (1976).

Christie, I., and Mitchell, A. R. *Int. J. Num. Meth. Eng.* **12**, 1764–1771 (1978).

Christie, I., Griffiths, D. F., Mitchell, A. R., and Sanz-Serna, J. M. "Product Approximation for Nonlinear Problems in the Finite Element Method", *Inst. Math. Appl. J. Num. Anal.* **1**, 253–266 (1981).

Cole, J. D. *Quart. Appl. Math.* **9**, 225–236 (1951).

Gresho, P. M., and Lee, R. L. *Comp. and Fluids* **9**, 223–253 (1981).

Griffiths, D. F. A Petrov–Galerkin Method for Hyperbolic Equations, *Conference on Mathematics of Finite Elements and Their Applications*, Brunel University, May 1981.

Griffiths, D. F., and Lorenz, J. *Comp. Meth. App. Mech. Eng.* **14**, 39–64 (1978).

Griffiths, D. F., and Mitchell, A. R. In *Finite Elements for Convection Dominated Flows* (ed. T. J. R. Hughes), AMD 34, pp. 91–104, ASME, New York (1979).

Heinrich, J. C., Huyakorn, P. S., Mitchell, A. R., and Zienkiewicz, O. C. *Int. J. Num. Meth. Eng.* **11**, 131–143 (1977).

Heinrich J. C., and Zienkiewicz, O. C. *Int. J. Num. Meth. Eng.* **11**, 1831–1844 (1977).

Hughes, T. J. R. *Int. J. Num. Meth. Eng.* **12**, 1359–1365 (1978).

Hughes, T. J. R., and Brooks, A. In *Finite Element Methods for Convection Dominated Flows* (ed. T. J. R. Hughes), AMD 34, pp. 19–35, ASME, New York (1979).

Huyakorn, P. S. *Appl. Math. Modelling* **1**, 187–195 (1977).

Kelly, D. W., Nakazawa, S., and Zienkiewicz, O. C. *Int. J. Num. Meth. Eng.* **15**, 1705–1711 (1980).

Leschziner, M. A. *Comp. Meth. Appl. Mech. Eng.* **23**, 293–312 (1980).

Mitchell, A. R., and Griffiths, D. F. *The Finite Difference Method in Partial Differential Equations*, Wiley, Chichester (1980).

Morton, K. W., and Parrott, A. K. *J. Comp. Phys.* **36**, 249–270 (1980).

Morton, K. W., and Stokes, A. Generalised Galerkin Methods of Hyperbolic Equations, *Conference on Mathematics of Finite Elements and Their Applications*, Brunel University, May 1981.

Richtmyer, R. D., and Morton, K. W. *Difference Methods for Initial-Value Problems*, Wiley, New York, 2nd Edn. (1967).

Roache, P. J. *Computational Fluid Dynamics*, Hermosa, Albuquerque, NM (1972).

Program BURG1

Program BURG1 contains the necessary code, written in FORTRAN, to solve the propagating shock problem governed by Burgers' equation; both by the traditional Galerkin (section 1.2.5) and by the spectral (section 5.2.2) method. Program BURG1, and BURG4 described in Appendix 2, were written in single precision on a CYBER 72.

The subroutines used by BURG1 are briefly described in Table A1.1. The main program is called BURG1. The main program, BURG1 performs the integration of eqs. (1.2.67) and (5.2.16). This is indicated in the flow diagram, Fig. A.1.1. It may be noted that Fig. A.1.1 contains the flow diagram for program BURG4.

The integration loop consists of four calls to GRAD to evaluate intermediate values of $\dot{a}$, required by the fourth-other Runge-Kutta scheme. Each time the array A (see Table A1.2 for description) is upgraded, subroutine BC is called to adjust $A(NN + 1)$ and $A(NN + 2)$ to ensure that the boundary conditions, eqs. (1.2.60) and (5.2.14), are satisfied.

After $\dot{a}_j$ is calculated the maximum value of $|\dot{a}_j \Delta t|$ is compared with the maximum value of $|a_j|$ to decide if Δt should be changed. No change is made if $0.1 \, DEL < |a_j \Delta t|/|a_j| < DEL$.

The main purpose of FORM1 is to set up the matrices C, AA, and E, and to evaluate the initial values of A. For the traditional Galerkin method $(IQL = 1)$ this involves inverting C, in MATINV, and solving eq. (1.2.72). For the spectral method the matrix inversion is not necessary.

In the present program C, AA, and E have been evaluated by Gauss quad-

Table A1.1. Subroutine Description for Program BURG1

Subroutine	Description
BURG1	Reads in control parameters; sets up Gauss quadrature parameters XI, H; integrates eqs. (1.2.67) and (5.2.16); controls stepsize HH; evaluates solutions U and UEX and rms and $L_{2,d}$ errors
FORM1	Evaluates test and trial functions; calculates C, AA, E, and the starting values of A
MATINV	Inverts C
GRAD1	Calculates the local gradients DK1, DK2, DK3 and DK4
BC	Adjusts $A(NN + 1)$ and $A(NN + 2)$ to satisfy boundary conditions
ERFC	Calculates complementary error function; needed for UEX

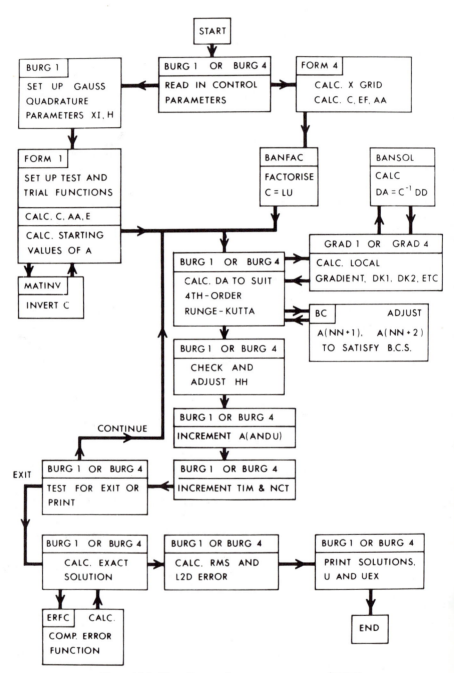

Figure A1.1. Flow diagram for programs BURG1 and BURG4

Table A1.2. Variable Description for Program BURG1

A	Unknown coefficients in trial solution, eqs. (1.2.64) and (5.2.15)
AA	Dissipation matrix $[\equiv \mathbf{C}$ in eqs. (1.2.67) and (5.2.16)$]$
AL	x location
B	Right-hand side of initial value matrix $[\equiv \mathbf{D}$ in eqs. (1.2.72) and (5.2.20)$]$
C	Mass matrix $[\equiv \mathbf{M}$ in eqs. (1.2.67) and (5.2.16)$]$
DA	$\dot{a}$ in BURG1; equivalent to DK1 etc. in GRAD1
DD	Local value of $\dot{a}$ (in GRAD1)
DEL, APS	Parameters controlling adjustment of timestep
DK1 etc.	Intermediate values of $\dot{a}$
E	Convection matrix $[\equiv \mathbf{B}$ in eqs. (1.2.67) and (5.2.16)$]$
HH	Current timestep
HHM	Maximum timestep
IPR	Print control
IQL	Control parameter; chooses between traditional and spectral methods
LST	Parameter controlling full or boundary calculation of UEX
NCT	Timestep counter
NINC	Print increment
NN	Number of unknown coefficients, A, directly calculated: NN = NOR − 2
NOR	Number of unknown coefficients in trial solution, eqs. (1.2.64) and (5.2.15)
NPR	Printer control
NST, NFIN	Do loop limits for upgrading components of A
NSTOP	Maximum number of timesteps
RE	Reynolds number
SUMD	rms error
SUML	Error in discrete L_2 norm
T	Trial function; Chebyshev polynomial (IQL = 1) or Legendre polynomial (IQL = 2)
TB	Trial function scaled to suit the initial conditions
TIM	Time
TMAX	Maximum time
TXI	dT/dx
U	Dependent variable being solved for; eqs. (1.2.64) and (5.2.15)
UEX	Exact solution
UL, UR	Boundary values of U; eqs. (1.2.60) and (5.2.14)

rature. This is a general procedure which, for the particular case considered here, could be avoided; i.e., the integral like eq. (1.2.68) could be evaluated analytically.

Once the maximum time, TMAX, or the maximum number of steps, NSTOP, is exceeded the program calculates the exact solution, UEX, and the rms and discrete L_2 errors. Subroutine ERFC is required to compute the exact solution.

A description of the major variables that are used in BURG1 is given in Table A1.2. A brief description of the input parameters is given at the beginning of the listing of BURG1. The listings of the subroutines are followed by some typical output for the traditional Galerkin method.

```
      PROGRAM BURG1(INPUT,OUTPUT,TAPE3,TAPE6=OUTPUT)
C
C     APPLIES TRADITIONAL OR SPECTRAL GALERKIN METHOD
C     TO THE PROPAGATING SHOCK PROBLEM GOVERNED BY
C     BURGERS EQUATION .  TIME INTEGRATION IS BY A VARIABLE
C     STEP 4TH-ORDER RUNGE-KUTTA SCHEME.
C
C
C     THE INPUT PARAMETERS ARE:
C
C     HH    IS THE INITIAL TIME STEP
C     HHM   IS THE MAXIMUM TIME STEP
C     TMAX  IS THE MAXIMUM TIME
C     RE    IS THE REYNOLDS NUMBER
C     NSTOP IS THE MAXIMUM NUMBER OF TIME STEPS
C     NOR   IS THE NUMBER OF UNKNOWN COEFFICIENTS, A
C     IQL = 1, TRADITIONAL GALERKIN METHOD
C     IQL = 2, SPECTRAL METHOD
C     NINC  IS THE PRINT INCREMENT
C     IPR = 1, FULL PRINT
C     IPR = 0, ESSENTIAL PRINT
C
      DIMENSION C(15,15),AA(15,15),A(15),E(15,15,15),XI(17),H(15)
     1,T(11,21),TXI(11,21),DA(15),U(21),UEX(21),DK3(15),DK4(15)
      DIMENSION XD(7),HD(7),A1(15),A2(15),A3(15),DK1(15),DK2(15)
      COMMON C,AA,E,RE,HH
      WRITE(6,10)
   10 FORMAT(26H OUTPUT FROM PROGRAM BURG1,//)
C
C     GAUSS QUADRATURE PARAMETERS
C
      DATA XD/0.2011940940,0.3941513471,0.5709721726,0.7244177314
     1,0.8482065834,0.9372733924,0.9879925180/
      DATA HD/0.1984314853,0.1861610001,0.1662692058,0.1395706779
     1,0.1071592205,0.0703660475,0.0307532420/
C
C     READ IN CONTROL PAPAMETERS
C
      READ(3,20)HH,HHM,TMAX,RE
   20 FORMAT(4E10.3)
      WRITE(6,30)HH,HHM,TMAX
   30 FORMAT(16H INIT. T/STEP = ,E10.3,14H MAX T/STEP = ,E10.3,
     112H MAX TIME = ,E10.3)
      WRITE(6,40)RE
   40 FORMAT(16H REYNOLDS NO. = ,E10.3)
C
      READ(3,50)NSTOP,NOR,IQL,NINC,IPR
   50 FORMAT(5I5)
      WRITE(6,60)NSTOP,NOR,IQL,NINC,IPR
   60 FORMAT(9H NSTOP = ,I5,7H NOR = ,I5,7H IQL = ,I5,
     18H NINC = ,I5,7H IPR = ,I5,//)
C
      DEL = 0.10
      APS = 2.00
      PI = 3.1415926536
      UL = 1.
      UR = 0.
      NN = NOR - 2
      NST = 1
      NFIN = NN
      NH = NN/2
```

```
C     XI AND H ARE GAUSS QUADRATURE SAMPLE-POINTS AND WEIGHTS
C
      H(8) = 0.20257 82419
      XI(8) = 0.
      DO 70 I = 1,7
      IA = 8 - I
      XI(I+8) = XD(I)
      XI(I) = -XD(IA)
      H(I+8) = HD(I)
      H(I) = HD(IA)
   70 CONTINUE
      XI(16) = -1.
      XI(17) = 1.
      IF(IPR .EQ. 0)GOTO 100
      DO 80 I = 1,15
      IA = I
   80 WRITE(6,90)I,XI(I),H(I)
   90 FORMAT(4H I= ,I2,5X,4H X= ,F12.10,5X,4H H= ,F12.10)
  100 CONTINUE
C
C     SUBROUTINE FORM1 EVALUATES T,TB,TXI,C,AA,E AND STARTING
C     VALUES FOR A
C
      CALL FORM1(IPR,NOR,IQL,XI,T,A,H,UL,UR)
C
      IF(IPR .EQ. 0)GOTO 130
      DO 120 I = 1,NOR
      WRITE(6,110)I,A(I),(C(I,K),K=1,NN)
  110 FORMAT(3H I=,I2, 3H A=,E12.5,3H C=,9E10.3)
  120 CONTINUE
  130 CONTINUE
C
C     INTEGRATE A USING A 4TH-ORDER RUNGE-KUTTA SCHEME
C
      HHMD = HHM
      TIM = 0.
      NCT = 1
      NPR = NINC
      LST = 1
C
  140 CALL GRAD1(DK1,A,NOR,IQL)
      DO 150 K = NST,NFIN
  150 A1(K) = A(K) + 0.5*DK1(K)
      CALL BC(NN,A1,UL,UR)
C
      CALL GRAD1(DK2,A1,NOR,IQL)
      DO 160 K = NST,NFIN
  160 A2(K) = A(K) + 0.5*DK2(K)
      CALL BC(NN,A2,UL,UR)
C
      CALL GRAD1(DK3,A2,NOR,IQL)
      DO 170 K = NST,NFIN
  170 A3(K) = A(K) + DK3(K)
      CALL BC(NN,A3,UL,UR)
C
      CALL GRAD1(DK4,A3,NOR,IQL)
      DO 180 K = NST,NFIN
      DA(K) =         (DK1(K)+2.*(DK2(K)+DK3(K))+DK4(K))/6.
  180 CONTINUE
```

```
C      CHECK APPROPRIATE STEPSIZE
C
       ACT = NCT + 1
       IF(NCT .LE. 9)HHM=HHMD*(ACT/10.)**2
       DAM = 0.
       AM = 0.
       DO 200 K = NST,NFIN
       IF(ABS(DA(K)) .LT. DAM)GOTO 190
       DAM= ABS(DA(K))
   190 IF(ABS(A(K)) .LT. AM)GOTO 200
       AM = ABS(A(K))
   200 CONTINUE
C
       IF(DAM/AM .GT. DEL)GOTO 220
       IF(DAM/AM .GE. 0.1*DEL)GOTO 240
       IF(APS*HH .GE. HHM)GOTO 240
       DO 210 K = NST,NFIN
   210 DA(K) = DA(K)*APS
       HH = HH*APS
       GOTO 240
   220 HHO = HH
       HH = HH*DEL*AM/DAM
       IF(HH .GT. HHM)HH=HHM
       IF(HH .LT. 0.01*HHMD)HH=0.01*HHMD
       DO 230 K = NST,NFIN
   230 DA(K) = DA(K)*HH/HHO
C
C      INCREMENT A
C
   240 DO 250 K = NST,NFIN
   250 A(K) = A(K) + DA(K)
C
       CALL BC(NN,A,UL,UR)
C
C      COMPUTE U
C
       DO 270 L = 1,21
       AL = L - 1
       AL = -1. + 0.1*AL
       U(L) = 0.
       DO 260 J = 1,NOR
       U(L) = U(L) + A(J)*T(J,L)
   260 CONTINUE
   270 CONTINUE
C
C      TEST FOR EXIT OR PRINT
C
       TIM = TIM + HH
       NCT = NCT + 1
       IF(TIM .GE. TMAX)GOTO 280
       IF(NCT .GE. NSTOP)GOTO 280
       IF(NCT .LT. NPR)GOTO 140
       NPR = NPR + NINC
       LST = 21
```

```
      IF(IPR .EQ. 0)GOTO 320
280   WRITE(6,290)NCT,TIM,HH,RE
290   FORMAT(7H NCT = ,I3,7H TIM = ,E12.5,6H HH = ,E12.5
     1,6H RE = ,E12.5)
      WRITE(6,300)U
300   FORMAT(3H U=,11F5.2)
      WRITE(6,310)(A(I),I=1,NOR)
310   FORMAT(3H A=,6E11.4)
320   CONTINUE
C
      SUMD = 0.
      DO 340 L = 1,21
      IF(LST .EQ. 1)GOTO 330
      IF(L .GT. 1 .AND. L .LT. 21)GOTO 340
330   CONTINUE
C
C     CALCULATE EXACT SOLUTION
C
      AL = L-1
      AL = -1.0 + 0.1*AL
      XB = -AL
      XA = AL - TIM
      PQ = SQRT(0.25*PI)
      PP = SQRT(PI*TIM/RE)
      XA= XA*PQ/PP
      XB = XB*PQ/PP
C
C     SUBROUTINE ERFC CALCULATES COMPLEMENTARY ERROR FUNCTION
C
      CALL ERFC(XA,EXA)
C
      SUMA = PP*EXA
C
      CALL ERFC(XB,EXB)
C
      SUMB = PP*EXB
      SUMC = EXP(0.5*RE*(AL-0.5*TIM))
      UEX(L) = SUMA/(SUMA+SUMC*SUMB)
      SUMD = SUMD + (U(L)-UEX(L))**2
340   CONTINUE
C
      IF(LST .EQ. 1)GOTO 350
      UR = UEX(21)
      UL = UEX(1)
      LST = 1
      GOTO 140
350   CONTINUE
C
C     COMPARE SOLUTION WITH EXACT
C
      SUML = SQRT(SUMD)
      SUMD = SQRT(SUMD/19.)
      DO 360 L = 1,21
      DU = U(L) - UEX(L)
360   WRITE(6,370)L,UEX(L),U(L),DU
370   FORMAT(3H L=,I3,6H UEX =,E11.4,5H  U =,E11.4,6H  DU =,E11.4)
      WRITE(6,380)SUMD,SUML
380   FORMAT(9H RMS/ERR=,E11.4,21H DISCRETE L2 ERROR = ,E11.4)
      STOP
      END
```

```
      SUBROUTINE GRAD1(DA,A,NOR,IQL)
C
C     COMPUTE LOCAL GRADIENT AS REQUIRED BY O.D.E. INTEGRATOR
C
      DIMENSION DA(15),C(15,15),AA(15,15),DD(15),A(15),E(15,15,15)
      COMMON C,AA,E,RE,HH
      NN = NOR - 2
C
      DO 30 K = 1,NN
      DD(K) = 0.
      DO 20 I = 1,NOR
      DKI = 0.
      DO 10 J = 1,NOR
   10 DKI = DKI + E(I,J,K)*A(J)
      DD(K) = DD(K) + (AA(I,K)/RE-DKI)*A(I)
   20 CONTINUE
   30 CONTINUE
C
      DO 50 I = 1,NN
      DA(I) = 0.
      IF(IQL .EQ. 2)DA(I) = DD(I)/C(I,I)
      IF(IQL .EQ. 2)GOTO 50
      DO 40 J = 1,NN
   40 DA(I) = DA(I) + C(J,I)*DD(J)
   50 DA(I) = DA(I)*HH
      RETURN
      END




      SUBROUTINE BC(NN,A,UL,UR)
C
C     COMPUTE A(NN+1),A(NN+2) TO SATISFY BOUNDARY CONDITIONS
C
      DIMENSION A(15)
      NH = NN/2
      SE = 0.
      SO = 0.
      DO 10 I = 1,NH
      IA = 2*I-1
      SO = SO + A(IA)
   10 SE = SE + A(IA+1)
      SO = SO + A(NN)
      A(NN+1) = -0.5*UL - SE + 0.5*UR
      A(NN+2) = 0.5*UL - SO  + 0.5*UR
      RETURN
      END
```

```
      SUBROUTINE FORM1(IPR,NOR,IQL,XI,T,A,H,UL,UR)
C
C     CALCULATE T,TB,TXI,C,AA,E AND STARTING VALUES OF A
C
      DIMENSION C(15,15),B(15),A(15),TB(11,21),XI(17)
      DIMENSION E(15,15,15),AA(15,15),T(11,21),TXI(11,21),H(15)
      COMMON C,AA,E,RE,HH
      NN = NOR - 2
      XIO = 0.
C
C     IQL = 1  USE CHEBYSHEV FUNCTIONS AS TEST AND TRIAL FUNCTIONS
C     IQL = 2  USE LEGENDRE FUNCTIONS AS TEST AND TRIAL FUNCTIONS
C
      DO 40 I = 1,17
      XET = 0.5*(XI(I)*(1.+XIO) - (1.-XIO))
      T(1,I) = 1.
      T(2,I) = XI(I)
      TB(1,I) = 1.
      TB(2,I) = XET
      TXI(1,I) = 0.
      TXI(2,I) = 1.
      DO 30 J = 3,NOR
      IF(IQL .EQ. 2)GOTO 10
      BJ = 2.
      CJ = 1.
      GOTO 20
   10 AJ = J - 2
      BJ = (2.*AJ+1.)/(AJ+1.)
      CJ = AJ/(AJ+1.)
   20 CONTINUE
      T(J,I) = BJ*XI(I)*T(J-1,I)   - T(J-2,I)*CJ
      TB(J,I) = BJ*XET*TB(J-1,I) - TB(J-2,I)*CJ
   30 TXI(J,I) = BJ*T(J-1,I) + BJ*XI(I)*TXI(J-1,I) - TXI(J-2,I)*CJ
   40 CONTINUE
C
      IF(IPR .EQ. 0)GOTO 80
      DO 50 I = 1,15
      WRITE(6,60)(I,(T(J,I),J=1,NOR))
   50 WRITE(6,70)I,(TXI(J,I),J=1,NOR)
   60 FORMAT(3H I=,I3,5X,6H    T= ,11E10.3)
   70 FORMAT(3H I=,I3,5X,6H  TXI= ,11E10.3)
   80 CONTINUE
C
C     EVALUATE C,AA,E
C
      DO 160 I = 1,NOR
      DO 120 K = 1,NOR
      DO 90 J = 1,NOR
   90 E(I,J,K) = 0.
      C(I,K) = 0.
      AA(I,K) = T(K,17)*TXI(I,17) - T(K,16)*TXI(I,16)
      IF(I .EQ. K)B(K) = 0.
C
      DO 110 L=1,15
      C(I,K) = C(I,K) + T(I,L)*T(K,L)*H(L)
      AA(I,K) = AA(I,K) - TXI(I,L)*TXI(K,L)*H(L)
      DO 100 J = 1,NOR
  100 E(I,J,K) = E(I,J,K) + T(I,L)*TXI(J,L)*T(K,L)*H(L)
      IF(I .EQ. K)B(K) = B(K) + TB(K,L)*H(L)
  110 CONTINUE
```

```
         IF(IPR .EQ. 0)GOTO 120
         WRITE(6,130)I,K,(E(I,J,K),J=1,NOR)
     120 CONTINUE
  C
         B(I) = B(I)*0.5*(1.+XIO)*UL
         IF(IPR .EQ. 0)GOTO 160
         WRITE(6,140)I,B(I),(C(I,K),K=1,NOR)
         WRITE(6,150)I,B(I),(AA(I,K),K=1,NOR)
     130 FORMAT(5H I,K=,2I5,5H  E= ,11E10.3)
     140 FORMAT(3H I=,I2,3H B=,E10.3,4H  C=,11E10.3)
     150 FORMAT(3H I=,I2,3H B=,E10.3,4H AA=,11E10.3)
     160 CONTINUE
  C
         IF(IQL .EQ. 1)GOTO 180
         DO 170 K = 1,NN
         AK = K
         C(K,K) = 1./(AK-0.5)
     170 A(K) = B(K)/C(K,K)
         GOTO 210
     180 CONTINUE
  C
  C      MODIFY B AND C TO ACCOUNT FOR BOUNDARY CONDITIONS
  C
         NH = NN/2
         DO 200 K = 1,NN
         DO 190 I = 1,NH
         IA = 2*I-1
         C(IA,K) = C(IA,K) - C(NN+2,K)
     190 C(IA+1,K) = C(IA+1,K) - C(NN+1,K)
         B(K) = B(K) + 0.5*(C(NN+1,K)-C(NN+2,K))*UL-0.5*UR*(C(NN+1,K)
        1 + C(NN+2,K))
     200 C(NN,K)= C(NN,K) - C(NN+2,K)
  C
  C      REDEFINE T ON A UNIFORM GRID
  C
     210 DO 250 L = 1,21
         AL = L-1
         AL = -1. + 0.1*AL
         T(1,L) = 1.
         T(2,L) = AL
         DO 240 J = 3,NOR
         IF(IQL .EQ. 2)GOTO 220
         BJ  = 2.
         CJ = 1.
         GOTO 230
     220 AJ = J - 2
         BJ = (2.*AJ+1.)/(AJ+1.)
         CJ = AJ/(AJ+1.)
     230 CONTINUE
         T(J,L) = BJ*AL*T(J-1,L) - T(J-2,L)*CJ
     240 CONTINUE
     250 CONTINUE
         IF(IQL .EQ. 2)GOTO 280
  C
  C      OBTAIN STARTING VALUES FOR A
  C
         CALL MATINV(C,NN)
  C
         DO 270 I = 1,NN
         A(I) = 0.
         DO 260 K = 1,NN
     260 A(I) = A(I) + C(K,I)*B(K)
     270 CONTINUE
  C
     280 CALL BC(NN,A,UL,UR)
  C
         RETURN
         END
```

```
      SUBROUTINE MATINV(AMAK,N)
C
C     PROGRAM CALCULATES INVERSE OF MATRIX AMAK GIVEN N
C     J = NUMBER OF ROW = 1,...,N
C     K = NUMBER OF COLUMN = 1,...,N
      DIMENSION AMAK(15,15),IPIVOT(15),INDEX(15,2),PIVOT(15)
      DETERM = 1.0
      DO 10 J=1,N
 10   IPIVOT(J) = 0
      DO 90 ID = 1,N
      AMAX = 0.0
      DO 30 J=1,N
      IF(IPIVOT(J) .EQ. 1)GOTO 30
      DO 20 K = 1,N
      IF(IPIVOT(K) .EQ. 1)GOTO 20
      IF(IPIVOT(K) .GT. 1)GOTO 120
      IF(ABS(AMAX) .GE. ABS(AMAK(J,K)))GOTO 20
      IROW = J
      ICOLUM = K
      AMAX = AMAK(J,K)
 20   CONTINUE
 30   CONTINUE
      IPIVOT(ICOLUM) = IPIVOT(ICOLUM) + 1
      IF(IROW .EQ. ICOLUM)GOTO 50
      DETERM = -DETERM
      DO 40 L=1,N
      SWAP = AMAK(IROW,L)
      AMAK(IROW,L) = AMAK(ICOLUM,L)
 40   AMAK(ICOLUM,L) = SWAP
 50   INDEX(ID,1) = IROW
      INDEX(ID,2) = ICOLUM
      PIVOT(ID) = AMAK(ICOLUM,ICOLUM)
      DETERM = DETERM*PIVOT(ID)
      AMAK(ICOLUM,ICOLUM) = 1.0
      DO 60 L=1,N
 60   AMAK(ICOLUM,L) = AMAK(ICOLUM,L)/PIVOT(ID)
      DO 80 L1=1,N
      IF(L1 .EQ. ICOLUM)GOTO 80
      T=AMAK(L1,ICOLUM)
      AMAK(L1,ICOLUM) = 0.0
      DO 70 L=1,N
 70   AMAK(L1,L) = AMAK(L1,L) - AMAK(ICOLUM,L)*T
 80   CONTINUE
 90   CONTINUE
C
      DO 110 ID=1,N
      L=N+1-ID
      IF(INDEX(L,1) .EQ. INDEX(L,2))GOTO 110
      JROW = INDEX(L,1)
      JCOLUM = INDEX(L,2)
      DO 100 K=1,N
      SWAP = AMAK(K,JROW)
      AMAK(K,JROW) = AMAK(K,JCOLUM)
      AMAK(K,JCOLUM) = SWAP
 100  CONTINUE
 110  CONTINUE
 120  CONTINUE
      RETURN
      END
```

```
      SUBROUTINE ERFC(S,ERC)
C     COMPLEMENTARY ERROR FUNCTION
C
      BB = ABS(S)
      IF(BB .LT. 4.)GOTO 10
      D = 1.
      GOTO 20
   10 CC = EXP(-BB*BB)
      T = 1./(1.+0.3275911*BB)
      D = 0.254829592*T - 0.284496736*T*T + 1.421413741*T*T*T
     1 - 1.453152027*T*T*T*T + 1.061405429*T*T*T*T*T
      D = 1. - D*CC
   20 IF(S .LT. 0.)D = -D
      ERF = D
      ERC = 1. - ERF
      RETURN
      END
```

```
OUTPUT FROM PROGRAM BURG1


INIT. T/STEP =    .100E-02 MAX T/STEP =    .200E-01 MAX TIME =    .400E+01
REYNOLDS NO. =    .100E+02
NSTOP =     65 NOR =     11 IQL =     1 NINC =     1 IPR =     0


NCT =  65 TIM =   .94000E+00 HH =   .16000E-01 RE =   .10000E+02
U= 1.00 1.00 1.00 1.00 1.00 1.00 1.00  .99  .99  .98  .96
U=  .92  .85  .74  .61  .46  .31  .19  .11  .06  .03
A=  .6667E+00 -.5197E+00 -.2122E+00  .1668E-01  .7232E-01  .2941E-01
A= -.9934E-02 -.1529E-01 -.4354E-02  .3169E-02  .1739E-02
L=  1 UEX =   .1000E+01  U =   .1000E+01  DU =   .8626E-07
L=  2 UEX =   .1000E+01  U =   .1000E+01  DU =   .3284E-03
L=  3 UEX =   .9999E+00  U =   .9987E+00  DU = -.1286E-02
L=  4 UEX =   .9998E+00  U =   .9987E+00  DU = -.1104E-02
L=  5 UEX =   .9996E+00  U =   .1000E+01  DU =   .6208E-03
L=  6 UEX =   .9990E+00  U =   .1001E+01  DU =   .1626E-02
L=  7 UEX =   .9977E+00  U =   .9984E+00  DU =   .7417E-03
L=  8 UEX =   .9948E+00  U =   .9937E+00  DU = -.1076E-02
L=  9 UEX =   .9888E+00  U =   .9869E+00  DU = -.1905E-02
L= 10 UEX =   .9767E+00  U =   .9759E+00  DU = -.7907E-03
L= 11 UEX =   .9538E+00  U =   .9551E+00  DU =   .1256E-02
L= 12 UEX =   .9128E+00  U =   .9151E+00  DU =   .2213E-02
L= 13 UEX =   .8448E+00  U =   .8461E+00  DU =   .1341E-02
L= 14 UEX =   .7423E+00  U =   .7426E+00  DU =   .3393E-03
L= 15 UEX =   .6067E+00  U =   .6077E+00  DU =   .9995E-03
L= 16 UEX =   .4537E+00  U =   .4554E+00  DU =   .1695E-02
L= 17 UEX =   .3085E+00  U =   .3080E+00  DU = -.4858E-03
L= 18 UEX =   .1919E+00  U =   .1884E+00  DU = -.3449E-02
L= 19 UEX =   .1106E+00  U =   .1087E+00  DU = -.1914E-02
L= 20 UEX =   .5991E-01  U =   .6168E-01  DU =   .1766E-02
L= 21 UEX =   .3079E-01  U =   .2863E-01  DU = -.2166E-02
RMS/ERR=  .1587E-02 DISCRETE L2 ERROR =    .6916E-02
```

Program BURG4

Program BURG4 contains the necessary code to solve the propagating shock problem governed by Burgers' equation (section 3.2.5). Options (IQL) are available to use a linear finite element, a quadratic finite element, or a three-point finite difference method.

The subroutines used by BURG4 are described in Table A2.1. A flow diagram for BURG4 is given in Fig A1.1. The structure of BURG4 is very similar to that of BURG1. For the finite element schemes the mass matrix C is either tridiagonal or pentadiagonal. This structure is utilized by BANFAC which factorizes C into $L \cdot U$ and by BANSOL which evaluates

$$DA = U^{-1}L^{-1}DD. \tag{A.2.1}$$

A description of the variables in BURG4, which have not already been described in Table A1.2, are given in Table A2.2. A brief description of the input parameters is given at the beginning of the listing of BURG4.

The listings of the subroutines of BURG4 are followed by some typical output produced by the linear finite element option. ERFC is listed in Appendix 1 and is not repeated here.

Table A2.1. Subroutine Description for Program BURG4

Subroutine	Description
BURG4	Reads in control parameters; integrates eqs. (3.2.59); controls step-size HH; evaluates solutions U and UEX and rms and $L_{2,d}$ errors.
FORM4	Sets up the x gridpoints; calculates C, EF, and AA
BANFAC	Factorizes C into $L \cdot U$
GRAD4	Calculates local gradients DK1, DK2, DK3, and DK4
BANSOL	Evaluates $DA = U^{-1}L^{-1}DD$
ERFC	Calculates complementary error function, needed for UEX

Table A2.2 Variable Description for Program BURG4

Variable	Description
A	Nodal unknowns in trial solution, $[\equiv \bar{u}_j$ in eq. (3.2.58)$]$
AA	Dissipation matrix $[\equiv \mathbf{C}$ in eq. (3.2.59)$]$
A1 . . . A3	Intermediate values of A
C	Mass matrix $[\equiv \mathbf{M}$ in eq. (3.2.59)$]$
EC, ED	Convection matrix for quadratic elements $[\equiv \mathbf{B}$ in eq. (3.2.59)$]$
EF	Convection matrix for 3 pt. $f.d.m$ $[\equiv \mathbf{B}$ in eq. (3.2.59)$]$
EY, EZ	Data used in FORM4 to evaluate EC, ED
IQL	Control parameter; chooses between linear and quadratic finite element methods and a 3 pt. finite difference method
NN	Number of nodal unknowns solved for; NN = NOR − 2
NOR	Number of nodes
STEP	x stepsize (stepsize adjacent to x = XIO on a variable grid)
U	$= A$, nodal unknowns
X	x coordinate
XL, XR	Boundary values of x

```
      PROGRAM BURG4(INPUT,OUTPUT,TAPE3,TAPE6=OUTPUT)
C
C     APPLIES LINEAR OR QUADRATIC GALERKIN FINITE ELEMENT
C     METHOD OR 3 PT. FINITE DIFFERENCE METHOD TO THE
C     PROPAGATING SHOCK PROBLEM GOVERNED BY BURGERS EQUATION.
C     TIME INTEGRATION IS BY A VARIABLE STEP 4TH-ORDER
C     RUNGE-KUTTA SCHEME.
C
C     THE INPUT PARAMETERS ARE:
C
C     HH     IS THE INITIAL TIME STEP
C     HHM    IS THE MAXIMUM TIME STEP
C     TMAX   IS THE MAXIMUM TIME
C     RE     IS THE REYNOLDS NUMBER
C     STEP   IS THE X-STEPSIZE
C     XL     IS THE L.H. X BOUNDARY
C     XR     IS THE R.H. X BOUNDARY
C     NSTOP  IS THE MAXIMUM NUMBER OF TIME STEPS
C     NOR    IS THE NUMBER OF GRIDPOINTS
C     IQL = 1, LINEAR ELEMENT
C     IQL = 2, QUADRATIC ELEMENT
C     IQL = 3, 3 PT. FINITE DIFFERENCE SCHEME
C     NINC   IS THE PRINT INCREMENT
C     IPR = 1, FULL PRINT
C     IPR = 0, ESSENTIAL PRINT
C
      DIMENSION C(7,201),AA(7,201),EF(7,201),U(203),UEX(203),X(203)
      DIMENSION A(203),A1(203),A2(203),DK1(203),DK2(203),DK3(203),
     1A3(203),DA(203),DK4(203),EC(5,5),ED(3,3)
      COMMON C,AA,RE,HH,EC,ED,EF,X,XL,XR
      WRITE(6,10)
   10 FORMAT(26H OUTPUT FROM PROGRAM BURG4,//)
C
C     READ IN CONTROL PARAMETERS
C
      READ(3,20)HH,HHM,TMAX,RE,STEP,XL,XR
   20 FORMAT(7E10.3)
      WRITE(6,30)HH,HHM,TMAX
   30 FORMAT(16H INIT. T/STEP = ,E10.3,14H MAX T/STEP = ,E10.3,
     112H MAX TIME = ,E10.3)
      WRITE(6,40)XL,XR,STEP,RE
   40 FORMAT(6H XL = ,E10.3,6H XR = ,E10.3,
     18H STEP = ,E10.3,16H REYNOLDS NO. = ,E10.3)
C
      READ(3,50)NSTOP,NOR,IQL,NINC,IPR
   50 FORMAT(5I5)
      WRITE(6,60)NSTOP,NOR,IQL,NINC,IPR
   60 FORMAT(9H NSTOP = ,I5,7H NOR = ,I5,7H IQL = ,I5
     1,8H NINC = ,I5,7H IPR = ,I5,//)
C
      DEL = 0.1
      APS = 2.0
      NN = NOR - 2
      ANN = NN
      PI = 3.1415926536
C
C     SUBROUTINE FORM4 CALCULATES X GRIDPOINTS AND C,EF,AA
C     WHICH ARE REQUIRED BY SUBROUTINE GRAD4
C
      CALL FORM4(IPR,NOR,STEP,IQL,U,A)
C
```

```
C     BANFAC FACTORISES C INTO L.U
C
      IF(IQL .NE. 3)CALL BANFAC(C,NN,IQL)
C
C     INTEGRATE A USING A 4TH-ORDER RUNGE-KUTTA SCHEME
C
      HHMD = HHM
      TIM = 0.
      NCT = 1
      NPR = NINC
      LST = 1
C
   70 CALL GRAD4(DK1,A,NOR,IQL)
      DO 80 K = 1,NOR
   80 A1(K) = A(K) + 0.5*DK1(K)
C
      CALL GRAD4(DK2,A1,NOR,IQL)
      DO 90 K = 1,NOR
   90 A2(K) = A(K) + 0.5*DK2(K)
C
      CALL GRAD4(DK3,A2,NOR,IQL)
      DO 100 K = 1,NOR
  100 A3(K) = A(K) + DK3(K)
C
      CALL GRAD4(DK4,A3,NOR,IQL)
      DO 110 K = 1,NOR
      DA(K) =        (DK1(K)+2.*(DK2(K)+DK3(K))+DK4(K))/6.
  110 CONTINUE
C
C     CHECK APPROPRIATE STEPSIZE
C
      ACT = NCT + 1
      IF(NCT .LE. 9)HHM=HHMD*(ACT/10.)**2
      DAM = 0.
      AM = 0.
      DO 130 K = 1,NOR
      IF(ABS(DA(K)) .LT. DAM)GOTO 120
      DAM= ABS(DA(K))
  120 IF(ABS(A(K)) .LT. AM)GOTO 130
      AM = ABS(A(K))
  130 CONTINUE
C
      IF(DAM/AM .GT. DEL)GOTO 150
      IF(DAM/AM .GE. 0.1*DEL)GOTO 170
      IF(APS*HH .GE. HHM)GOTO 170
      DO 140 K = 1,NOR
  140 DA(K) = DA(K)*APS
      HH = APS*HH
      GOTO 170
  150 HHO = HH
      HH = HH*DEL*AM/DAM
      IF(HH .GT. HHM)HH=HHM
      IF(HH .LT. 0.01*HHMD)HH=0.01*HHMD
      DO 160 K = 1,NOR
  160 DA(K) = DA(K)*HH/HHO
C
  170 DO 180 K = 1,NOR
```

```
C
C      INCREMENT A (AND U)
C
       A(K) = A(K) + DA(K)
       U(K) = A(K)
  180 CONTINUE
       TIM = TIM + HH
       NCT = NCT + 1
C
C      TEST FOR EXIT OR PRINT
C
       IF(TIM .GE. TMAX)GOTO 190
       IF(NCT .GE. NSTOP)GOTO 190
       IF(NCT .LT. NPR)GOTO 70
       NPR = NPR + NINC
       LST = NOR
C
       IF(IPR .EQ. 0)GOTO 220
  190 WRITE(6,200)NCT,TIM,HH,RE
  200 FORMAT(7H NCT = ,I3,7H TIM = ,E12.5,6H HH = ,E12.5,
      16H RE = ,E12.5)
       WRITE(6,210)(A(K),K=1,NOR)
  210 FORMAT(5H A = ,6E12.5)
  220 CONTINUE
C
       SUMD = 0.
       DO 240 L = 1,NOR
       IF(LST .EQ. 1)GOTO 230
       IF(L .GT. 1 .AND. L .LT.NOR)GOTO 240
  230 CONTINUE
C
C      CALCULATE EXACT SOLUTION
C
       AL = X(L)
       XB = -AL
       XA = AL - TIM
       PQ = SQRT(0.25*PI)
       PP = SQRT(PI*TIM/RE)
       XB = XB*PQ/PP
       XA = XA*PQ/PP
C
C      SUBROUTINE ERFC CALCULATES COMPLEMENTARY ERROR FUNCTION
C
       CALL ERFC(XA,EXA)
C
       SUMA = PP*EXA
C
       CALL ERFC(XB,EXB)
C
       SUMB = PP*EXB
       SUMC = EXP(0.5*RE*(AL-0.5*TIM))
       UEX(L) = SUMA/(SUMA+SUMC*SUMB)
       SUMD = SUMD + (U(L)-UEX(L))**2
  240 CONTINUE
```

```
C
      IF(LST .EQ. 1)GOTO 250
      A(1) = UEX(1)
      A(NOR) = UEX(NOR)
      LST = 1
      GOTO 70
  250 CONTINUE
C
C     COMPARE SOLUTION WITH EXACT
C
      SUML = SQRT(SUMD)
      SUMD = SQRT(SUMD/ANN)
      DO 260 L = 1,NOR
      DU = U(L) - UEX(L)
  260 WRITE(6,270)L,UEX(L),U(L),DU
  270 FORMAT(3H L=,I3,6H UEX =,E11.4,4H   U=,E11.4,6H   DU= , E11.4)
      WRITE(6,280)SUMD,SUML
  280 FORMAT(9H RMS/ERR=,E11.4,21H DISCRETE L2 ERROR = ,E11.4)
      STOP
      END
```

```
      SUBROUTINE FORM4(IPR,NOR,STEP,IQL,U,A)
C
C     CALCULATE GRID AND C,EF,AA
C
      DIMENSION EF(7,201),AA(7,201),C(7,201),A(203),U(203),X(203)
      DIMENSION EY(5,5),EZ(3,3),EC(5,5),ED(3,3)
      COMMON C,AA,RE,HH,EC,ED,EF,X,XL,XR
      DATA EY/1.,0.,-1.,0.,0.,2.,-8.,6.,0.,0.,2.,-12.,0.,12.
     1,-2.,0.,0.,-6.,8.,-2.,0.,0.,1.,0.,-1./
      DATA EZ/-3.,4.,-1.,-8.,0.,8.,1.,-4.,3./
      XIO = 0.
C
C     CALCULATE X GRIDPOINTS
C
      NH = NOR/2
      ANR = NH
      DXI = 1./ANR
      BS2 = ((XR-XIO)-STEP/DXI)/(1.-DXI*DXI)
      BS1 = XR - XIO - BS2
      NCO = NOR - NH
      NCP = NCO + 1
      DO 10 I = NCP,NOR
      AI = I - NCO
      XI = AI*DXI
      X(I)  = XIO + BS1*XI + BS2*XI*XI*XI
   10 CONTINUE
C
      ANL = NCO - 1
      DXL = 1./ANL
      AS2 = ((XIO-XL)-STEP/DXL)/(1.-DXL*DXL)
      AS1 = XIO - XL - AS2
      DO 20 I = 1,NCO
      AI = NCO - I
      XI = -AI*DXL
      X(I) = XIO + AS1*XI+AS2*XI*XI*XI
   20 CONTINUE
C
      IF(IQL .NE. 2)GOTO 40
      DO 30 I = 1,NH
      IA = 2*I
   30 X(IA) = 0.5*(X(IA-1)+X(IA+1))
   40 CONTINUE
C
      IF(IPR .EQ. 0)GOTO 60
      WRITE(6,50)(X(I),I=1,NOR)
   50 FORMAT(4H X= ,10E12.5)
   60 CONTINUE
C
C     CALCULATE AA,EF,C
C
      NN = NOR - 2
      IF(IQL .EQ. 2)GOTO 130
      IF(IQL .EQ. 1)GOTO  90
```

```
C
C        3 PT F.D.M.
C
         DO   80 K = 1,NN
         DO   70 I = 1,7
         EF(I,K) = 0.
  70 AA(I,K) = 0.
         DM1 = X(K+1)-X(K)
         DP1 = X(K+2)-X(K+1)
         DUM = 2./(DM1+DP1)
         AA(2,K) = DUM/DM1
         AA(4,K) = DUM/DP1
         AA(3,K) = -AA(2,K)-AA(4,K)
         DIM = DM1/DP1
         EF(4,K) = 0.5*DUM*DIM
         EF(2,K) = -0.5*DUM/DIM
         EF(3,K) = -EF(2,K)-EF(4,K)
  80 CONTINUE
         GOTO 180
  90 CONTINUE
C
C        LINEAR F.E.M.
C
         DO  120 K = 1,NN
         DO  100 I = 1,7
         C(I,K) = 0.
         AA(I,K) = 0.
 100 CONTINUE
         AA(3,K) = +1./(X(K+1)-X(K))
         AA(4,K) =-(X(K+2)-X(K))/(X(K+1)-X(K))/(X(K+2)-X(K+1))
         AA(5,K) = +1./(X(K+2)-X(K+1))
         IF(K .EQ. 1)GOTO 110
         C(3,K)     = (X(K+1)-X(K))/6.
 110 C(4,K) = (X(K+2)-X(K))/3.
         IF(K .EQ. NN)GOTO 120
         C(5,K) = (X(K+2)-X(K+1))/6.
 120 CONTINUE
         GOTO 180
C
C        QUADRATIC F.E.M.
C
 130 DO 150 KA = 1,NH
         K = 2*KA - 1
         DO 140 I = 1,7
         C(I,K) = 0.
         C(I,K+1) = 0.
         AA(I,K+1) = 0.
         AA(I,K) = 0.
 140 CONTINUE
         XKM = X(K+2)-X(K)
         AA(3,K) = 8./3./XKM
         AA(4,K) = -2.*AA(3,K)
         AA(5,K) = AA(3,K)
         AA(2,K+1) =-1./3./XKM
         AA(3,K+1) = -8.*AA(2,K+1)
         AA(4,K+1) = AA(2,K+1)*7.
         C(3,K)     =     XKM/15.
         C(4,K) = 8.*C(3,K)
         C(5,K) = C(3,K)
         C(2,K+1)   =     -XKM/15./2.
         C(3,K+1) = -2.*C(2,K+1)
         C(4,K+1) = 2.*C(3,K+1)
```

```
      IF(K+2 .GT. NN)GOTO 150
      XKP = X(K+4)-X(K+2)
      AA(6,K+1) =-1./3./XKP
      AA(5,K+1) = -8.*AA(6,K+1)
      AA(4,K+1) = AA(4,K+1) + AA(6,K+1)*7.
      C(6,K+1)  =      -XKP/15./2.
      C(5,K+1) = -2.*C(6,K+1)
      C(4,K+1) = C(4,K+1) + 2.*C(5,K+1)
  150 CONTINUE
C
      DO 160 I = 1,5
      DO 160 J = 1,5
  160 EC(I,J) = EY(J,I)/30.
      DO 170 I = 1,3
      DO 170 J = 1,3
  170 ED(I,J) = EZ(J,I)/15.
  180 CONTINUE
C
C     SET THE INITIAL VALUES OF A (AND U)
C
      DO 190 I = 1,NOR
      IF(X(I) .LT. 0.)A(I) = 1.
      IF(X(I) .LT. 0.)IH = I
      IF(X(I) .GE. 0.)A(I) = 0.
  190 CONTINUE
      A(IH+1) = 0.5
      DO 200 I = 1,NOR
  200 U(I) = A(I)
      IF(IPR .EQ. 0)RETURN
C
C     WRITE AA,EF,C,EC AND ED IF REQUIRED
C
      DO 210 I = 1,NOR
      WRITE(6,220)I,(AA(K,I),K=1,7)
       IF(IQL .EQ. 3)WRITE(6,240)I,(EF(K,I),K=1,7)
       IF(IQL .NE. 3)WRITE(6,230)I,(C(K,I),K=1,7)
  210 CONTINUE
  220 FORMAT(3H I=,I2,4H AA=,7E15.8)
  230 FORMAT(3H I=,I2,4H  C=,7E15.8)
  240 FORMAT(3H I=,I2,4H EF=,7E15.8)
      IF(IQL .NE. 2)RETURN
C
      DO 250 I = 1,5
  250 WRITE(6,260)(EC(I,J),J=1,5)
  260 FORMAT(4H EC=,10E11.4)
      WRITE(6,270)((ED(I,J),J=1,3),I=1,3)
  270 FORMAT(4H ED=,10E11.4)
      RETURN
      END
```

```
      SUBROUTINE GRAD4(DA,A,NOR,IQL)
C
C     COMPUTE LOCAL GRADIENT AS REQUIRED BY O.D.E. INTEGRATOR
C
      DIMENSION DA(203),C(7,201),AA(7,201),DD(203),A(203)
     1,EC(5,5),ED(3,3),EF(7,201),X(203)
      COMMON C,AA,RE,HH,EC,ED,EF,X,XL,XR
      NN = NOR-2
      NH = NOR/2
C
      DO 120 KA = 1,NH
      KB = 2*KA - 1
C
      GOTO(10,20,80),IQL
   10 K = KB
C
C     LINEAR F.E.M.
C
      DEM = (A(K)+A(K+1)+A(K+2))/3.
      DKJ = DEM*(A(K+2)-A(K))/2.
      DUM = AA(3,K)*A(K)+AA(4,K)*A(K+1)+AA(5,K)*A(K+2)
      K = KB +1
      IF(K .GT. NN)GOTO 110
      DEM = (A(K)+A(K+1)+A(K+2))/3.
      DKI = DEM*(A(K+2)-A(K))/2.
      DIM = AA(3,K)*A(K)+AA(4,K)*A(K+1)+AA(5,K)*A(K+2)
      GOTO 110
C
C     QUADRATIC F.E.M.
C
   20 DKI = 0.
      DKJ = 0.
      DO 50 I = 1,5
      IA = KB - 2 + I
      DO 40 J = 1,5
      JA = KB - 2 + J
      IF(KB .GE. NN)GOTO 30
      DKI = DKI + EC(I,J)*A(IA+1)*A(JA+1)
   30 IF(I .EQ. 1 .OR. I .EQ. 5)GOTO 40
      IF(J .EQ. 1 .OR. J .EQ. 5)GOTO 40
      DKJ = DKJ + ED(I-1,J-1)*A(IA)*A(JA)
   40 CONTINUE
   50 CONTINUE
C
      DIM = 0.
      DUM = 0.
      K = KB
      DO 70 I = 2,6
      IA = I + K - 3
      IF(I .EQ. 2 .OR. I .EQ. 6)GOTO 60
      DUM = DUM + AA(I,K)*A(IA)
   60 IF(KB .GE. NN)GOTO 70
      DIM = DIM + AA(I,K+1)*A(IA+1)
   70 CONTINUE
      GOTO 110
```

```
C
C       3 PT F.D.M.
C
   80 DKI = 0.
      DKJ = 0.
      DIM = 0.
      DUM = 0.
      DO 100 I = 2,4
      IA = KB + I - 2
      DKJ=DKJ+EF(I,KB)*A(IA)
      IF(KB .GE. NN)GOTO 90
      DKI=DKI+EF(I,KB+1)*A(IA+1)
   90 DUM = DUM + AA(I,KB)*A(IA)
      IF(KB .GE. NN)GOTO 100
      DIM = DIM + AA(I,KB+1)*A(IA+1)
  100 CONTINUE
      DKJ = DKJ*A(KB+1)
      DKI = DKI*A(KB+2)
C
  110 DD(KB+1)=DUM/RE - DKJ
      DD(KB+2) = DIM/RE - DKI
  120 CONTINUE
C
      DD(1) = 0.
      DD(NOR) = 0.
      IF(IQL .NE. 3)GOTO 140
      DO 130 I = 1,NN
  130 DA(I+1) = DD(I+1)*HH
      DA(1) = 0.
      DA(NOR) = 0.
      RETURN
C
  140 CONTINUE
      DD(1) = 0.
      DD(NOR) = 0.
C
C       BANSOL EVALUATES DA = (C**-1)*DD
C
      CALL BANSOL(DD,DA,C,NN,IQL)
C
      DO 150 I = 1,NN
  150 DA(I+1) = DA(I+1)*HH
      DA(1) = 0.
      DA(NOR) = 0.
      RETURN
      END
```

```
      SUBROUTINE BANFAC(B,NN,NQL)
C
C     FACTORISES BAND MATRIX ARISING FROM LINEAR OR QUADRATIC
C     ELEMENTS INTO L.U .  ENTRIES STORED IN BAND.. D,B,A,C,E
C     NQL = 1 TRIDIAGONAL SYSTEM.. LINEAR ELEMENTS
C     NQL = 2 PENTADIAGONAL SYSTEM.. QUADRATIC ELEMENTS
C
      DIMENSION B(7,201)
      IF(NQL .EQ. 2)GOTO 20
      NP = NN - 1
      DO 10 J = 1,NP
      B(3,J+1) = B(3,J+1)/B(4,J)
      B(4,J+1) = B(4,J+1) - B(3,J+1)*B(5,J)
   10 CONTINUE
      RETURN
C
   20 NH = NN/2
      DO 50 I = 1,2
      DO 40 J = 1,NH
      JA = 2*(J-1)
      IF(I .EQ. 2)GOTO 30
C
C     I=1, FIRST PASS, REDUCE TO TRIDIAGONAL
C
      JB = JA + 2
      B(2,JB) = B(2,JB)/B(3,JB-1)
      B(3,JB) = B(3,JB) - B(2,JB)*B(4,JB-1)
      B(4,JB) = B(4,JB) - B(2,JB)*B(5,JB-1)
      GOTO 40
C
C     I=2, SECOND PASS, REDUCE TO UPPER TRIANGULAR
C
   30 JB = JA + 3
      B(3,JB-1) = B(3,JB-1)/B(4,JB-2)
      B(4,JB-1) = B(4,JB-1) - B(3,JB-1)*B(5,JB-2)
      B(3,JB) = B(3,JB)/B(4,JB-1)
      B(4,JB) = B(4,JB) - B(3,JB)*B(5,JB-1)
      B(5,JB) = B(5,JB) - B(3,JB)*B(6,JB-1)
   40 CONTINUE
   50 CONTINUE
      RETURN
      END
```

```
      SUBROUTINE BANSOL(R,X,B,NN,NQL)
C
C     USES L.U FACTORISATION TO SOLVE FOR X GIVEN R
C
      DIMENSION R(203),X(203),B(7,201)
      X(1) = 0.
      IF(NQL .EQ. 2)GOTO 30
C
C     NQL = 1,  TRIDIAGONAL SYSTEM
C
      NP = NN-1
      DO 10 J = 1,NP
      JA = J + 1
   10 R(JA+1) = R(JA+1) - B(3,JA)*R(JA)
      X(NN+1) = R(NN+1)/B(4,NN)
      DO 20 J = 1,NP
      JA = NN - J
      X(JA+1) = (R(JA+1)-B(5,JA)*X(JA+2))/B(4,JA)
   20 CONTINUE
      RETURN
C
C     NQL = 2,  PENTADIAGONAL SYSTEM
C
   30 NH = NN/2
      DO 60 I = 1,2
      DO 50 J = 1,NH
      JA = 2*J
      DO 40 K = 1,I
      JB = JA - 1 + K
   40 R(JB+1) = R(JB+1) - B(I+1,JB)*R(JB)
   50 CONTINUE
   60 CONTINUE
      X(NN+2) = 0.
      X(NN+1) = R(NN+1)/B(4,NN)
      DO 70 J = 1,NH
      JA = NN - 2*(J-1) - 1
      X(JA+1) = (R(JA+1)-B(5,JA)*X(JA+2)-B(6,JA)*X(JA+3))/B(4,JA)
      X(JA) = (R(JA) - B(5,JA-1)*X(JA+1))/B(4,JA-1)
   70 CONTINUE
      RETURN
      END
```

```
OUTPUT FROM PROGRAM BURG4


INIT. T/STEP =   .100E-02 MAX T/STEP =   .100E-01 MAX TIME =   .800E+00
XL  = -.100E+01 XR =   .100E+01 STEP =   .200E+00 REYNOLDS NO. =   .100E+02
NSTOP =     65 NOR =     11 IQL =     1 NINC =     5 IPR =     0


NCT =    65 TIM =   .47100E+00 HH =   .80000E-02 RE =   .10000E+02
A =   .10000E+01  .10000E+01  .10003E+01  .99888E+00  .97455E+00  .85142E+00
A =   .56079E+00  .23387E+00  .53978E-01  .48665E-02  .52062E-03
L=  1 UEX =   .1000E+01  U=   .1000E+01  DU=   .4077E-06
L=  2 UEX =   .1000E+01  U=   .1000E+01  DU=   .2164E-04
L=  3 UEX =   .9996E+00  U=   .1000E+01  DU=   .6558E-03
L=  4 UEX =   .9960E+00  U=   .9989E+00  DU=   .2883E-02
L=  5 UEX =   .9713E+00  U=   .9745E+00  DU=   .3283E-02
L=  6 UEX =   .8589E+00  U=   .8514E+00  DU= -.7482E-02
L=  7 UEX =   .5661E+00  U=   .5608E+00  DU= -.5314E-02
L=  8 UEX =   .2233E+00  U=   .2339E+00  DU=   .1056E-01
L=  9 UEX =   .5294E-01  U=   .5398E-01  DU=   .1035E-02
L= 10 UEX =   .8403E-02  U=   .4867E-02  DU= -.3536E-02
L= 11 UEX =   .9269E-03  U=   .5206E-03  DU= -.4063E-03
RMS/ERR=   .5044E-02 DISCRETE L2 ERROR =   .1513E-01
```

Index